GENETICS, GENOMICS AND BREEDING OF PEANUTS

Genetics, Genomics and Breeding of Crop Plants

Series Editor

Chittaranjan Kole
Vice-Chancellor
Bidhan Chandra Agricultural University
Mohanpur, West Bengal
India

Books in this Series:

Published or in Press:

- Jinguo Hu, Gerald Seiler & Chittaranjan Kole: *Sunflower*
- Kristin D. Bilyeu, Milind B. Ratnaparkhe & Chittaranjan Kole: *Soybean*
- Robert Henry & Chittaranjan Kole: *Sugarcane*
- Kevin Folta & Chittaranjan Kole: *Berries*
- Jan Sadowsky & Chittaranjan Kole: *Vegetable Brassicas*
- James M. Bradeen & Chittaranjan Kole: *Potato*
- C.P. Joshi, Stephen DiFazio & Chittaranjan Kole: *Poplar*
- Anne-Françoise Adam-Blondon, José M. Martínez-Zapater & Chittaranjan Kole: *Grapes*
- Christophe Plomion, Jean Bousquet & Chittaranjan Kole: *Conifers*
- Dave Edwards, Jacqueline Batley, Isobel Parkin & Chittaranjan Kole: *Oilseed Brassicas*
- Marcelino Pérez de la Vega, Ana María Torres, José Ignacio Cubero & Chittaranjan Kole: *Cool Season Grain Legumes*
- Yi-Hong Wang, Tusar Kanti Behera & Chittaranjan Kole: *Cucurbits*
- Albert G. Abbott & Chittaranjan Kole: *Stone Fruits*
- Barbara E. Liedl, Joanne A. Labate, John R. Stommel, Ann Slade & Chittaranjan Kole: *Tomato*
- Byoung-Cheorl Kang & Chittaranjan Kole: *Peppers and Eggplants*
- Hongwei Cai, Toshihiko Yamada & Chittaranjan Kole: *Forage Crops*

GENETICS, GENOMICS AND BREEDING OF PEANUTS

Editors

Nalini Mallikarjuna
International Crop Research Institute for
the Semi-Arid Tropics (ICRISAT)
Patancheru
Andhra Pradesh
India

Rajeev K Varshney
International Crops Research Institute for the
Semi-Arid Tropics (ICRISAT)
Patancheru
Andhra Pradesh
India

CRC Press
Taylor & Francis Group
Boca Raton London New York

CRC Press is an imprint of the
Taylor & Francis Group, an **informa** business
A SCIENCE PUBLISHERS BOOK

CRC Press
Taylor & Francis Group
6000 Broken Sound Parkway NW, Suite 300
Boca Raton, FL 33487-2742

Printed in the United States of America on acid-free paper

International Standard Book Number: 978-1-4822-3835-8 (Hardback)

Visit the Taylor & Francis Web site at
http://www.taylorandfrancis.com

CRC Press Web site at
http://www.crcpress.com

Science Publishers Web site at
http://www.scipub.net

Preface to the Series

Genetics, genomics and breeding has emerged as three overlapping and complimentary disciplines for comprehensive and fine-scale analysis of plant genomes and their precise and rapid improvement. While genetics and plant breeding have contributed enormously towards several new concepts and strategies for elucidation of plant genes and genomes as well as development of a huge number of crop varieties with desirable traits, genomics has depicted the chemical nature of genes, gene products and genomes and also provided additional resources for crop improvement.

In today's world, teaching, research, funding, regulation and utilization of plant genetics, genomics and breeding essentially require thorough understanding of their components including classical, biochemical, cytological and molecular genetics; and traditional, molecular, transgenic and genomics-assisted breeding. There are several book volumes and reviews available that cover individually or in combination of a few of these components for the major plants or plant groups; and also on the concepts and strategies for these individual components with examples drawn mainly from the major plants. Therefore, we planned to fill an existing gap with individual book volumes dedicated to the leading crop and model plants with comprehensive deliberations on all the classical, advanced and modern concepts of depiction and improvement of genomes. The success stories and limitations in the different plant species, crop or model, must vary; however, we have tried to include a more or less general outline of the contents of the chapters of the volumes to maintain uniformity as far as possible.

Often genetics, genomics and plant breeding and particularly their complimentary and supplementary disciplines are studied and practiced by people who do not have, and reasonably so, the basic understanding of biology of the plants for which they are contributing. A general description of the plants and their botany would surely instill more interest among them on the plant species they are working for and therefore we presented lucid details on the economic and/or academic importance of the plant(s); historical information on geographical origin and distribution; botanical origin and evolution; available germplasms and gene pools, and genetic and cytogenetic stocks as genetic, genomic and breeding resources; and basic information on taxonomy, habit, habitat, morphology, karyotype, ploidy level and genome size, etc.

Classical genetics and traditional breeding have contributed enormously even by employing the phenotype-to-genotype approach. We included detailed descriptions on these classical efforts such as genetic mapping using morphological, cytological and isozyme markers; and achievements of conventional breeding for desirable and against undesirable traits. Employment of the *in vitro* culture techniques such as micro- and megaspore culture, and somatic mutation and hybridization, has also been enumerated. In addition, an assessment of the achievements and limitations of the basic genetics and conventional breeding efforts has been presented.

It is a hard truth that in many instances we depend too much on a few advanced technologies, we are trained in, for creating and using novel or alien genes but forget the infinite wealth of desirable genes in the indigenous cultivars and wild allied species besides the available germplasms in national and international institutes or centers. Exploring as broad as possible natural genetic diversity not only provides information on availability of target donor genes but also on genetically divergent genotypes, botanical varieties, subspecies, species and even genera to be used as potential parents in crosses to realize optimum genetic polymorphism required for mapping and breeding. Genetic divergence has been evaluated using the available tools at a particular point of time. We included discussions on phenotype-based strategies employing morphological markers, genotype-based strategies employing molecular markers; the statistical procedures utilized; their utilities for evaluation of genetic divergence among genotypes, local landraces, species and genera; and also on the effects of breeding pedigrees and geographical locations on the degree of genetic diversity.

Association mapping using molecular markers is a recent strategy to utilize the natural genetic variability to detect marker-trait association and to validate the genomic locations of genes, particularly those controlling the quantitative traits. Association mapping has been employed effectively in genetic studies in human and other animal models and those have inspired the plant scientists to take advantage of this tool. We included examples of its use and implication in some of the volumes that devote to the plants for which this technique has been successfully employed for assessment of the degree of linkage disequilibrium related to a particular gene or genome, and for germplasm enhancement.

Genetic linkage mapping using molecular markers have been discussed in many books, reviews and book series. However, in this series, genetic mapping has been discussed at length with more elaborations and examples on diverse markers including the anonymous type 2 markers such as RFLPs, RAPDs, AFLPs, etc. and the gene-specific type 1 markers such as EST-SSRs, SNPs, etc.; various mapping populations including F_2, backcross, recombinant inbred, doubled haploid, near-isogenic and pseudotestcross; computer software including MapMaker, JoinMap, etc. used; and different types of genetic maps including preliminary, high-resolution, high-density, saturated, reference, consensus and integrated developed so far.

Mapping of simply inherited traits and quantitative traits controlled by oligogenes and polygenes, respectively has been deliberated in the earlier literature crop-wise or crop group-wise. However, more detailed information on mapping or tagging oligogenes by linkage mapping or bulked segregant analysis, mapping polygenes by QTL analysis, and different computer software employed such as MapMaker, JoinMap, QTL Cartographer, Map Manager, etc. for these purposes have been discussed at more depth in the present volumes.

The strategies and achievements of marker-assisted or molecular breeding have been discussed in a few books and reviews earlier. However, those mostly deliberated on the general aspects with examples drawn mainly from major plants. In this series, we included comprehensive descriptions on the use of molecular markers for germplasm characterization, detection and maintenance of distinctiveness, uniformity and stability of genotypes, introgression and pyramiding of genes. We have also included elucidations on the strategies and achievements of transgenic breeding for developing genotypes particularly with resistance to herbicide, biotic and abiotic stresses; for biofuel production, biopharming, phytoremediation; and also for producing resources for functional genomics.

A number of desirable genes and QTLs have been cloned in plants since 1992 and 2000, respectively using different strategies, mainly positional cloning and transposon tagging. We included enumeration of these and other strategies for isolation of genes and QTLs, testing of their expression and their effective utilization in the relevant volumes.

Physical maps and integrated physical-genetic maps are now available in most of the leading crop and model plants owing mainly to the BAC, YAC, EST and cDNA libraries. Similar libraries and other required genomic resources have also been developed for the remaining crops. We have devoted a section on the library development and sequencing of these resources; detection, validation and utilization of gene-based molecular markers; and impact of new generation sequencing technologies on structural genomics.

As mentioned earlier, whole genome sequencing has been completed in one model plant (Arabidopsis) and seven economic plants (rice, poplar, peach, papaya, grapes, soybean and sorghum) and is progressing in an array of model and economic plants. Advent of massively parallel DNA sequencing using 454-pyrosequencing, Solexa Genome Analyzer, SOLiD system, Heliscope and SMRT have facilitated whole genome sequencing in many other plants more rapidly, cheaply and precisely. We have included extensive coverage on the level (national or international) of collaboration and the strategies and status of whole genome sequencing in plants for which sequencing efforts have been completed or are progressing currently. We have also included critical assessment of the impact of these genome initiatives in the respective volumes.

Comparative genome mapping based on molecular markers and map positions of genes and QTLs practiced during the last two decades of the last century provided answers to many basic questions related to evolution, origin and phylogenetic relationship of close plant taxa. Enrichment of genomic resources has reinforced the study of genome homology and synteny of genes among plants not only in the same family but also of taxonomically distant families. Comparative genomics is not only delivering answers to the questions of academic interest but also providing many candidate genes for plant genetic improvement.

The 'central dogma' enunciated in 1958 provided a simple picture of gene function—gene to mRNA to transcripts to proteins (enzymes) to metabolites. The enormous amount of information generated on characterization of transcripts, proteins and metabolites now have led to the emergence of individual disciplines including functional genomics, transcriptomics, proteomics and metabolomics. Although all of them ultimately strengthen the analysis and improvement of a genome, they deserve individual deliberations for each plant species. For example, microarrays, SAGE, MPSS for transcriptome analysis; and 2D gel electrophoresis, MALDI, NMR, MS for proteomics and metabolomics studies require elaboration. Besides transcriptome, proteome or metabolome QTL mapping and application of transcriptomics, proteomics and metabolomics in genomics-assisted breeding are frontier fields now. We included discussions on them in the relevant volumes.

The databases for storage, search and utilization on the genomes, genes, gene products and their sequences are growing enormously in each second and they require robust bioinformatics tools plant-wise and purpose-wise. We included a section on databases on the gene and genomes, gene expression, comparative genomes, molecular marker and genetic maps, protein and metabolomes, and their integration.

Notwithstanding the progress made so far, each crop or model plant species requires more pragmatic retrospect. For the model plants we need to answer how much they have been utilized to answer the basic questions of genetics and genomics as compared to other wild and domesticated species. For the economic plants we need to answer as to whether they have been genetically tailored perfectly for expanded geographical regions and current requirements for green fuel, plant-based bioproducts and for improvements of ecology and environment. These futuristic explanations have been addressed finally in the volumes.

We are aware of exclusions of some plants for which we have comprehensive compilations on genetics, genomics and breeding in hard copy or digital format and also some other plants which will have enough achievements to claim for individual book volume only in distant future. However, we feel satisfied that we could present comprehensive deliberations on genetics, genomics and breeding of 30 model and economic plants, and their groups in a few cases, in this series. I personally feel also happy that I could work with many internationally celebrated scientists who edited the book volumes on the leading plants and plant groups and included chapters authored by many scientists reputed globally for their contributions on the concerned plant or plant group.

We paid serious attention to reviewing, revising and updating of the manuscripts of all the chapters of this book series, but some technical and formatting mistakes will remain for sure. As the series editor, I take complete responsibility for all these mistakes and will look forward to the readers for corrections of these mistakes and also for their suggestions for further improvement of the volumes and the series so that future editions can serve better the purposes of the students, scientists, industries, and the society of this and future generations.

Science publishers, Inc. has been serving the requirements of science and society for a long time with publications of books devoted to advanced concepts, strategies, tools, methodologies and achievements of various science disciplines. Myself as the editor and also on behalf of the volume editors, chapter authors and the ultimate beneficiaries of the volumes take this opportunity to acknowledge the publisher for presenting these books that could be useful for teaching, research and extension of genetics, genomics and breeding.

Chittaranjan Kole

Preface

Peanut (*Arachis hypogaea* L. Millsp), a grain legume crop, which originated in South America, has become an important crop worldwide. Especially in the context of the developing world, where the crop is grown in a marginal environment by resource-poor farmers, peanut is either a crop for food security or for income generation. Due to exposure of the crop to a range of biotic and abiotic stresses, the crop productivity in developing countries is about 1 ton/ha. Therefore, it is imperative for peanut researchers across the world not only to understand peanut biology indepth but to use this information for crop improvement that can help in improving the livelihood of the poor in the developing countries.

Although peanut researchers have made great progress during the last 5–6 years, many of the latest findings are in the form of publications in various peer-reviewed journals. Much information has been generated on ways to get to the germplasm of interest and utilize it in breeding programs. The generation and utilization of a wide array of new sources of tetraploid peanut (also called synthetics) are expected to broaden the genetic base of peanut and to introduce useful traits. While transferring superior alleles from wild species and unadapted germplasm in elite varieties, there is an inherent issue of linkage drag. However, with the availability of large-scale molecular markers, dense genetic maps, and the information on the QTLs for traits of interest, issues like linkage drags can be overcome. Furthermore, recent advances in genomics, proteomics and bioinformatics are expected to enhance precision and efficiency in peanut breeding.

In view of above and with an objective of compiling information at one place, we planned to have a book dedicated to peanut. We, indeed, are privileged to have a panel of eminent scientists who are authorities in their fields, to write chapters for the book. The most important aspect of these chapters is that they don't just provide a compilation but also present a critical appraisal and future direction in the particular areas. In summary, the volume documents the latest advances in research on germplasm, molecular cytogenetics, genetic maps, trait mapping, transcriptomics, proteomics and bioinformatics.

We would like to avail this opportunity to extend our sincere thanks to all the authors (Annexure I) who accepted our invitation and wrote excellent articles. Sincere thanks are also due to the reviewers (Annexure II) who spent their quality time, for the sake of high-quality science, for providing useful suggestions to further improve the quality of chapters. The editors are also thankful to Dr. William Dar, Director General, ICRISAT for his encouragement to do and share high quality science and Dr. C.L.L. Gowda, Deputy Director General-Research, ICRISAT for his support. We are also thankful to several colleagues from ICRISAT for useful discussions and support during the preparation of the book. The editors thank Prof. C. Kole, Series Editor for his invitation and help in editing this volume. The editors, also would like to thank Dr. Manish K Pandey, Dr. Manish Roorkiwal, Dr. Reyazul Rouf Mir and Ms. Anu Chitikineni for their help in editing this book.

The editors also recognize that the editorial work for this book volume took away precious time that they could have spent with their respective families. Nalini Mallikarjuna thanks her husband P Mallikarjuna for his unstinted support and encouragement. Rajeev K. Varshney also appreciates the help, support and understanding of his wife Monika and his children Prakhar and Preksha who allowed their time to be taken away to fulfill RKV's editorial responsibilities for this book volume in addition to research, managerial and other institutional duties at ICRISAT and Generation Challenge Program (GCP).

Hyderabad
India

Nalini Mallikarjuna
Rajeev K. Varshney

Contents

List of Contributors

Amareshwari, P
Osmania University, Hyderabad, India.

Araujo, ACG
Embrapa Genetic Resources and Biotechnology, Brasília-DF, Brazil.

Bertioli, David J
Universidade de Brasília, Brasília, Brazil.

Bhatnagar-Mathur, Pooja
International Crops Research Institute for the Semi-Arid Tropics (ICRISAT), Hyderabad, India.
Ph: +91 (40) 30713485

Chen, Xiaoping
Crops Research Institute, Guangdong Academy of Agricultural Sciences, Guangzhou 510640, China.
Ph: +862087597315

Brewster, George
Florida A&M University, Tallahassee, USA.

Dwivedi, Sangam L
International Crops Research Institute for the Semi-Arid Tropics (ICRISAT), Patancheru 502324, AP, India.
Ph: +91 (40) 30713617

Gottschalk, Virginia
Florida A&M University, Tallahassee, USA.

Guimarães, PM
Embrapa Genetic Resources and Biotechnology, Brasília-DF, Brazil.

Guo, Baozhu
USDA-ARS, Crop Protection and Management Research Unit, Tifton, USA.

Heslop-Harrison, P
University of Leicester, Leicester, UK.

Holbrook, Corley C
USDA-ARS, Crop Genetics and Breeding Research Unit, Tifton, USA.

Isobe, Sachiko
Kazusa DNA Research Institute, Chiba, Japan.

Janila, P
International Crops Research Institute for the Semi-Arid Tropics (ICRISAT), Hyderabad, India.
Ph +91 (40) 30713583

Katam, Ramesh
Florida A&M University, Tallahassee, USA.
Ph: 850-559-3134

Latinwo, Lekan M
Florida A&M University, Tallahassee, USA.

Lavia, Graciela I
Instituto de Botánica del Nordeste (CONICET-UNNE) and Facultad de Ciencias Exactas y Naturales y Agrimensura, Universidad Nacional del Nordeste, Corrientes, Argentina.

Leal-Bertioli, SCM
Embrapa Genetic Resources and Biotechnology, Brasília-DF, Brazil.

Lee Ching
University of Wisconsin, Madison, WI USA.

Liang, Xuanqiang
Crops Research Institute, Guandong Academy of Agricultural Sciences, Guangzhou, 510640, China.
Ph: +862087597315

Liao, Boshou
CAAS Oil Crops Research Institute, Wuhan, China.

Mallikarjuna, Nalini
International Crops Research Institute for the Semi-Arid Tropics (ICRISAT), Hyderabad, India.
Ph: +918212412494

Nielen, S
International Atomic Energy Agency, Vienna, Austria.

Ortíz, Alejandra
Instituto de Botánica del Nordeste (CONICET-UNNE) and Facultad de Ciencias Exactas y Naturales y Agrimensura, Universidad Nacional del Nordeste, Corrientes, Argentina.

Pandey, Manish K
International Crops Research Institute for the Semi-Arid Tropics (ICRISAT), Hyderabad, India.
Ph: +91 40 3071 3397

Paxton, Payton
USDA-ARS Cropping Systems Research Laboratory, Lubbock, TX, USA.

Rani, A Roja
Osmania University, Hyderabad, India.

Robledo, Germán
Instituto de Botánica del Nordeste (CONICET-UNNE) and Facultad de Ciencias Exactas y Naturales y Agrimensura, Universidad Nacional del Nordeste, Corrientes, Argentina.

Sakata, Katsumi
Maebashi Institute of Technology, Maebashi, Japan.

Schwarzacher, T
University of Leicester, Leicester, UK.

Seijo, J Guillermo
Instituto de Botánica del Nordeste (CONICET-UNNE) and Facultad de Ciencias Exactas y Naturales y Agrimensura, Universidad Nacional del Nordeste, Corrientes, Argentina.

Sharma, Kiran K
International Crops Research Institute for the Semi-Arid Tropics (ICRISAT), Hyderabad, India.
Ph: +91 (40) 30713300

Sharma, Shivali
International Crops Research Institute for the Semi-Arid Tropics (ICRISAT), Patancheru 502324, India.
Ph: +91 (40) 307133570

Shirasawa, Kenta
Kazusa DNA Research Institute, Chiba, Japan.

Sunkara, Sowmini
International Crops Research Institute for the Semi-Arid Tropics (ICRISAT), Hyderabad, India.
Ph: +91 (40) 30713485

Survajahala, Prashanth
Bioclues.org.

Upadhyaya, Hari D
International Crops Research Institute for the Semi-Arid Tropics (ICRISAT), Patancheru 502324, AP, India.
Ph: +91 (40) 30713333

Varshney, Rajeev K
International Crops Research Institute for the Semi-Arid Tropics (ICRISAT), Hyderabad, India.
Ph: +914030713305

Venkatesh, K
Department of Genetics and Biotechnology, Osmania University, Hyderabad, India.

Williams Crayger
Department of Biological Sciences, Florida A&M University, Tallahassee, FL USA

Zhang, Xinyou
Henan Academy of Agricultural Sciences, Zhengzhou, China.

Abbreviations

AFLP	:	Amplified fragment length polymorphism
BSA	:	Bulked segregant analysis
CAAS (China)	:	Chinese Academy of Agricultural Sciences
CAGE	:	Cap-analysis gene expression
CaMV	:	Cauliflower mosaic virus
CENARGEN	:	National Research Center for Genetic Resources and Biotechnology
CGIAR	:	Consultative Group on International Agricultural Research
CIMMYT	:	International Maize and Wheat Improvement Center
cM	:	CentiMorghan
CRI (China)	:	Crops Research Institute
DArT	:	Diversity array technology
DDRT-PCR	:	Differential display reverse transcription polymerase chain reaction
DGR (India)	:	Directorate of Groundnut Research
ELS	:	Early leaf spot
EMBRAPA	:	Brazilian Agricultural Research Corporation
EST	:	Express sequence tag
FAD	:	Fatty acid desaturase
FISH	:	Fluorescent *in situ* hybridization
GAAS (China)	:	Guangdong Academy of Agricultural Sciences
GAB	:	Genomics-assisted breeding
GISH	:	Genomic *in situ* hybridization
GRAV	:	Groundnut rosette assistor virus
GRV	:	Groundnut rosette virus
GS	:	Genomic selection
HiCEP	:	High coverage expression profiling
IBONE (Argentina)	:	Instituto de Botánica the Northeast
ICAR (India)	:	Indian Council of Agricultural Research
ICPV	:	Indian peanut clump virus

ICRISAT (India) :	International Crops Research Institute for the Semi-Arid Tropics
IITA :	International Institute for Tropical Agriculture
IL :	Introgression line
INTA :	InstitutoNacional de TechnologiaAgropecuaria
IRRI :	International Rice Research Institute
ISSR :	Inter-simple sequence repeats
LD :	Linkage disequilibrium
LLS :	Late leaf spot
LTR :	Long terminal repeat
MABC :	Marker-assisted backcrossing
MARS :	Marker-assisted recurrent selection
MAS :	Marker-assisted slection
MITEs :	Miniature inverted-repeat transposable elements
MPSS :	Massively parallel signature sequencing
NBPGR (India) :	National Bureau of Plant Genetic Resources
NCSU (USA) :	North Carolina State University
NGS :	Next-generation sequencing
OCRI (China) :	Oil Crops Research Institute
PAC :	Preharvest aflatoxin contamination
PBND :	Peanut bud necrosis disease
PDR :	Pathogen-derived resistance
PGC :	Peanut Genome Consortium
PGP :	Peanut Genome Project
PMAGE :	Polony multiplex analysis of gene expression
PMV :	Peanut mottle virus
PStV :	Peanut stripe virus
PTGS :	Post-transcriptional gene silencing
PVE :	Phenotypic variance explained
QTL :	Quantitative trait locus
RAPD :	Random apmplified polymorphic DNA
RFLP :	Restriction fragment length polymorphism
RIL :	Recombinant inbred line
RLD :	Root length density
SAGE :	Serial analysis of gene expression
SCAR :	Sequence-characteized amplified region
SCMR :	SPAD chlorophyll meter reading
SLA :	Specific leaf area
SNP :	Single nucleotide polymorphism
SSH :	Suppressive subtractive hybridization
TAMU (USA) :	Texas A & M University
TOG :	Tentative orthologous gene

TSV	:	Tobacco streak virus
TSWV	:	Tomato spotted wilt virus
USDA (USA)	:	U.S. Department of Agriculture

List of Reviewers

Dutta, Manoranjan: National Bureau of Plant Genetic Resources, New Delhi, India

Guimarães, Patricia M: Embrapa Genetic Resources and Biotechnology, Brasília DF, Brazil

Guo, Baozhu: USDA-ARS, Tifton, USA

Iwata, Aiko: University of Georgia, Athens, USA

Katam, Ramesh: Florida A&M University, Florida, USA

Nadaf, Hajisaheb: University of Agricultural Sciences, Dharwad, India

Naveen, Puppala: New Mexico State University, New Mexico, USA

Pandey, Manish K: ICRISAT, Hyderabad, India

Sastry, Sameera: ICRISAT, Hyderabad, India

Schwarchzacher, Trude: University of Leicester, Leicester, UK

Suravajhala, Prashanth: WWW.Bioclues.org

1

Genetics, Genomics and Breeding of Peanut: An Introduction

Nalini Mallikarjuna[1] and *Rajeev K Varshney*[1,2,3,4,*]

ABSTRACT

Peanut stands second to soybean in both area and production in the world among legume oilseeds crops and is grown in >100 countries. Genetic barriers have not allowed sharing of useful alleles from wild relatives leaving the primary gene pool with a very narrow genetic base. Improving pod yield and oil content have been the main focus, along with providing resistance/tolerance against important biotic/abiotic stresses. Realizing the ever increasing demand among consumers, productivity needs to be increased significantly without compromising the oil quality and providing defense shield against biotic and abiotic stress. It is very difficult to achieve the above milestones without integrating the modern genomics tools with conventional breeding programs. The last decade witnessed significant progress in terms of genomic resources and molecular breeding activities. The objective of this book is to critically review the current updates on different aspects of peanut such as germplasm collections, genetics, genomics, transcriptomics, bioinformatics together with traditional and molecular breeding. The book also summarizes the success stories achieved through trait mapping and application of molecular markers

[1]International Crops Research Institute for the Semi-Arid Tropics (ICRISAT), Hyderabad, India.
[2]Crops Research Institute, Guandong Academy of Agricultural Sciences, Guangzhou, China.
[3]CGIAR Generation Challenge Programme, c/o CIMMYT, Mexico DF, Mexico.
[4]The University of Western Australia, Crawley, Australia.
*Corresponding author: *r.k.varshney@cgiar.org*

in improving important traits. This chapter provides highlights of different chapters which are expected to be a good resource for young researchers, breeders and policy makers for employing better strategies towards food security.

Keywords: Peanut, Groundnut, *Arachis*, Allotetraploid, Amphidiploid, Synthetic, Genepool, Peanut improvement

1.1 Introduction

Peanut (*Arachis hypogaea* L. Mill sp.) with a postfix of nut in the name is not a nut in the true sense, but is a leguminous crop, and because of a nutty cover, the pod wall, is called peanut. On the other hand, because of its growth and ripening of seed inside the ground, it is also known as groundnut. It has many characteristics of nuts such as high amount of fat (46g/100g) and other important constituents such as vitamins, protein, minerals and phytochemicals. During expeditions on foot to the South and North Poles by *Discovery* and *Terra Nova* expeditions in the early 19th century, consuming peanuts was the deciding factor between life and death. Peanut butter was the ideal foodstuff, freeing explorers from the transport and kindling of cooking fuel (a near-impossibility in the frigid polar winds), and high enough in protein and calories to fuel the party and keep them from freezing to death in the harsh weather and freezing night-time temperatures.

Peanuts are rich in nutrients, providing over 30 essential nutrients and phytonutrients such as niacin, folate, fiber, magnesium, vitamin E, manganese and phosphorus, etc. (Savage and Keenan 1994; Whitley et al. 2011). Plumpy'nut a ready-to-use therapeutic food made from peanut is a popular source of nutrient for malnutritioned children in Africa due to the presence of about 25% protein, a higher proportion than in any true nut. Peanuts are found to contain high concentration of antioxidant polyphenols than other nuts and other antioxidant sources such as blackberries, strawberries, carrots or beets (Craft et al. 2010). Furthermore, peanuts are a significant source of resveratrol equivalent to that present in red grapes (Sanders et al. 2000), a chemical associated with reduction in risk of cardiovascular disease (Fraser et al. 1992; Hu et al. 1998; Prineas et al. 1993), cancer (Awad et al. 2000) and anti-aging properties, hence would have a high impact in both health and cosmetic industry. In addition, peanuts are also a source of coenzyme Q10 (Pravst et al. 2010), as are oily fish, beef, soybeans and spinach.

Peanut is believed to have originated in South America and was first domesticated in the Brazilian-Paraguayan region (Vavilov 1951). The area of the valleys of Paraguay and Parana rivers is the most likely center of origin. Excavation in coastal Peru dating back to 800 BC evidenced the cultivation of peanut. From South America, peanut spread to other parts of the world. It was commonly found in the West Indies but not in the United States in pre-Columbian times. Peanut was introduced to the Old World in the 16th century when the Portuguese took the seeds from America to Africa. The Spaniards introduced it into the Philippines. It then spread to China, India, Japan, Malaysia and other parts of the world. Human interaction through selection of most suited lines over the centuries resulted in loss of much of the genetic diversity/desirable alleles and genes whose importance is now been realized. Further, we are still unaware of future preferences for the so-called "lost" genetic diversity including genes for resistance/tolerance to biotic and abiotic stresses, as well as taste and nutritional composition along with yield. The domesticated peanut is an amphidiploid or allotetraploid, meaning that it has two sets of chromosomes from two different species, thought to be *A. duranensis* and *A. ipaensis* (Kochert et al. 1991, 1996; Seijo et al. 2007). These species combined in the wild to form the tetraploid species, which gave rise to the domesticated peanut. Cultivated peanut contains a fraction of the genetic diversity, which is not more than 13% (Varshney et al. 2009a), found in their closest wild relatives in section *Arachis* (Kochert et al. 1991), a legacy of the "domestication bottleneck".

The first domestication bottleneck was the combination of two species *A. duranensis* and *A. ipaensis* amongst 26 species from section *Arachis*. Crossing experiments between *A. duranensis* and *A. ipaensis* have shown that the diploid hybrid is highly sterile (Mallikarjuna et al. 2011a). Had the hybrid been fertile, there would not have been the need to double its chromosome number to form the fertile amphidiploid. It would probably have remained a diploid than a tetraploid as it is today. So the second bottleneck was in the formation of a diploid sterile hybrid. The third bottleneck was in the process of chromosome duplication to form the allotetraploid as it is known in literature that polyploidy causes genetic bottlenecks (Sanford 1983). Ancient farmers would have selected relatively few plants from the progenitors of modern crops, in a limited number of places and a similar situation would have existed for peanut in South America. This can be visualized as the fourth bottleneck. The early Portuguese and Spanish traders during their expeditions spread a few selected genotypes to the rest of the world thus giving rise to yet another, the fifth, bottleneck, superimposed by the sixth bottleneck which is the self-pollinating nature of the crop. To conclude peanut is the product of evolution after a series of six bottlenecks.

Traditionally wild relatives of peanut were directly used in a crossing program producing triploids as *Arachis* species in the compatible gene pool are diploid and cultivated peanut is a tetraploid. Triploids are cumbersome to use for peanut improvement, but such efforts have not gone without dividends (Mallikarjuna et al. 2004a,b). More recently, development of amphidiploid and autotetraploids utilizing *Arachis* species has been encouraging. With the development of synthetic peanut by combining putative A and B genome donors as well as many other A and B genome *Arachis* species, a range of tetraploid peanut was synthesized at ICRISAT (Mallikarjuna et al. 2011a). Screening some of these has shown that they have not lost the traits present in the diploid species (Mallikarjuna et al. 2011b). Newly synthesized tetraploids possessing several good traits are fairly easy to use in the crossing program as minimum segregation distortion (Fonceka et al. 2009) and minimal disturbance in meiosis has been observed (Mallikarjuna et al. 2012) with high pollen fertility in the hybrids. Fonceka et al. (2012) used a synthetic amphidiploid (Favero et al. 2006) and applied conventional breeding technique to capture the genetic diversity in peanut wild relatives. In their study, a set of 122 Introgression Lines (ILs) that offered an extensive coverage of the cultivated peanut genome with generally a unique fragment per line and overlapping fragments between contiguous lines were developed and thus these newly developed synthetics have opened new avenues for peanut improvement using new sources of tetraploid/synthetic peanuts.

1.2 Pre-breeding Efforts

Although pre-breeding does not produce new varieties, but it does turn up intermediate lines that breeders find easier to use. It throws up enough variation to sustain breeding activities especially with the assistance of molecular markers. Many public agricultural research institutions, such as the International Rice Research Institute (IRRI), International Maize and Wheat Improvement Center (CIMMYT), International Institute for Tropical Agriculture (IITA) and International Crops Research Institute for Semi-Arid Tropics (ICRISAT) have active pre-breeding programs in their mandate crops. Pre-breeding is the link between conservation and use of wild crop relatives. Out of all the raw materials at the breeder's disposal, the diversity of wild crop relatives has been relatively neglected. The conserved germplasm in the gene banks is for present and future use. Urbanization, explosion in population growth, dwindling water resources and in a 2°C-warmer world, we may not have a choice but to bring in new sources of variation that the new sources of synthetic tetraploids are offering. Some of the recent success at ICRISAT in utilizing wild *Arachis* species for tackling those diseases/pests for which high levels of resistance is not present in cultivated peanut germplasm has been achieved. These are stable tetraploid introgression or aptly called the pre-breeding lines with trait(s) of interest.

i. *Aflatoxin* resistance: Inadequate levels of resistance in peanut germplasm are one of the important factors for not having resistance to *A. flavus*-

aflatoxin resistance in peanut. This means sources of resistances have to be scouted beyond the cultivated/primary gene pool. The report from Xue et al. (2005) showed that *Arachis* species *A. duranensis* (eight accessions) and *A. cardenasii* (two accessions) from section *Arachis* had high levels of resistance to aflatoxin production and interspecific derivatives derived from them continued to show the trait. ICRISAT screened advance generation lines derived from *A. cardenasii* in aflatoxin sick plot for three consecutive years and found many of the lines with low aflatoxin production. This opens up new avenues for aflatoxin resistance breeding in peanut (Mallikarjuna N and Sudini H, unpubl. data).

ii. *Late leaf spot*: Sources of resistance to Late Leaf Spot (LLS) caused by *Cercosporidium personatum* Berk. & M.A. Curtis is higher in wild *Arachis* species compared to moderate levels of resistance in cultivated germplasm (Subhramanyam et al. 1989). *Arachis cardenasii* derived lines showed resistance to LLS when screened under unprotected field conditions in different locations (Mallikarjuna N and Sudini H, unpubl. data).

iii. *Peanut bud necrosis disease*: Peanut Bud Necrosis Disease (PBND) is an economically important virus disease of peanut in many Asian countries where peanut is grown. The disease causes crop losses exceeding US$ 89 million in India alone (Anon 1992). Sources of resistance is absent in cultivated germplasm (Reddy 1998). Many of the *Arachis* species have been found to be resistant to the disease (Reddy et al. 2000). Stable lines derived from *Arachis* species were screened for PBND in disease hot spot location and a few resistant lines were identified (Sunkad and Mallikarjuna N, unpubl. data).

iv. *Root rot*: Among the soil-borne fungal diseases of peanut, stem rot caused by *S. rolfsii* is a potential threat to groundnut production throughout the world. The disease causes severe damage during any stage of crop growth, and yield losses over 25% have been reported by Mayee and Datar (1988). Sources of resistance to the constraint are not up to the desired level in cultivated germplasm. Stable lines derived from *Arachis* species were screened for *S. rolfsii* in the disease hot spot location at Dharwad, Karnataka state, India, and a few lines with resistance were obtained (Kenchanagowda R and Mallikarjuna N, unpubl. data).

v. *Spodoptera litura: Spodoptera litura* also called fall army worm, a polyphagous insect, is becoming an important insect pest of groundnut with sources of resistance to the pest absent in cultivated germplasm. Yield losses of groundnuts have been directly associated with higher density of larvae of *S. litura*, and the intensity of defoliation (Panchbhavi and Nethradani 1987).

Currently, no peanut cultivar is known to express resistance to *S. litura,* however, some wild relatives of peanut were found resistant to *S. litura*. Neonate larvae suffer high levels of mortality and the development of older larvae on resistant wild species is severely inhibited (Stevenson et al. 1993a). Flavonoids chlorogenic acid, quercetin and rutin present in *Arachis kempff mercadoi* responsible for resistance to *S. litura* were identified (Stevenson et al. 1993b). Mallikarjuna et al. (2004a) developed lines utilizing *A. kempff mercadoi* and screened the lines for *S. litura* resistance. Resistant derivatives were found to have high levels of flavonoids and antibiosis mechanism prevented larval growth. Susceptible derivatives and the female parent, *A. hypogaea* had low levels of flavonoids (Mallikarjuna et al. 2004b).

1.3 Germplasm Resources and Cytogenetics

Chapter 2 entitled "***Genetic Resources, Diversity and Association Mapping in Peanut***" deals with the conservation of a large collection of peanut germplasm including wild *Arachis* species, which is the key to the success in crop improvement. Much of the diversity of wild crop relatives that is available in gene banks is not actively used because most crop-breeding programs are generally not set up to best use it and wild relatives are viewed as too unwieldy to use with sufficient ease and speed. Further, all the closely related wild relatives in section *Arachis* are diploid whereas cultivated peanut is a tetraploid and, hence, crossing diploids with tetraploids or vice versa is a not a straight-forward process. Therefore, objective oriented manageable germplasm sets possessing agronomically important traits such as reference set, core collection, mini-core collection and mini-mini core collections were structured in order to use these sets judiciously and more efficiently. In addition to these sets, amphidiploids originating from distant wild *Arachis* species were also developed in order to overcome bottlenecks associated with peanut domestication and are being currently used in alien introgressions and for several other genetical and breeding applications which will enrich existing variability of primary gene pool.

Chapter 3 entitled "***Classical and Molecular Cytogenetics in Arachis***" deals with the recent progress made in understanding the chromosome complements of peanut and related wild species. In order to conduct genetical studies, proper understanding on chromosome number, size and structure play a significant role especially during integration of genomics tools with conventional breeding programs. Cytogenetics has played an important role through classical cytological studies in revealing important information about the complexity of the peanut genome. Comparative cytological mapping studies helped in defining chromosome numbers and their karyotype features to establish the relationships among species and the taxonomic sections. Further, efforts made to understand chromosome structure and genome evolution within the genus by using chromosome specific markers developed by fluorescent *in situ* hybridization (FISH). These studies have also revealed variation in karyotype structure, which represents different genomes. This chapter also addresses the integration of genomic *in situ* hybridization (GISH) approaches with FISH analysis to differentiate chromosomes of the two progenitors of cultivated tetraploid peanut, i.e., *A. duranensis* and *A. ipaënsis* along with identification of center of origin.

1.4 Conventional and Molecular Breeding

Chapter 4 on "***Peanut Breeding***" reviews the recent progress in peanut breeding worldwide. Basically genetic enhancement through conventional approaches has been achieved for few qualitative traits such as resistance to *Sclerotinia* blight, root-knot nematode and Tomato Spotted Wilt Virus (TSWV), which is benefitting US peanut producers >$200 million annually. Similar trends have also been observed in China as efforts led to at least 30% yield increase during the past two decades. However, the conventional approaches are not able to address the further increased yield demand as well as existing and future breeding challenges. In such a scenario, integration of genomics with conventional breeding approaches has become mandatory for developing superior cultivars with higher yield, better quality and enhanced resilience.

Chapter 5 on "***Molecular Markers, Genetic Maps, and QTLs for Molecular Breeding in Peanut***" deals with development and use of genomic resources and their utilization in peanut improvement. The international research community neglected development of genomic resources in peanut, which left this crop in the group of "Orphan Crops". Nevertheless, recent efforts resulted in the development of limited genomic resources and it was mostly in the last decade, which witnessed a speedy development due to collaborative effort among several research partners. Large scale of molecular markers such as Simple Sequence Repeat (SSR) and Diversity Array Technology (DArT) markers have been developed recently, which paved the way for construction of genetic maps, initially for diploids and then for tetraploids (Pandey et al. 2012; Varshney et al. 2013). This chapter provides detailed development of genomics resources such as markers, genetic maps and Quantitative Trait Loci (QTL) analysis. SSR markers assembled from public domain or collaborators were screened on parental genotypes and several genetic maps were constructed based on cultivated × cultivated genotypes. These genetic maps were then used for marker-trait associations for drought tolerance and foliar diseases. The identified markers/gene specific markers were successfully deployed through Marker-

Assisted Backcrossing (MABC) for improving elite peanut varieties. With international collaborations, dense reference consensus genetic map was successfully improved from 897 (Gautami et al. 2012b) to 3,693 marker loci (Shirasawa et al. 2013) and these dense consensus maps will set the platform for several other genetic and molecular breeding activities in peanut. Efforts to sequencing the peanut genome have been initiated with the formation of a Peanut Genome Consortium (PGC) (http://www.peanutbioscience.com/peanutgenomeproject.html).

1.5 Genome Structure and Proteomics

Chapter 6 of this book entitled "***An Overview of Peanut Genome Structure***" is focused in defining peanut genome structure. The studies in this area have revealed that the A and B genomes are of similar size and are composed mostly of metacentric chromosomes. With the genome of about 2.8 Gb for tetraploid peanut seems to have high repetitive DNA content. *A. duranensis* and *A. ipaënsis* are considered as the most probable diploid ancestors and donors of the A and B genomes, respectively. The cytogenetical studies integrated with genomic approaches revealed the possible ancestors, period of origin of different subgenomes. The genetic similarity between A and B genomes at sequence level is very low. Most importantly, cultivated peanut genetically behaves as a diploid, the two subgenomes have a very high genetic synteny, and do not appear to have undergone major structural rearrangements after polyploidization. The structural genomics revealed significant genetic similarity of peanut subgenomes with other legumes that diverged during evolution.

This crop is witnessing a transition phase wherein the efforts are continued towards development of genomic resources along with a large amount of sequencing data, which need efficient data storage and retrieval system along with statistical analysis support. The transcriptome represents messenger RNA (mRNA) expressed in a particular cell/tissue/organ/organism, and their quantity at a particular growth stage. These resources are the primary source from where DNA markers are being developed to use in several genetical studies. Transcriptomics also improves understanding the genetic mechanism underlying important agronomic traits for peanut improvement. Transcriptomics or genomewide transcriptional profiling allows simultaneous examination of transcriptome that is the term designated to the specific subset or complete set of mRNA expressed in a particular cell, tissue, organ or organism, and their quantity for a given developmental stage or physiological condition. It has been increasingly used to understand transcriptomes of a range of peanut tissues at different developmental stages under various environmental stresses. Chapter 7 entitled "***Peanut Transcriptomics***" reviews commonly used transcriptomic technologies, the definition of the transcriptomes for three principal tissues (pod/seed, root and leaf), the transcriptomics of stress response in peanut, as well as the use of transcriptome for marker development. Peanut transcriptomics will make great contributions to the understanding of genetic mechanism underlying important agronomic traits for peanut improvement.

Peanut improvement by conventional or molecular approach relies on an understanding of the biology of the plant, particularly interactions occurring across hierarchical scales of organization. In this regard, the application of metabolomics and proteomics is poised to deliver large volumes of data on protein and metabolite fluctuations associated with developmental and environmental cues. The challenge for the peanut research community will be to ensure that similar data is generated, interpreted and integrated towards crop improvement. Chapter 8 entitled "***Advances in Proteomics Research in Peanut Genetics and Breeding***" gives an insight into this field of research.

1.6 Transgenic Breeding

Chapter 9 entitled "***Transgenic Interventions in Peanut Crop Improvement: Progress and Prospects***" deals with the strength of transgenic technology specially when there is no source of variation for a trait in peanut genepool. Advances in tissue culture and genetic engineering comprising important areas of biotechnology have provided alternative pest control strategies. The development and standardization of protocols for genetic transformation in peanut for several genes are briefly discussed. Although this technology has not been received open heartedly in many countries, it has shown great potential in peanut improvement. Let us hope that once the regulatory issues are solved in certain countries of the world, the material will be available for peanut improvement across the world.

1.7 Bioinformatics Tools

Even though peanut lagged behind in generation of sequences still with the limited genomic resources studies on genetic diversity, genetic mapping and QTL analysis could be conducted. Expressed Sequence Tags (ESTs) generated provided raw material/sequence to apply some bioinformatics tools for extracting information to use them in genetic and breeding applications. Chapter 10 entitled "***Applications of Bioinformatics Tools to Genetic Mapping and Diversity Analysis in Peanut***" presents an overview of available resources on peanut bioinformatics and their role in elucidating biological and genomic information on peanut.

1.8 Summary

In summary, developments in last decade in several research areas promise to fill the research gaps required for handling the genetic bottlenecks in a better and precise way. Much needed diversification using genomic tools will facilitate the use of a diverse source for improving existing cultivars to equip them with genes for high resilience and new cultivars with desired traits. Although such concerns are raised at many of the scientific gatherings, however not many initiatives have been taken even for very important food crops. Hence, this is the prime time to retrieve desirable alleles not only to address existing problems but also for the future as well in order to sustain food production. Therefore, this is an effort to update the peanut research community on developments at different aspects in peanut by compiling all the developments made till date. It is also foreseen that compilation of updates will encourage more inter-displinary collaborations to tackle existing as well as advance initiatives to address future problems.

References

Anonymous (1992) The Medium Term Plan, 1994–1998, vol I, Main Report. ICRISAT, Patancheru, Andhra Pradesh, India, p 80.

Awad AB, Chan KC, Downie AC, Fink CS (2000) Peanuts as a source of beta-sitosterol, a sterol with anticancer properties. Nutr Cancer 36: 238–241.

Craft BD, Kosińska A, Amarowicz R, Pegg RB (2010) Antioxidant properties of extracts obtained from raw, dry-roasted, and oil-roasted US peanuts of commercial importance. Plant Foods Hum Nutr 65: 309–310.

Favero AP, Simpson CE, Valls JFM, Yuksel B (2006) Study of the evolution of cultivated peanut through crossability studies among *Arachis ipaensis*, *A. duranensis*, and *A. hypogaea*. Crop Sci 46: 1546–1552.

Fonceka D, Hodo-Abalo T, Rivallan R, Faye I, Sall MN, Ndoye O, Fávero AP, Bertioli DJ, Glaszmann JC, Courtois B, Rami J-F (2009) Genetic mapping of wild introgressions into cultivated peanut: a way toward enlarging the genetic basis of a recent allotetraploid. BMC Plant Biol 9: 103 doi:10.1186/1471-2229-9-103.

Fonceka D, Tossim HA, Rivallan R, Vignes H, Faye I, Ndoye O, Moretzsohn MC, Bertioli DJ, Glaszmann JC, Courtois B, Rami J-F (2012) Fostered and left behind alleles in peanut: interspecific QTL mapping reveals footprints of domestication and useful natural variation for breeding. BMC Plant Biol 12: 26.

Fraser G, Sabate J, Beeson LW, Strahan LW (1992) Possible effect of nut consumption and risk of coronary heart disease. Arch Internet Med 152: 1416–1424.

Hu FB, Tampler MJ, Manson JE, Rimm E, Colditz GA, Rosner BA, Speizer FE, Henneckens CH, Willett WC (1998) Frequent nut consumption and risk of coronary heart disease in women: prospective cohort study. Brit Med J 317: 341–345.

Kochert G, Stalker HT, Gimenes M, Galgaro L, Lopes CR, Moore K (1996) RFLP and cytogenetic evidence on the origin and evolution of allotetraploid domesticated peanut, *Arachis hypogaea* (Leguminosae). Am J Bot 83: 1282–1291.

Kochert G, Halward T, Branch WD, Simpson CE (1991) RFLP variability in peanut (*Arachis hypogaea* L.) cultivars and wild species. Theor Appl Genet 81: 565–570.

Mallikarjuna N, Pande S, Jadhav DR, Sastri DC, Rao JN (2004a) Introgression of disease resistance genes from *Arachis kempff-mercadoi* into cultivated groundnut. Plant Breed 123: 573–576.

Mallikarjuna N, Jadhav DR, Kranthi KR, Kranthi S (2004b) Influence of foliar chemical compounds on the development of *Spodoptera litura* (Fab.) on interspecific derivatives of groundnut. J Appl Entomol 128: 321–328.

Mallikarjuna N, Senthilvel S, Hoisington D (2011) Development of synthetic groundnuts (*Arachis hypogaea* L.) to broaden the genetic base of cultivated groundnut. Genet Resour Crop Evol 58: 889–907.

Mallikarjuna N, Jadhav DR, Reddy K, Husain F, Das K (2012) Screening new *Arachis* amphidiploids, and autotetraploids for resistance to late leaf spot by detached leaf technique. Eur J Plant Pathol 132: 17–21.

Mayee CD, Datar VV (1988) Diseases of groundnut in the tropics. Rev Trop Plant Pathol 5: 169–198.

Panchbhavi KS, Nethradani CR (1987) Yield of groundnut as affected by varying larval density of *Spodoptera litura* (Fabricius) (Lepidoptera: Noctuidae). Indian J Agric Sci 57: 525–527.

Pandey MK, Monyo E, Ozias-Akins P, Liang X, Guimarães P, Nigam SN, Upadhyaya HD, Janila P, Zhang X, Guo B, Cook DR, Bertioli DJ, Michelmore R, Varshney RK (2012) Advances in *Arachis* genomics for peanut improvement. Biotechnol Adv 30: 639–651.

Pravst I, Katja Z, Janko Z (2010) Coenzyme Q10 contents in foods and fortification strategies. Crit Rev Food Sci Nutr 50: 269–280.

Prineas RJ, Kushi LH, Folsom AR, Bostick RM (1993) Letter to the editor. New England J Med 329–359.

Reddy DVR (1998) Control measures for the economically important peanut viruses. In: Hadidi A, Khetarpal RK, Koganezawo A (eds) Plant Virus Disease Control. APS Press, American Phytopathological Society, St. Paul, Minnesota, USA, pp 541–546.

Reddy AS, Reddy LJ, Mallikarjuna N, Abdurahman MD, Reddy YV, Bramel PJ, Reddy DVR (2000) Identification of resistance to peanut bud necrosis virus (PBNV) in wild *Arachis* germplasm. Ann Appl Biol 37: 135–139.

Sanders TH, McMichael Jr RW, Keith HW (2000) Occurrence of resveratrol in edible peanuts. J Agric Food Chem 48: 1243–1246.

Sanford JC (1983) Ploidy manipulations. In: Moore JN, Janick J (eds) Methods in Fruit Breeding. Purdue University Press, West Lafayette, IN, USA, pp 100–123.

Savage GP, Keenan JI (1994) The composition and nutritive value of groundnut kernels. In: Smartt J (ed) The Groundnut Crop: A Scientific Basis for Improvement. Chapman and Hall, London, UK, pp 173–213.

Seijo JG, Lavia GI, Fernández A, Krapovickas A, Ducasse DA, Bertioli DJ, Moscone EA (2007) Genomic relationships between the cultivated peanut (*Arachis hypogaea*, Leguminosae) and its close relatives revealed by double GISH. Am J Bot 94: 1963–1971.

Shirasawa K, Bertioli DJ, Varshney RK, Moretzsohn MC, Leal-Bertioli SCM, Thudi M, Pandey MK, Rami J-F, Foncéka D, Gowda MVC, Qin H, Guo B, Hong Y, Liang X, Hirakawa H, Tabata S, Isobe S (2013) Integrated consensus map of cultivated peanut and wild relatives reveals structures of the A and B genomes of *Arachis* and divergence of the legume genomes. DNA Res pp 1–12, doi:10.1093/dnares/dss042.

Stevenson PC, Anderson JC, Blaney WM, Simmonds MJS (1993a) Developmental inhibition of *Spodoptera litura* (Fab.) larvae by a novel caffeoylquinic acid from the wild groundnut *Arachis paraguariensis* (Chod et Hassl.). J Chem Ecol 19: 2917–2933.

Stevenson PC, Blaney WM, Simmonds MJS, Wightman JA (1993b) The identification and characterization of resistance in wild species of *Arachis* to *Spodoptera litura* (Lepidoptera: Noctuidae). Bull Entomol Res 83: 421–429.

Subrahmanyam P, Moss JP, McDonald D, Rao PVS, Rao VR (1985) Resistance to leaf spot caused by *Cercosporidium personatum* in wild *Arachis* species. Disease 69: 951–954.

Subrahmanyam P, Rao VR, Mcdonald D, Moss JP, Gibbons RW (1989) Origins of resistance to rust and late leaf spot in peanut (*Arachis hypogaea* Fabaceae). Econ Bot 43: 444–455.

Varshney RK, Hoisington DA, Nayak SN, Graner A (2009b) Molecular plant breeding: Methodology and achievements. In: Daryl J Somers, Peter Langridge, J Perry Gustafson (eds) Methods in Molecular Biology, Plant Genomics, vol 513. Humana Press, a part of Springer Science and Business Media, LLC DOI: 10.1007/978-1-59745-427-8_15.

Varshney RK, Bertioli DJ, Moretzsohn MC, Vadez V, Krishnamurthy L, Aruna R, Nigam SN, Moss BJ, Seetha K, Ravi K, He G, Knapp SJ, Hoisington DA (2009b) The first SSR-based genetic linkage map for cultivated groundnut (*Arachis hypogaea* L.). Theor Appl Genet 118: 729–739.

Varshney RK, Mohan SM, Gaur PM, Gangarao NVPR, Pandey MK, Bohra A, Sawargaonkar S, Kimurto PK, Janila P, Saxena KB, Fikre A, Sharma M, Pratap A, Tripathi S, Datta S, Chaturvedi SK, Anuradha G, Babbar A, Chaudhary RG, Mhase MB, Bharadwaj CH, Mannur DM, Harer PN, Guo B, Liang X, Nadarajan N, Gowda CLL (2013) Achievements and prospects of genomics-assisted breeding in three legume crops of the semi-arid tropics. Biotechnol Adv http://dx.doi.org/10.1016/j.biotechadv.2013.01.001.

Vavilov NI (1951) Phytogeographic basis of plant breeding. The origin, variation, immunity and breeding of cultivated plants. Chron Bot 13: 1–366.

Whitley ML, Isleib TG, Hendrix KW, Sanders TH, Dean LO (2011) Environmental and varietal effects on niacin content of raw and roasted peanuts. Peanut Sci 38: 20–25.

Xue HQ, Isleib TG, Stalker HT, Payne GA, Obrian G (2005) Evaluation of *Arachis* species and interspecific tetraploid lines for resistance to aflatoxin production by *Aspergillus flavus*. Peanut Sci 31: 134–141.

2

Genetic Resources, Diversity and Association Mapping in Peanut

Hari D Upadhyaya, Shivali Sharma* and *Sangam L Dwivedi*

ABSTRACT

Globally, peanut is an important crop, providing both oil and protein, and genebanks across the world have conserved a large collection of peanut germplasm including wild *Arachis* species. The key to the success in crop improvement depends on how effectively and efficiently the new genetic variation is introduced to broaden the genetic base of cultigens. The genus *Arachis* harbors considerable diversity for morpho-agronomic traits including resistance to abiotic and biotic stresses. Impressive progress have been made towards developing a large number of markers specific to peanut in addition to the technological breakthrough in developing high-throughput genotyping platforms for unlocking the genetic variation present in the germplasm collections. Using core and mini core collections and genomic tools, peanut researchers have identified a number of diverse germplasm possessing agronomically beneficial traits that are now being used in peanut breeding. Amphidiploids originating from distant wild *Arachis* species crosses are expected to unravel the variation not earlier available to peanut research community due to bottlenecks associated with peanut domestication.

Keywords: peanut, *Arachis*, germplasm, diversity, abiotic/biotic stresses, mapping

International Crops Research Institute for the Semi-Arid Tropics (ICRISAT), Patancheru 502324, AP, India.
*Corresponding author: *h.upadhyaya@cgiar.org*

2.1 Introduction

Peanut (groundnut) (*Arachis hypogaea* L.), an important food legume crop, grown in more than 108 countries representing tropical, subtropical and warm temperate regions of the world, extending cultivation between 40°N to 40°S. It ranks 6th among the oilseed crops and 13th among food crops of the world. Globally it is cultivated on 24.1 million ha area, with a total production of 37.6 million tons and average productivity of 1.6 t ha^{-1} (FAO 2010: http://www.faostat.fao.org). Developing countries account for 96% of the global peanut area and 92% of the global production. Asia accounts for 64% of the global peanut production, while Africa only 28%. In Asia, India and China together produce 57% of the global peanut production. Other countries in Asia that produce substantial peanut include Indonesia, Myanmar and Vietnam. Peanut in Africa is widely distributed, grown over 50 countries across the continent. Nigeria, Sudan, Senegal, Chad and Ghana together account for 54% of the peanut production in the continent, with Nigeria the leading producer. North America contributes ~5% of the global peanut production, 80% of which is in the USA, the world's fourth-largest producer. Europe and Oceania contribute less than 1% peanut production.

Peanut seeds, rich in oil, protein, minerals and vitamins, are consumed in a variety of forms. About two-thirds of global peanut production is crushed for extracting vegetable oil, while the remaining is used in the form of edible products and seeds. Peanut cake obtained after oil extraction is used as protein-rich meal for livestock or for making other food products. The haulms are an important source of good quality animal fodder, especially in developing countries. Peanut helps improve soil fertility through biological nitrogen fixation

Peanut production is adversely affected by both biotic and abiotic stresses. Rust, early leaf spot and late leaf spot are the most common and widely distributed foliar diseases, while peanut bud necrosis virus in South Asia, rosette disease in Africa and bacterial wilt in Southeast Asia are the major diseases of peanut, impacting yield and quality. The pests are of localized importance, for example, leaf miner and *Spodoptera* in South and Southeast Asia, termites in Africa and corn earworm, lesser corn stock borer and southern corn rootworm in North America. Drought is one of the major abiotic stresses, potentially limiting the peanut productivity worldwide. Peanut quality is adversely affected by aflatoxin contamination. All these factors either alone or in combination cause substantial yield losses worldwide, which necessitates the utilization of host-plant resistance to ameliorate losses to peanut production caused by biotic and abiotic stresses (Dwivedi et al. 2003 and references therein).

Legumes, including peanut, has a narrow genetic base, particularly due to bottleneck associated with their evolution. Various studies have shown that the cultivated peanut originated by a single hybridization event between two wild diploid species with distinct genome giving rise to a sterile hybrid followed by a spontaneous duplication of chromosomes producing fertile tetraploid (peanut) that remain reproductively isolated from its wild ancestors (Kochert et al. 1991; Jung et al. 2003; Seijo et al. 2004). Both pre- and post-zygotic hybridization barriers have been shown to restrict crossing between cultivated peanut and wild *Arachis* species (Halward and Stalker 1987). Crop genetic resources are the reservoir of many useful genes but general reluctance of the breeders to use exotic germplasm has severely restricted the introgression of useful variation present in exotic germplasm including wild *Arachis* species. The main reason for such low use of germplasm is due to the difficulties in evaluating large sets of germplasm across multilocations to get reliable information about the traits of economic importance. This chapter provides information about the nature and extent of peanut genetic resources preserved across genebanks globally, the pattern of diversity unearthed in cultivated and wild *Arachis* species and various approaches including genomic tools to promote utilization of genetic resources to broaden the genetic base for sustainable peanut production.

2.2 Origin, Dissemination and Gene Pools

The genus *Arachis* contains nine sections comprising 80 species. Most of these species are diploid with $2n = 2x = 20$ and $2n = 2x = 18$ (*A. praecox, A. palustris* and *A. decora* in section *Arachis* and *A. porphyrocalyx* in section *Erectoides*) except *A. pseudovillosa, A. glabrata* and *A. nitida* in section *Rhizomatosae* and *A. hypogaea* (cultivated peanut) and *A. monticola* in section *Arachis,* which are tetraploid with $2n = 4x = 40$ (reviewed in Upadhyaya et al. 2011a). The genus *Arachis* originated in South America, where it is widely distributed, mainly in Argentina, Brazil, Paraguay and Uruguay. The cultivated *A. hypogaea* probably originated in the region of southern Bolivia and northern Argentina, since its progenitor *A. monticola,* the only wild allotetraploid species that crosses with *A. hypogaea* is found in this area (Krapovickas 1969). The diploid species *A. duranensis* and *A. ipaensis,* the most likely donors of A and B genomes, are restricted to northwest Argentina and southeast Bolivia (Krapovickas and Gregory 1994, 2007) that overlap to the segmental allotetraploid, *A. monticola*/*A. hypogaea* and *A. monticola* (Seijo et al. 2004). This, together with evidence on archeological and morphological diversity, indicate that this region may be the center of origin and the primary center of diversity for *A. hypogaea* (Krapovickas and Rigoni 1957; Hammons 1994; Singh and Simpson 1994; Kochert et al. 1996).

Archeological evidences suggest that peanut has been cultivated for over 3,500 years. Domestication probably first took place in northern Argentina and southern Bolivia and was subsequently introduced to Africa, India and the Far East by the Portuguese and from the west coast of South America to the western pacific to Indonesia and China by the Spaniards in the early 16th century; and later on from Asian countries to east Africa. By the middle of the 16th century, peanut was introduced to North America and other parts of the world. Subsequent spread of the crop to different agroclimatic zones brought further diversification and variability in growth habit and seed and pod characteristics, which resulted into evolution of a number of morphologically distinct botanical varieties that predominate and show high levels of diversity in some geographical regions (Singh 1995; Singh and Nigam 1997).

The four gene pools in genus *Arachis* include i) primary gene pool (*A. hypogaea* landraces and its wild form *A. monticola*), ii) secondary gene pool (diploid species from section *Arachis* that are cross-compatible with *A. hypogaea*), iii) tertiary gene pool (species of section *Procumbentes*, weakly cross-compatible with *A. hypogaea*), and iv) the remaining *Arachis* species from other seven sections, not easily cross-compatible with *A. hypogaea* (Singh and Simpson 1994).

2.3 Conserving *Arachis* Species Diversity

Crop genetic diversity is threatened by several factors such as replacement of traditional varieties and landraces with genetically uniform high yielding cultivars, changes in dietary habits, habitat loss, natural calamities, land and crop conversion, introduction of exotic crops, environmental pollution and above all global warming. These genetically uniform modern cultivars could become vulnerable to new pests and diseases resulting into epidemics as have been seen in the past (Tatum 1971). Such experiences necessitate the use of diverse sources in plant breeding programs with a view to broaden the genetic base of crop cultigens. The landraces, exotic germplasm and wild relatives are the repository of many useful genes/alleles and can be utilized in breeding programs to develop new high yielding climate resilient cultivars.

The CGIAR comprising 11 genebanks, represents the largest concerted effort toward collecting, preserving and utilizing global agricultural resources and holds nearly 7,60,000 samples of the estimated 7.4 million accessions of different crops preserved globally (FAO 2009). There are a number of germplasm banks, which are conserving the peanut germplasm worldwide. International Crops Research Institute for the Semi-Arid Tropics (ICRISAT), Patancheru, India has the global responsibility of collecting, conserving and distributing the peanut germplasm comprising landraces, modern cultivars, genetic stocks, mutants and wild *Arachis* species. It contains 14,968 accessions of cultivated peanut and 477 accessions of 48 wild *Arachis* species representing eight sections from 92 countries. These accessions came from donations as well as from collecting missions launched in different countries. Other major genebanks holding peanut germplasm include the National Bureau of Plant Genetic Resources, New Delhi (14,593 accessions), India; the Directorate of Groundnut Research, Junagadh (8,934 accessions), India; the United States Department of Agriculture (9,964 accessions), USA; the Instituto Nacional de Tecnologia Agropecuaria, Argentina (8,347 accessions); the Institute of Crop Germplasm Resources, Beijing (6,565 accessions), China; Institute of Oil Crops Research, Wuhan (5,688 accessions), China; Crop Science Department, North Carolina State University, Raleigh, USA (3,788 accessions) and Estación Experimental Agropecuaria Manfredi, EEA INTA Manfredi, Argentina (2,158 accessions) (http://apps3.fao.org). Other genebanks specialized in conservation and maintenance of wild *Arachis* species are Texas A&M University, USA and Centro de Investigaciones de Nataima, Instituto Colombiano Agropecuario, Colombia (http://apps3.fao.org).

2.4 Assessing Species Diversity

2.4.1 Wild Relatives

Besides resistance to biotic and abiotic stresses (Dwivedi et al. 2003, 2008), wild *Arachis* species also harbor novel sources of variation for morpho-agronomic and nutritional traits, which can be used to enhance the genetic base of cultivated peanut. Stalker (1990) reported substantial variation in leaflet size and shape, branching habit and flower size among 73 wild *Arachis* species accessions, while Singh et al. (1996) found that the height of the main axis, length of apical leaflet on the main stem, length of isthmus between pods, seed width, and reaction to rust accounted for the greater part of variation amongst 42 *A. duranensis* accessions. The *A. pintoi* accessions from section *Caulorrhizae* also showed greater variability for most of the morphological traits (Carvalho and Quesenberry 2009). Nautiyal et al. (2008) based on relative injury identified NRCG 11824 (*A. glabrata*) as heat tolerant and NRCG 12042 (*A. paraguariensis*) as cold tolerant, while NRCG 11786 (*A. appresipila*) susceptible both to heat and cold. Their study indicated that the plants with thicker leaves were better protected from heat injuries and epi-cuticular wax load helped in maintaining stomatal regulation and leaf water relations, thus enabling these species to thrive under water-limited environments. More recently, Upadhyaya et al. (2011b) evaluated the largest collections of wild *Arachis* species (269 accessions) from 20 wild *Arachis* species representing six sections for 41 morpho-agronomic traits and 89 accessions for nutritional traits. The species accessions showed large variability for days to flowering, pod and seed characteristics, specific leaf area (SLA) and for SPAD chlorophyll meter reading (SCMR). For example, *A. pusilla* accessions, ICG 14898 and ICG 14906 flowered in 13–14 days, the earliest flowering accessions among the 20 wild *Arachis* species studied and were a week earlier than the earliest flowering cultivated groundnut germplasm, Chico. Further, *A. duranensis* had maximum intraspecific variability for 23 of the 41 traits. The other species with desirable traits was *A. villosa* (high SCMR at 60 and 80 days after sowing). The best wild *Arachis* accessions, possessing one to five desirable agronomic, nutritional and drought related traits identified in this study were ICG 8144 (*A. villosa*) high in SCMR, low SLA, high sugar content; ICG 13223, 13244, 14868, 14872, 14874, 14884 (*A. stenosperma*) superior in pod length and width and/or seed length and width; ICG 13211 (*A. pusilla*) earliest to flower; ICG 13178 (*A. monticola*) and ICG 13189 (*A. duranensis*) high in sugar; ICG 15142 (*A. pusilla*) and ICG 13227 (*A. dardani*) high in protein), which may be exploited to broaden the genetic base of cultivated peanut (Upadhyaya et al. 2011b).

2.4.2 *Cultigens*

2.4.2.1 Geographical Diversity

Geographical pattern of diversity, involving 13,342 accessions from 92 countries and 14 regions conserved in ICRISAT genebank, revealed that South America (where primary and seven secondary centers of diversity are located) and China (an important center of diversity) are under-represented, which necessitates the exploration and collection of germplasm from these regions (Upadhyaya et al. 2002a). This study further revealed that though South America is under-represented in terms of number of accessions (10.06% of total accessions) but contained adequate diversity for morphological and agronomic traits. Further, the principal component analysis grouped the 14 regions into three clusters—accessions from North America, Middle East, and East Asia in the first cluster; South America in the second cluster; and West Africa, Europe, Central Africa, South Asia, Oceania, South Africa, East Africa, Southeast Asia, Central Asia, and Caribbean in the third cluster.

2.4.2.2 Biological Diversity

Several studies in the past involving 22 to 125 genotypes were used to quantify variability for morpho-agronomic traits (Vaddoria and Patel 1990; Reddy and Gupta 1992; Pathirana 1993; Senapati and Roy 1998; Singh and Chaubey 2003; John et al. 2005; Kotzamanidis et al. 2006; Sumathi and

Muralidharan 2007; Korat et al. 2009; Sumathi et al. 2009; Sadeghi et al. 2011; Vekariya et al. 2011; Nautiyal et al. 2012).

Upadhyaya (2003) used morpho-agronomic traits to study phenotypic diversity in peanut core collection consisting of 1,704 accessions of which 910 belong to subsp. *fastigiata* (var. *fastigiata, vulgaris, aequatoriana, peruviana*) and 794 to subsp. *hypogaea* (var. *hypogaea, hirsuta*). The two groups, subsp. *fastigiata* and *hypogaea*, differed significantly for most of the traits with the *hypogaea* accessions having significantly greater mean pod length, pod width, seed length, seed width, yield per plant and 100-seed weight. The *fastigiata* accessions had higher plant height, leaflet length, leaflet width and shelling percentage. He detected maximum diversity between ICG 13479 and ICG 8422 in the *fastigiata* group and between ICG 13723 and ICG 20016 in the *hypogaea* group. Further, he found that 12–15 morpho-agronomic traits explained most of the phenotypic variability. The ICRISAT peanut mini core (184 accessions) also showed sufficient variability for most of the morphological traits (Madhura et al. 2011). Swamy et al. (2003) evaluated Asia-specific peanut core collection (504 accessions, Upadhyaya et al. 2001) for 20 agronomic traits for two seasons that detected sufficient variability (except pod yield) explaining multivariate polymorphism.

2.4.2.3 Trait Diversity

2.4.2.3.1 Early maturity: Most breeding programs aim at developing early-maturing cultivars that matches with the available crop duration. Appropriate time to flowering is a major component of crop adaptation, particularly in the environments where the growing season is restricted by terminal drought and high temperature. In most breeding programs, Chico has been used as the source of early maturity, which resulted in a narrow genetic base of peanut cultivars of this early maturing source. In a study involving sources of early maturity germplasm revealed that 21 such germplasm had similar maturity as of Chico but produced ~12 and ~8% greater pod yield at 75 and 90 days harvest including the control cultivars such as Gangapuri and JL 24 (Upadhyaya et al. 2006). These new sources of early maturity grouped into three distinct clusters—cluster I comprises of ICG 9930, ICG 4558, Gangapuri and Chico, cluster II 12 landraces and JL 24, while cluster III had seven landraces.

2.4.2.3.2 Yield and component traits: Evaluation of a peanut core collection for Asia across multilocations resulted in identification of 15 *fastigiata*, 20 *vulgaris*, and 25 *hypogaea* type peanut accessions for pod yield, total pods, shelling percentage, 100-seed weight and oil content. Clustering of these accessions together with controls resulted in four clusters in *fastigiata* and three clusters each in *vulgaris* and *hypogaea*. The control cultivars, Gangapuri in *fastigiata*, ICGS 44 in *vulgaris* and ICGS 76 and S 230 in *hypogaea* clustered separately indicating that the selected accessions were diverse from the control cultivars (Upadhyaya et al. 2005).

2.4.2.3.3 Drought tolerance: Upadhyaya (2005) evaluated peanut mini core collection for SLA and SCMR, associated with drought tolerance. The five and 13 most promising *vulgaris* and *hypogaea* accessions identified in this study clustered into four groups, all *vulgaris* types (ICG 118, ICG 14985, ICG 2106, ICG 5236 and ICG 6654) clustering with controls (Gangapuri, ICGS 44 and ICGS 76) in Cluster I, most of the *hypogaea* accessions including control cultivar M 13 (*hypogaea*) in cluster II, while ICG 6766 and ICG 14523 forming separate clusters, indicating that these two are diverse and can be used in breeding to enhance drought tolerance in peanut cultivars. Ravindra et al. (1990) reported that GG 2 in comparison to J 11, JL 24 and TMV 10 has an inherent ability to produce more pods under drought stress occurring at any growth stage, while Ratnakumar and Vadez (2011) suggested that genotypes with lower leaf area may use water more sparingly under intermittent drought stress, which will have less damaging consequences for reproductive and pod development than genotypes having larger leaf area.

2.4.2.3.4 Cold tolerance: Upadhyaya et al. (2009) studied the phenotypic diversity in 158 cold tolerant peanut germplasm, which germinated at 12°C and had substantial diversity for agronomic traits. The clustering pattern grouped the cold tolerant accessions into four clusters. The accessions in these four clusters differed in mean, variance and range for agronomic traits. The cold tolerant accessions were superior to control cultivars for several agronomic traits indicating the potential of these accessions in developing genetically diverse cold tolerant peanut cultivars.

2.4.2.3.5 Salinity: Srivastava (2010) evaluated 275 accessions representing mini core, high yielding breeding lines and landraces from salinity prone areas and identified ICG 5195, ICG 442, ICG 7283, ICG 1711, ICG 2106 and ICG 1519 as good sources of salinity tolerance.

2.4.2.3.6 Resistance to diseases: ICG 11426, ICG 13787 and ICG 8760 were identified as resistant to rust and late leaf spot, and ICG 14985, ICG 3673, ICG 6025, ICG 12625, ICG 13787 and ICG 8760 for *Aspergillus flavus*. Of these, ICG 13787 and ICG 8760 were resistant to all the three diseases (Kusuma et al. 2007); ICG 875, ICG 928, ICG 1668, and ICG 14466 resistant to the bud necrosis disease (Ahmed 2008). In another study, accessions from ICRISAT peanut mini core were found more resistant to seed colonization by

Aspergillus flavus and aflatoxin production from ICRISAT peanut mini core than those from Chinese peanut mini core, with ICG 6813, ICG 12370, ICG 4750, ICG 4156, ICG 12625, ICG 12697, ICG 14482 combining resistance to both seed invasion and aflatoxin production (Jiang et al. 2010a). Molecular profiling study further revealed that ICG 12625 (resistance to aflatoxin production) and ICG 4750 (resistance to seed invasion) were diverse from the rest of the accessions (Jiang et al. 2010a). ICG 36, ICG 118, ICG 1448, ICG 434, ICG 1415, ICG 5745, ICG 76, ICG 1668, ICG 14710, ICG 6057, ICG 6201, ICG 1455, ICG 397 and ICG 7633 were identified as resistance to bacterial wilt (personal communication). US researchers identified a number of accessions tolerant to root-knot nematode, early leaf spot, pepper spot, tomato spotted wilt virus and soil borne fungal diseases, including pre-harvest aflatoxin contamination (Isleib et al. 1995; Anderson et al. 1996; Holbrook et al. 1998, 2000; Franke et al. 1999; Damicone et al. 2010; Chamberlin et al. 2010).

2.4.2.3.7 Seed quality traits: Upadhyaya et al. (2012) found sufficient variation for protein, oil and fatty acid composition including oleic (O), linoleic (L) fatty acids and O/L ratio in peanut mini core collection. Subsp. *fastigiata* as a group has shown relatively high variation for protein, while subsp. *hypogaea* for high O/L ratio. They identified accessions with high protein and oil, and better O/L ratio. Cluster analysis delineated these accessions into three clusters: cluster 1 those with high oleic acid, high pod yield, and high 100-seed weight, cluster 2 with those having high O/L and early flowering, while cluster 3 had accessions with high protein and high shelling percentage.

2.5 Genetic and Genomic Resources to Promote Utilization of Germplasm in Breeding

2.5.1 Core/mini core Subsets to Identifying New Sources of Variation

The low use of germplasm accessions in breeding programs is mainly due to the lack of reliable information on traits of economic importance such as yield, resistance to biotic and abiotic stresses and quality traits, which often show high genotype x environment interactions that require replicated multilocational evaluations. This is a costly and resource-demanding task owing to the large size of the germplasm collections. Thus, the collection needs to be sampled to get the size of the collections to a manageable level for meaningful evaluation. Frankel (1984) coined the term "core collection" to sample representative variability from the entire collection. A core collection contains 10% of the accessions from the entire collection that captures most of the available diversity in the species (Brown 1989a).

Frankel and Brown (1984) suggested that greater use of germplasm in crop improvement is possible if a small collection representing diversity is made available for characterization and utilization. Thus, core collection has a reduced size containing a diverse set of germplasm that represents the entire collection. Such a core collection can be evaluated extensively and the information derived could be used to guide more efficient utilization of the entire collection (Brown 1989b).

A number of reduced subsets in the form of core or mini core collections (Upadhyaya and Ortiz 2001) have been reported in peanut (Table 2-1), which researchers at ICRISAT and national programs have used to identify new sources of variation for important agronomic and nutritional traits (Table 2-2) or resistance to various biotic and abiotic stresses (Table 2-3). For example, 21 accessions combining early maturity with high yield (Upadhyaya et al. 2006), 60 accessions having greater pod yield, shelling percentage, 100-seed weight and oil content (Upadhyaya et al. 2005), or 12 accessions with 100-seed weight ≥60 g (Upadhyaya et al. 2010), Likewise, a number of accessions with high oil and protein contents or accessions with O/L ratio greater than 3.0 have been identified (Upadhyaya et al. 2012). Furthermore, accessions tolerant to drought (Upadhyaya 2005), low temperature (Upadhyaya et al. 2009), salinity (Srivastava 2010), aflatoxin (Kusuma et al. 2007; Jiang et al. 2010a) or resistant to bud necrosis disease (Ahmed 2008) have been reported. Few of these accessions have also shown

Table 2-1 Core and mini core collections as reported in peanut.

Reduced subset	# accessions used in forming reduced subset	# traits used in forming reduced subset	# accessions in constituted subset	% of accessions in reduced subset representing entire collection	Reference
US Valencia core	630	26	77	12.22	Dwivedi et al. 2008
US core	7,432	6	831	11.18	Holbrook et al. 1993
Chinese core	6,390	15	576	9.01	Jiang et al. 2008
Asian core	4738	15	504	10.64	Upadhyaya et al. 2001
Global core	14,310	14	1,704	11.91	Upadhyaya et al. 2003
USA mini core	831	16	111	13.36	Holbrook and Dong 2005
ICRISAT mini core	1,704	31	184	1.28	Upadhyaya et al. 2002b

Table 2-2 Promising germplasm accessions identified for agronomic and nutritional traits.

Traits	Few promising germplasm	Reference
Early maturity	ICG# 4558, 4890, 9930, and 11605 having early maturity and 3–4 seeds per pod	Upadhyaya et al. 2006
Large seed size	ICG# 2381, 5016, 5051, 5745, 5662, 6057, 6766, 8760, 11219, 11855, 11862, and 14482	Upadhyaya et al. 2010
Yield and component traits	60 accessions: ICG# 4, 29, 3443, 14161, 11188, 7140, 2918 and others	Upadhyaya et al. 2005
Protein content (>30%)	5 accessions: ICG# 36, 5779, 3421, 3584, and 2019	Upadhyaya et al. 2012
Oil content (>50%)	ICG 442	Upadhyaya et al. 2012
Oleic acid (≥60%)	6 accessions: ICG# 2381, 10185, 15419, 12276, 7243, and 11088	Upadhyaya et al. 2012
O/L ratio (>3.0)	12 accessions: ICG# 2381 (O/L ratio of 7.0), 10185, 6022, 1274, 7243, 6766, 12625, 12276, 15419, PI 274193, PI 290594, PI 468271	Upadhyaya et al. 2012; Dean et al. 2009

Table 2-3 Promising germplasm accessions having tolerance/resistance to abiotic/biotic stresses.

Traits	Promising accessions	Reference
Drought	18 accessions: ICG# 14523, 6766, 7243, 862, 6654, 14985 and others	Upadhyaya 2005
	30 accessions: ICG# 11088, 12697, 8751, 3140, 3584 and others	Hamidou et al. 2012
Low temperature	15 accessions with superior pod yield: ICG# 12625, 7898, 11130, 6148, 7013, 6022, 7905, 7884, 4992, 9515, 10915, 10567, 1710, 11088 and 10945	Upadhyaya et al. 2009
Salinity	6 accessions: ICG# 5195, 442, 7283, 1711, 2106, and 1519	Srivastava 2010
Rhizoctonia limb rot resistant	6 accessions: PI# 343398, 343361, 288178, 331326, 497351 and 274193	Franke et al. 1999
Late leaf spot	7 accessions: ICG# 12625, 11426, 12672, 13787, 14475, 2857, and 8760	Kusuma et al. 2007
Rust	5 accessions: ICG# 9809, 11088, 11426, 13787, and 8760	Kusuma et al. 2007
A. flavus	12 accessions: ICG# 14985, 3673, 6025, 12625, 13787, 8760, 6813, 12370, 4750, 4156, 12697, and 14482	Kusuma et al. 2007; Jiang et al. 2010a
Bud necrosis	4 accessions: ICG# 875, 928, 1668, and 14466	Ahmed 2008
Combined resistance to Sclerotinia blight, pepper spot and web blotch	5 accessions: PI# 274193, 497599, 458619, 468195, and 259796	Damicone et al. 2010
Sclerotinia blight resistant	39 accessions	Chamberlin et al. 2010

multiple resistances—ICG 11426, ICG 13787 and ICG 8760 resistant to late leaf spot and rust or ICG 13787 and ICG 8760 resistant to late leaf spot, rust and *A. flavus* (Kusuma et al. 2007).

2.5.2 Assessing Population Structure and Diversity in Germplasm

Of late, a number of publications have come out detailing the allelic richness and diversity amongst the cultivated peanut germplasm. Most of these studies reported on an average 3–15 alleles per locus (Table 2-4), with Barkley et al. (2007) and Kottapalli et al. (2011) detecting 13–15 alleles. Furthermore, a few of these studies clearly separated accessions based on botanical groups or accessions within species into different clusters, clearly indicating diversity among accessions (Ferguson et al. 2004; Moretzsohn et al. 2004; He et al. 2005; Mace et al. 2007; Kottapalli et al. 2011). For example, 10 simple sequence repeat (SSR) loci separated South American landraces from African and Asian landraces (Ferguson et al. 2004); *hypogaea* and *fastigiata* forming distinct clusters (Mace et al. 2007); Valencia accessions clustering into different groups (Kottapalli et al. 2011); SSR loci contributing more variation to rust and/or late leaf spot (Mace et al. 2006) and bacterial wilt (Mace et al. 2007) or SSR loci detecting more diversity in *fastigiata* than those of *hypogaea* accessions (Jiang et al. 2007); Chinese peanut mini core contributing more diversity than that of ICRISAT mini core with accessions L 2 Gangguo (a Chinese genotype) and ICG 12625 (an ICRISAT genotype) revealing the highest genetic dissimilarity (Jiang et al. 2010b). Likewise, the South American landraces showed high allelic diversity than those from Africa and Asia (Ferguson et al. 2004), while a new marker (Ah-041) differentiated AA-genome species accessions with those from non-AA genome species accessions (Moretzsohn et al. 2004). These examples clearly demonstrate that marker-based information provides breeders critical inputs to plan future breeding strategies in peanut.

2.5.3 Molecular Markers and Genetic Maps

Molecular markers are important for germplasm characterization, to assess variability for identifying genetically diverse traits-specific germplasm and marker-trait association in crop improvement programs. DNA-based markers provide a reliable means of estimating the genetic relationships between genotypes and taxonomic groups as compared to morphological markers. Restriction Fragment Length Polymorphism (RFLP), Random Apmplified Polymorphic DNA (RAPD) and Amplified Fragment Length Polymorphism (AFLP) markers detected little variation among *A. hypogaea* cultivars and germplasm lines whereas abundant polymorphism amongst

Table 2-4 Allelic richness and diversity, and grouping of accessions based on allelic diversity in peanut.

# genotypes	# SSR	Allelic richness	Summarized findings	Reference
22 accessions including wild species	6	38 fragments, averaged 6 fragments per primer	Effective in detecting molecular variation in cultivated peanut	Hopkins et al. 1999
24 accessions	56	79 alleles, averaged 4 alleles per locus	Highlighted a simple and reliable way in obtaining polymorphic SSR markers from cultivated peanut	He et al. 2003
188 landraces	10	89 alleles, averaged 7 alleles per locus	Discriminated South American landraces than those from Africa and Asia	Ferguson et al. 2004
60 accessions	8	74 alleles, averaged 8.4 alleles per locus	Highlighted the usefulness of SSR markers for genetic diversity analysis of cultivated peanut	Moretzsohn et al. 2004
60 cultivated and 36 wild species accessions	12	-	More diversity amongst the Brazilian germplasm; two major groups among *A. hypogaea* accessions; high marker transferability between *Arachis* species, Ah-041 showing 100% transferability across species	Moretzsohn et al. 2004
48 accessions representing six botanical varieties	38	-	Differentiated six botanical types (*fastigiata, vulgaris, hypogaea, hirsuta, peruviana* and *aequatoriana*) into separate groups	He et al. 2005
22 accessions	23	135 alleles, averaged 6 alleles per locus	Few SSRs differentiated rust and/or late leaf spot resistant germplasm	Mace et al. 2006
46 accessions representing six botanical types	32	107 alleles, averaged 3 alleles per locus	Two distinct groups corresponding to subspecies *hypogaea* and *fastigiata;* six alleles differentiated bacterial wilt resistant and susceptible accessions	Mace et al. 2007
31 bacterial wilt resistant accessions	SSR: 78; AFLP 126	91 loci (3.14 alleles per locus) with SSR and 72 loci (2.25 alleles per locus) with AFLP markers	14H06, 7G02, 3A8, 16C6, and P1M62 were effective in detecting polymorphism	Jiang et al. 2007
US peanut mini core and wild relatives	31	477 alleles, averaged 15.4 alleles per locus	13 transferable markers across all wild relative accessions	Barkley et al. 2007
114 Valencia peanut accessions	52	683 alleles, averaged 13 alleles per locus	Differentiated Valencia germplasm into five clusters with two distinct major groups	Kottapalli et al. 2011
466 accessions	26	-	Greater diversity among the Chinese mini core accessions; L 2 Gangguo and ICG 12625 the most genetically diverse; among six botanical types, accessions of *fastigiata* and *hypogaea* were more diverse than other types	Jiang et al. 2010b

wild *Arachis* species (Halward et al. 1991; Halward et al. 1992; Paik-Ro et al. 1992; He and Prakash 1997; Subramanian et al. 2000; Gimenes et al. 2002).

The SSR (also known as microsatellites) and Single Nucleotide Polymorphism (SNP) markers are becoming important in molecular breeding of most crops including peanut because of their codominant nature, high polymorphism and transferability among related species. Concerted effort during the past decade has led to the development of >6,000 SSR markers and about >2,000 SNPs at the University of Georgia, USA (Pandey et al. 2012). The other marker system, a Diversity Array Technology (DArT) platform comprising of about 15,000 DArT clones has been developed at DArT Ptv. Ltd (Australia) in collaboration with ICRISAT (India), CIRAD (France), Catholic University of Brasília and EMBRAPA (Brazil). Use of DArT arrays with a range of genotypes representing diploid (AA, BB) and tetraploid (AABB) genome species showed low polymorphism in tetraploids but more diversity among accessions from diploid species (Kilian 2008; Varshney et al. 2010), which indicate that DArT markers may not be very useful in peanut breeding and genetics; however, more useful in monitoring genome introgression from diploid to cultivated peanut (Pandey et al. 2012).

Moretzsohn et al. (2005) constructed A-genome based genetic map in *Arachis* by using an F_2 population derived from a cross between two diploid species with AA genome (*A. duranensis* x *A. stenosperma*). This genetic map placed 170 SSR loci on 11 linkage groups (LGs) covering 1,231 cM, with an average distance of 7.24 cM. Further, Gobbi et al. (2006) developed B-genome based F_2 map by crossing *A. ipaensis* and *A. magna*, which mapped 130 SSR loci into 10 LGs. However, these maps have limited value to cultivated peanut, a tetraploid, and hence there is a need to develop tetraploid-based genetic maps in cultivated peanut. Varshney et al. (2009) were probably the first to construct SSR-based genetic linkage map for the cultivated peanut that mapped 135 SSR loci into 22 LGs, which was further saturated with the current map having 191 SSR loci into 20 LGs and a total map distance of 1,785 cM (Ravi et al. 2011). Subsequently, several other genetic maps for cultivated peanut have become available (reviewed in Pandey et al. 2012).

2.5.4 Markers Associated with Agronomically Beneficial Traits

Using genetic mapping, a number of markers/QTLs (Quantitative Trait Loci) associated with useful traits have been reported and used to introgress beneficial traits to cultivated peanut. Introgression of nematode-resistant gene from *A. cardenasii* into *A. hypogaea* was reported in 10 of 11 linkage groups (Garcia et al. 1995), which were used to develop nematode-resistant germplasm (Garcia et al. 1996). Two dominant genes conferring resistance to root-knot nematode, *Meloidogyne arenaria* race 1 were mapped using RAPD and sequence-characterized amplified region (SCAR) markers (Garcia et al. 1996). One marker, Z3/265, closely linked with *M. arenaria* resistance, was mapped to a linkage group on a backcross map in an area known to contain *A. cardenasii* introgression. This marker was cloned to make SCAR and RFLP probes, which further confirmed the linkage with nematode resistance. Subsequently, the RFLP markers linked to a locus for resistance to *M. arenaria* race 1 has been identified by various workers (Choi et al. 1999; Church et al. 2000; Seib et al. 2003) that provided a useful selection method for identifying resistance to the peanut root-knot nematode. Likewise, RAPD markers associated with nematode resistance in BC_4F_2 of the cross involving Florunner and TxAg-6 has been identified: *RKN410* and *RKN440* closely linked with each other identified a resistance gene derived from either *A. cardenasii* or *A. diogoi*, while. *RKN229*, inherited from *A. cardenasii* or *A. diogoi* was 9 cM away from this locus (Burow et al. 1996). Herselman et al. (2004) identified 20 putative AFLP markers associated with aphid vector of peanut rosette disease, of which, 12 mapped to five linkage groups covering a map distance of 139.4 cM, while Varma et al. (2005) reported two to seven SSR alleles associated with rust resistance in two F_2 populations. A few SSR markers associated with yield and yield contributing traits were also reported (Liang et al. 2009; Selvaraj et al. 2009).

Molecular mapping of drought tolerance traits identified 153 main-effect and 25 epistatic QTLs (Varshney et al. 2009; Ravi et al. 2011; Gautami et al. 2012). A major QTL each for leaf rust (Khedikar et al. 2010; Sujay et al. 2012) and late leaf spot (Sujay et al. 2012) has been identified for use in peanut breeding. Likewise, QTLs associated with peanut nutritional traits such as oil and protein contents have been identified (Sarvamangala et al. 2011). Diagnostic markers for resistance to nematode (Nagy et al. 2010), leaf rust (Khedikar et al. 2010), late leaf spot (Sujay et al. 2012) and high-oleate trait (Chu et al. 2009; Chen et al. 2010) are available for use in molecular breeding of peanut.

2.5.5 Association Mapping

The phenotypic variation of agronomically important traits is influenced by multiple QTLs, their interaction, the environment and the QTL x environment interactions. Association mapping, also known as Linkage Disequilibrium (LD) mapping, is a relatively new and promising tool for dissecting complex traits. Association mapping in comparison to the traditional linkage mapping has major advantages due to increased mapping resolution through exploitation of historical and evolutionary recombination events at the population level (Risch and Merikangas 1996; Nordborg and Tavare 2002). The prerequisites to perform association mapping include a dense genetic linkage map, passport information and phenotypic data, an understanding of population structure, and contrasting genotypes for beneficial traits (Kresovich et al. 2002). The marker-trait association approach relies on the assumption that an allele responsible for a phenotype and the associated flanking markers are inherited as a block, and therefore neutral marker-based selection will be predictive of allelic content at critical genes determining favorable phenotype. Such marker-trait associations in a collection of plant genetic resources would allow the assessment of the genetic potential of specific genotypes prior to phenotypic evaluation and identification of superior trait alleles in germplasm collection (Gebhardt et al. 2004).

Belamkar et al. (2010) used 32 highly-polymorphic SSRs to study population structure and LD in 96 peanut genotypes comprising 92 US peanut mini core accessions, the diploid progenitors *A. duranensis* (AA) and *A. ipaensis* (BB) and synthetic amphidiploid accession TxAG-6 and a widely grown US peanut cultivar, Florunner. The population structure revealed that the diploid progenitors and their synthetic amphidiploid grouped separately from most mini core accessions. UPGMA and model-based clustering divided the population into four subgroups, two major subgroups representing subspecies *fastigiata* and *hypogaea*, a third group containing mixed individuals, while the fourth containing diploid progenitors and TxAG-6. Unified mixed linear model analysis incorporating population structure and kinship identified several SSR loci associated with drought tolerance traits. This study revealed the importance of LD mapping in exploiting the natural variation present in cultivated peanut. Wang et al. (2011) studied the population structure and marker-trait association by genotyping 94 accessions with 81 SSRs and two functional SNPs from Fatty Acid Desaturase 2 (FAD2), which identified four major subpopulations, related to four botanical varieties. Candidate-gene association analysis verified that one functional SNP from the *FAD2A* gene is significantly associated with oleic acid (C18:1), linoleic acid (C18:2), and oleic-to-linoleic (O/L) ratio across this diverse collection.

2.5.6 Amphidiploids as Source of Agronomically Beneficial Traits

Utilization of wild *Arachis* species following interspecific hybridization has resulted in the development of many elite germplasm lines and cultivars with improved level of resistance to diseases and insect-pests (Dwivedi et al. 2008 and references cited therein). Varieties such as Spancross (Hammons 1970), Tamnut 74 (Simpson and Smith 1975), Coan (Simpson and Starr 2001), NemaTAM (Simpson et al. 2003), having a genetic base from wild *Arachis* species, were released for cultivation in the USA. Likewise, ICGV-SM 85048 and ICGV-SM 86715 have been released for cultivation in Mauritius (Nigam et al. 1998; Moss et al. 1998).

The development and utilization of synthetic amphidiploids such as TxAG-6 with high genetic variations (Simpson et al. 1993) in breeding programs has made possible the transfer of resistance genes from wild species into cultivated peanut. This amphidiploid has been synthesized using species that are not in the direct lineage of the cultigen. However, it is crossable with the cultivated peanut and produced fertile progenies thus proved useful for introducing genetic variability into the cultigen. Crosses involving TxAG-6 with cultivated peanut has resulted in the release of two cultivars (Coan and NemaTAM) carrying genes for root-knot nematode (*M. arenaria*) resistance from *A. cardenasii* (Simpson and Starr 2001, Simpson et al. 2003). TxAG-6 is a small-seeded (~7–8 g 100 seed weight) and low-yielding (2–5 g plant^{-1}), which when crossed with TMV 2 (32 g 100 seed weight) at ICRISAT produced progenies with cryptic introgression and showed much higher seed weight and yielded 23 to 68% more than TMV 2 (3,343 kg ha^{-1}). These progenies also out yielded by 10 to 50% the highest yielding control cultivar ICGV 91114 (3,741 kg ha^{-1}, 49 g 100-seed weight^{-1}) (Upadhyaya 2008). This demonstrated that the novel alleles of wild relatives, that were considered to be lost in evolution to cultivated types, could be used to enhance the trait value in peanut cultivar development. Encouraged with this, the researchers at ICRISAT have developed a number of amphidiploids that are being assessed for releasing novel variation for use in peanut breeding (Mallikarjuna et al. 2012).

2.6 Conclusions

Natural genetic variation and means to exploit such variability is the key to the success of crop improvement programs. Large collections of peanut germplasm including wild *Arachis* species have been preserved in genebanks worldwide, representing a large spectrum of diversity in the genus *Arachis*. Development and evaluation of small-sized subsets such as core and mini core have resulted in the identification of trait-specific germplasm accessions for agronomic traits including resistance to abiotic and biotic stresses and nutritional traits, which would result in the enhanced utilization of genetic resources to broaden the genetic base to face new challenges to peanut production. Considerable variability for some traits of interest exists in wild *Arachis* gene pools, which can be brought, using wide hybridization and applying novel tools, in crop cultigens for sustainable production of peanut globally. Several elite germplasm lines and the cultivars carrying resistances from wild *Arachis* species have been released for use as a resource in crop breeding or even for direct cultivation. More importantly, amphidiploids are now being developed using species that were earlier not easily crossable, and the work so far revealed that these amphiploids have the potential to release hidden variability that was locked due to bottlenecks associated with the origin of cultivated peanut, thus making available more variability to peanut research community. Unlike in the 80s and 90s, the availability of a large number of PCR-based markers (SSRs and SNPs), high-throughput genotyping platforms and bioinformatics resources have enabling effects towards identifying and tracking allelic variants associated with beneficial traits and identifying desirable recombinant plants with the traits of interest, thus accelerating molecular breeding in peanut improvement.

References

Ahmed K (2008) Evaluation of groundnut (*Arachis hypogaea* L.) germplasm (Mini core collection). MSc Thesis, University of Agricultural Sciences, Dharwad, Bangalore, India.

Anderson WF, Holbrook CC, Culbreath AK (1996) Screening the peanut core collection for resistance to tomato spotted wilt virus. Peanut Sci 23: 57–61.

Barkley NA, Dean RE, Pittman RN, Wang ML, Holbrook CC, Pederson GA (2007) Genetic diversity of cultivated and wild-type peanuts evaluated with M13-tailed SSR markers and sequencing. Genet Res Camb 89: 93–106.

Belamkar V, Selvaraj MG, Ayers JL, Payton PR, Puppala N, Burow MD (2010) A first insight into population structure and linkage disequilibrium in peanut, and association mapping of drought tolerance-related traits in the US peanut minicore collection. In: Proceedings of Plant and Animal Genomes XVIII Conference, 9–13 January 2010, San Diego, USA, p 423.

Brown AHD (1989a) Core collections: a practical approach to genetic resources management. Genome 31: 818–824.

Brown AHD (1989b) The case for core collections. In: Brown AHD, Frankel OH, Marshall DR, Williams JT (eds) The Use of Plant Genetic Resources. Cambridge University Press, Cambridge, UK, pp 136–155.

Burow MD, Simpson CE, Paterson AH, Starr JL (1996) Identification of peanut (*Arachis hypogaea* L.) RAPD markers diagnostic of root-knot nematode (*Meloidogyne arenaria* (Neal) Chitwood) resistance. Mol Breed 2: 369–379.

Carvalho MA, Quesenberry KH (2009) Morphological characterization of the USA *Arachis pintoi* Krap. and Greg. collection. Plant Syst Evol 277: 1–11.

Chamberlin KDC, Melouk HA, Payton ME (2010) Evaluation of the US peanut mini core collection using a molecular marker for resistance to *Sclerotinia minor* Jagger. Euphytica 172: 109–115.

Chen CY, Guo B, Qin H, Feng S, Dang P, Culbreadth A, Wang ML, Holbrook C (2010) Morphological and molecular marker characterization of two RIL mapping populations in cultivated peanuts. In: Proceedings of Plant and Animal Genome XIX Conference, 9–13 January 2010, San Diego, USA, p 422.

Choi K, Burow MD, Church G, Burow G, Paterson AH, Simpson CE, Starr JL (1999) Genetics and mechanism of resistance to *Meloidogyne arenaria* in peanut germplasm. J Nematol 31: 283–290.

Chu Y, Holbrook CC, Ozias-Akins P (2009) Two alleles of ahFAD2B control the high oleic acid trait in cultivated peanut. Crop Sci 49: 2029–2036.

Church GT, Simpson CE, Burow MD, Paterson AH, Starr JL (2000) Use of RFLP markers for identification of individuals homozygous for resistance to *Meloidogyne arenaria* in peanut. Nematology 2: 575–580.

Damicone JP, Holbrook CC, Smith DL, Melouk HA, Chamberlin KD (2010) Reaction of the core collection of peanut germplasm to *Sclerotinia* blight and pepper spot. Peanut Sci 37: 1–11.

Dean LL, Hendrix KW, Holbrook CC, Sanders TH (2009) Content of some nutrients in the core of the peanut germplasm collection. Peanut Sci 36: 104–120.

Dwivedi SL, Crouch JH, Nigam SN, Ferguson ME, Paterson AH (2003) Molecular Breeding of groundnut for enhanced productivity and food security in the semi-arid tropics: Opportunities and challenges. Adv Agron 80: 153–221.

Dwivedi SL, Upadhyaya HD, Blair MW, Bertioli DJ, Nielen S, Ortiz RO (2008) Enhancing crop gene pools with beneficial traits using wild relatives. Plant Breed Rev 30: 179–230.

FAO (2009) The state of *ex situ* conservation. In: The Second State of the World's Plant Genetic Resources for Food and Agriculture. FAO, Rome, Italy, pp 51–87.

FAO (2010) FAO (http://apps.fao.org/page/collections?subset=agriculture), Rome, Italy.

Ferguson ME, Burow MD, Schultze SR, Bramel PJ, Paterson AH, Kresovich S, Mitchell S (2004) Microsatellite identification and characterization in peanut (*A. hypogaea* L.). Theor Appl Genet 108: 1064–1070.

Franke MD, Bennemen TB, Holbrook CC (1999) Identification of resistance to *Rhizoctonia* limb rot in a core collection of peanut germplasm. Plant Dis 83: 944–948.

Frankel OH (1984) Genetic perspective of germplasm conservation. In: Arber W, Limensee K, Peacock WJ, Stralinger P (eds) Genetic Manipulations: Impact on Man and Society. Cambridge University Press, Cambridge, UK, pp 161–170.

Frankel OH, Brown AHD (1984) Plant genetic resources today: a critical appraisal. In: Holden JHW, Williams JT (eds) Crop Genetic Resources: Conservation and Evaluation. George Allen and Unwin, London, UK, pp 249–257.

Garcia GM, Stalker HT, Kochert G (1995) Introgression analysis of an interspecific hybrid population in peanuts (*Arachis hypogaea* L.) using RFLP and RAPD markers. Genome 38: 166–176.

Garcia GM, Stalker HT, Shroeder E, Kochert G (1996) Identification of RAPD, SCAR and RFLP markers tightly linked to nematode resistance genes introgressed from *Arachis cardenasii* into *Arachis hypogaea*. Genome 39: 836–845.

Gautami B, Pandey MK, Vadez V, Nigam SN, Ratnakumar P, Krishnamurthy L, Radhakrishnan T, Gowda MVC, Narasu ML, Hoisington DA, Knapp SJ, Varshney RK (2012) QTL analysis and consensus genetic map for drought tolerance traits based on three RIL populations of cultivated groundnut (*Arachis hypogaea* L.). Mol Breed 32: 757–772.

Gebhardt C, Ballvora A, Walkemeier B, Oberhagemann P, Schuler K (2004) Assessing genetic potential in germplasm collections of crop plants by marker-trait association: a case study for potatoes with quantitative variation of resistance to late blight and maturity type. Mol Breed 13: 93–102.

Gimenes MA, Lopes CR, Galgaro ML, Valls JFM, Kochert G (2002) RFLP analysis of genetic variation in species of section *Arachis*, genus *Arachis* (Leguminosae). Euphytica 123: 421–429.

Gobbi A, Teixeira C, Moretzsohn M, Guimaraes P, Leal-Bertioli D, Lopes CR, Gimenes M (2006) Development of a linkage map to species of B genome related to the peanut (*Arachis hypogaea*-AABB). In: Plant and Animal Genome XIV Conference, 14–18th January 2006, San Diego, CA, USA, p 679.

Halward T, Stalker HT (1987) Incompatibility mechanism in interspecific peanut hybrids. Crop Sci 27: 456–460.

Halward T, Stalker HT, Larue E, Kochert G (1991) Genetic variation detectable with molecular markers among unadapted germplasm resources of cultivated peanut and related wild species. Genome 34: 1013–1020.

Halward T, Stalker HT, Larue E, Kochert G (1992) Use of single-primer DNA amplification in genetic studies of peanut (*Arachis hypogaea* L.). Plant Mol Biol 18: 315–325.

Hamidou F, Ratnakumar P, Halilou O, Mponda O, Kapewa T, Monyo E, Faye I, Ntare BR, Nigam SN, Upadhyaya HD, Vadez V (2012) Selection of intermittent drought tolerant lines across years and locations in the reference collection of groundnut (*Arachis hypogaea* L.). Field Crops Res 126: 189–199.

Hammons RO (1970) Registration of Spancross peanuts (Reg. No. 3). Crop Sci 10: 459–460.

Hammons RO (1994) The origin and history of the groundnut. In: Smartt J (ed) The Groundnut Crop: A Scientific Basis for Improvement. Chapman Hall, London, UK, pp 24–39.
He G, Prakash CS (1997) Identification of polymorphic DNA markers in cultivated peanut (*Arachis hypogaea* L.). Euphytica 97: 143–149.
He G, Meng R, Newman M, Gao G, Pittman RN, Prakash CS (2003) Microsatellites as DNA markers in cultivated peanut (*Arachis hypogaea* L.). BMC Plant Biol 3: 3.
He G, Meng R, Gao H, Gou B, Gao G, Newman M, Pittman RN, Prakash SH (2005) Simple sequence repeat markers for botanical varieties of cultivated peanut (*Arachis hypogaea* L.). Euphytica 142: 131–136.
Herselman LR, Thwaites FM, Kimmins B, Courtois PJA, Merwe VD, Seal SE (2004) Identification and mapping of AFLP markers linked to peanut (*Arachis hypogaea* L.) resistance to the aphid vector of groundnut rosette disease. Theor Appl Genet 109: 1426–1433.
Holbrook CC, Dong W (2005) Development and evaluation of a mini core collection for the U.S. peanut germplasm collection. Crop Sci 45: 1540–1544.
Holbrook CC, Anderson WF, Pittman RN (1993) Selection of a core collection from the US germplasm collection of peanut. Crop Sci 33: 859–861.
Holbrook CC, Bruniard J, Moore KM, Knauft DA (1998) Evaluation of the peanut core collection for oil content. In: Agronomy Abstracts 1998. American Society of Agronomy, Madison, WI, USA, p 159.
Holbrook CC, Stephenson MG, Johnson AW (2000) Level and geographical distribution of resistance to *Meloidogyne arenaria* in the US peanut germplasm collection. Crop Sci 40: 1168–1171.
Hopkins MS, Casa AM, Wang T, Michell SE, Dean RE, Kochert GD, Kresovich S (1999) Discovery and characterization of polymorphic simple sequence repeats (SSRs) in peanut. Crop Sci 39: 1243–1247.
Isleib TG, Beute MK, Rice PW, Hollowell JE (1995) Screening the core collection for resistance to *Cylindrocladium* black rot and early leaf spot. Proc Am Peanut Res Edu Soc 27: 25.
Jiang H, Liao B, Ren X, Lei Y, Mace E, Fu T, Crouch JH (2007) Comparative assessment of genetic diversity of peanut (*Arachis hypogaea* L.) genotypes with various levels of resistance to bacterial wilt through SSR and AFLP analyses. J Genet Genom 34: 544–554.
Jiang HF, Ren XP, Liao BS, Huang JQ, Lei Y, Chen BY, Guo B, Holbrook CC, Upadhyaya HD (2008) Peanut core collection established in china and compared with ICRISAT mini core collection. Acta Agron Sin 34: 25–30.
Jiang HF, Ren XP, Wang SY, Zhang XJ, Huang JQ, Liao BS, Holbrook CC, Upadhyaya HD (2010a) Development and evaluation of peanut germplasm with resistance to *Aspergillu flavus* from core collection. Acta Agron Sin 36: 428–434.
Jiang HF, Ren XP, Zhang XJ, Huang JQ, Lei Y, Yan LY, Liao BS, Upadhyaya HD, Holbrook CC (2010b) Comparison of genetic diversity between peanut mini core collections from China and ICRISAT by SSR markers. Acta Agron Sin 36: 1084–1091.
John K, Vasanthi RP, Venkateswarlu O, Naidu PH (2005) Variability and correlation studies for quantitative traits in Spanish bunch groundnut (*Arachis hypogaea* L.) genotypes. Legume Res 28: 189–193.
Jung S, Tate PL, Horn R, Kochert G, Moore K, Abbott AG (2003) The phylogenetic relationship of possible progenitors of the cultivated peanut. J Hered 94: 334–340.
Khedikar YP, Gowda MVC, Sarvamangala C, Patgar KV, Upadhyaya HD, Varshney RK (2010) A QTL study on late leaf spot and rust revealed one major QTL for molecular breeding for rust resistance in groundnut (*Arachis hypogaea* L.). Theor Appl Genet 121: 971–984.
Kilian A (2008) DArT-based whole genome profiling and novel information technologies in support system of modern breeding of groundnut. In: Proceedings of 3rd International Conference for Peanut Genomics and Biotechnology on Advances in *Arachis* through Genomics and Biotechnology (AAGB), 4–8 November 2008, Hyderabad, India, p 02.

Kochert G, Halward T, Branch WD, Simpson CE (1991) RFLP variability in peanut (*Arachis hypogaea* L.) cultivars and wild species. Theor Appl Genet 81: 565–570.

Kochert G, Stalker HM, Gimense M, Galgaro L, Lopes CR, Moore K (1996) RFLP and cytogenetics evidence on the origin and evolution of allotetraploid domesticated peanut, *Arachis hypogaea* (Leguminosae). Am J Bot 83: 1282–1291.

Korat VP, Pithia MS, Savaliya JJ, Pansuriya AG, Sodavadiya PR (2009) Studies on genetic variability in different genotypes of groundnut (*Arachis hypogaea* L.). Legume Res 32: 224–226.

Kottapalli P, Upadhyaya HD, Kottapalli KR, Payton P, Dwivedi SL, Burow M, David KO, Sanogo S, Puppala N (2011) Population structure and diversity in Valencia peanut germplasm collection. Crop Sci 51: 1089–1100.

Kotzamanidis ST, Stavropoulos N, Ipsilandis CG (2006) A systematic analysis for grouping Greek peanut (*Arachis hypogaea* L.) germplasm based on eight important phenotypic traits. Asian J Plant Sci 5: 481–486.

Krapovickas A (1969) The origin, variability and spread of groundnut (*Arachis hypogaea* L.). In: Ucko PJ, Dimbley G (eds) The Domestication and Exploitation of Plants and Animals. Duckworth, London, UK, pp 135–151.

Krapovickas A, Rigoni UA (1957) Nuevas especies de *Arachis* vinculadas al problem del origen del mani. Darwinians 11: 431–455.

Krapovickas A, Gregory WC (1994) Taxonomia del genero *Arachis* (Leguminosae). Bonplandia 8: 1–186 (in Spanish with English abstr).

Krapovickas A, Gregory WC (2007) Taxonomy of the genus *Arachis* (Leguminosae) (Translated by DE Williams and CE Simpson). Bonplandia 16: 1205.

Kresovich S, Luongo AJ, Schloss SJ (2002) Mining the gold: finding allelic variants for improved crop conservation and use. In: Engels JMM, Rao VR, Brown AHD, Jackson MT (eds) Managing Plant Genetic Diversity. CAB International, Wallingford, UK, pp 379–386.

Kusuma VP, Yugandhar G, Ajay BC, Gowda MVC, Upadhyaya HD (2007) Identification of sources of multiple disease resistance in groundnut (*Arachis hypogaea* L.) mini core. In: Proceedings of National Seminar by Indian Society of Oilseeds Research (ISOR), 29–31 January 2007, Directorate of Oilseeds Research, Rajendranagar, Hyderabad, India, pp 31–32.

Liang X, Chen X, Hong Y, Liu H, Zhou G, Li S, Guo B (2009) Utility of EST-derived SSR in cultivated peanut (*Arachis hypogaea* L.) and *Arachis* wild species. BMC Plant Biol 9: 35.

Mace ES, Phong DT, Upadhyaya HD, Chandra S, Crouch JH (2006) SSR analysis of cultivated groundnut (*Arachis hypogaea* L.) germplasm resistant to rust and late leaf spot diseases. Euphytica 152: 317–330.

Mace ES, Yuejin W, Boshou L, Upadhyaya HD, Chandra S, Crouch JH (2007) Simple sequence repeat (SSR)-based diversity analysis of groundnut (*Arachis hypogaea* L.) germplasm resistant to bacterial wilt. Plant Genet Resour Characteriz Utiliz 5: 27–36.

Madhura C, Kenchanagoudar, Basamma PV, Sujay V (2011) Morphological characterization of a minicore subset in different botanical types of groundnut (*Arachis hypogaea* L.). Int J Plant Sci 6: 280–284.

Mallikarjuna N, Senthilvel S, Hoisington D (2012) Development of new sources of tetraploid *Arachis* to broaden the genetic base of cultivated groundnut (*Arachis hypogaea* L.). Genet Resour Crop Evol 58: 889–907.

Moretzsohn MC, Hopkins MS, Mitchell SE, Kresovich S, Valls JFM, Ferreira ME (2004) Genetic diversity of peanut (*Arachis hypogaea* L.) and its wild relatives based on the analysis of hyper variable regions of the genome. BMC Plant Biol 4: 11.

Moretzsohn MC, Leoi L, Proite K, Guimaraes PM, Leal-Bertioli SCM, Gimenes MA, Martins WS, Valls JFM, Grattapaglia D, Bertioli DJ (2005) A microsatellite-based, gene-rich linkage map for the AA genome of *Arachis* (Fabaceae). Theor Appl Genet 111: 1060–1071.

Moss JP, Singh AK, Nigam SN, Hilderbrand GL, Goviden N, Ismael FM (1998) Registration of ICGV-SM 87165 peanut germplasm. Crop Sci 38: 572.

Nagy E, Chu Y, Guo Y, Khanal S, Tang S, Li Y, Dong W, Timper P, Taylor C, Ozias-Akins P, Corley HC, Beilinson V, Nielsen N, Stalker H, Knapp S (2010) Recombination is suppressed in an alien introgression in peanut harboring *Rma*, a dominant root-knot nematode resistance gene. Mol Breed 26: 357–370.

Nautiyal PC, Rajgopal K, Zala PV, Pujari DS, Basu M, Dhadhal BA, Nandre BM (2008) Evaluation of wild *Arachis* species for abiotic stress tolerance: I. Thermal stress and leaf water retentions. Euphytica 159: 43–57.

Nautiyal PC, Ravindra V, Rathnakumar AL, Ajay BC, Zala PV (2012) Genetic variations in photosynthetic rate, pod yield and yield components in Spanish groundnut cultivars during three cropping seasons. Field Crops Res 125: 83–91.

Nigam SN, Hildebrand GH, Bock KR, Ismael FM, Govinden N, Subrahmanyam P, Reddy LJ (1998) Registration of ICGV-SM 85048 peanut germplasm. Crop Sci 38: 572–573.

Nordborg M, Tavare S (2002) Linkage disequilibrium: What history has to tell us. Trends Genet 18: 83–90.

Paik-Ro OG, Smith RL, Knauft DA (1992) Restriction fragment length polymorphism evaluation of six peanut species within the *Arachis* section. Theor Appl Genet 84: 201–208.

Pandey MK, Monyo E, Ozias-Akins P, Liang X, Guimarães P, Nigam SN, Upadhyaya HD, Janila P, Zhang X, Guo B, Cook DR, Bertioli DJ, Michelmore R, Varshney RK (2012) Advances in *Arachis* genomics for peanut improvement. Biotechnol Adv 30: 639–651.

Pathirana R (1993) Yield component analysis of bunch groundnut (*Arachis hypogaea* L. ssp. *fastigiata*) germplasm in Sri Lanka. Trop Agric 70: 256–259.

Ratnakumar P, Vadez V (2011) Groundnut (*Arachis hypogaea*) genotypes tolerant to intermittent drought maintain a high harvest index and have small leaf canopy under stress. Funct Plant Biol 38: 1016–1023.

Ravi K, Vadez V, Isobe S, Mir RR, Guo Y, Nigam SN, Gowda MV, Radhakrishnan T, Bertioli DJ, Knapp SJ, Varshney RK (2011) Identification of several small effect main QTLs and large number of epistatic QTLs for drought tolerance in groundnut (*Arachis hypogaea* L.). Theor Appl Genet 122: 1119–1132.

Ravindra V, Nautiyal PC, Joshi YC (1990) Physiological analysis of drought resistance and yield in groundnut (*Arachis hypogaea* L.). Trop Agri 67: 290–296.

Reddy KR, Gupta RVS (1992) Variability and inter-relationship of yield and its component characters in groundnut. J Maharashtra Agric Univ 17: 224–226.

Risch N, Merikangas K (1996) The future of genetic studies of complex human diseases. Science 273: 1516–1517.

Sadeghi SM, Javid F, Niyaki SAN (2011) Assessment of genetic diversity in peanut (*Arachis hypogaea* L.) genotypes using quantitative traits by cluster analysis method. Res J Biol Sci 6: 293–297.

Sarvamangala C, Gowda CLL, Varshney RK (2011) Identification of quantitative trait loci for protein content, oil content and oil quality for groundnut (*Arachis hypogaea* L.). Field Crops Res 122: 49–59.

Seib JC, Wunder L, Gallo-Meagher M, Carpentieri-Pipolo V, Gobert DW, Dickson DW (2003) Marker-assisted selection in screening peanut for resistance to root-knot nematode. In: Proc American Peanut Research Education Society, July 8–11, 2003, Clearwater, Florida, 35: 90.

Seijo JG, Lavia GI, Fernandez A, Krapovickas A, Ducasse D, Moscone EA (2004) Physical mapping of the 5S and 18S-25S rRNA genes by fish as evidence that *Arachis duranensis* and *A. ipaensis* are the wild diploid progenitors of *A. hypogaea*. Am J Bot 91: 1294–1303.

Selvaraj MG, Narayana M, Schubert AM, Ayers JL, Baring MR, Burow MD (2009) Identification of QTLs for pod and kernel traits in cultivated peanut by bulked segregant analysis. Elec J Biotechnol 12: DOI: 10.2225/vol12.

Senapati BK, Roy K (1998) Correlation coefficients among stability parameters of yield and yield contributing characters in groundnut (*Arachis hypogaea* L.). Legume Res 21: 37–40.

Simpson CE, Smith OD (1975) Registration of Tamnut 74 peanut (Reg. No. 19). Crop Sci 15: 603–604.

Simpson CE, Starr JL (2001) Registration of COAN peanut. Crop Sci 41: 918.
Simpson CE, Starr JL, Nelson SC, Woodard KE, Smith OD (1993) Registration of T*x*AG-6 and T*x*AG-7 peanut germplasm. Crop Sci 33: 1418.
Simpson CE, Starr JL, Church GT, Burow MD, Paterson AH (2003) Registration of 'Nema TAM' peanut. Crop Sci 43: 1561.
Singh AK (1995) Groundnut, *Arachis hypogaea* Lin. (Leguminosae-Papilionoideae). In: Simmonds NW, Smartt J (eds) Evolution of Crop Plants, 2nd ed. Longman Group Ltd., London, pp 246–250.
Singh AK, Nigam SN (1997) Groundnut. In: Fuccillo D, Sears L, Stapleton P (eds) Biodiversity in Trust. Conservation and Use of Plant Genetic Resources on CGIAR Centres. Cambridge University Press, UK, pp 114–127.
Singh AK, Simpson CE (1994) Biosystematics and genetic resources. In: Smartt J (ed) The Groundnut Crop: A Scientific Basis for Improvement. Chapman and Hall, London, UK, pp 96–137.
Singh AK, Subrahmanyam P, Gurtu S (1996) Variation in a wild groundnut species, *Arachis duranensis* Krapov, Gregory WC. Genet Resour Crop Evol 43: 135–142.
Singh SB, Chaubey AK (2003) Variability and association amongst quantitative characters in peanut (*Arachis hypogaea* L.). New Agri 14: 97–101.
Srivastava N (2010) Identification of contrasting parental material for salinity tolerance based on morphological, physiological, and biochemical study and molecular diversity by SSR molecular markers in groundnut (*Arachis hypogaea*). PhD Thesis, Jawaharlal Nehru Technological University, Hyderabad, India.
Stalker HT (1990) A morphological appraisal of wild species in section *Arachis* of peanuts. Peanut Sci 17: 117–122.
Subramanian V, Gurtu S, Nageswara Rao RC, Nigam SN (2000) Identification of DNA polymorphism in cultivated groundnut using random amplified polymorphic DNA (RAPD) assay. Genome 43: 656–660.
Sujay V, Gowda MVC, Pandey MK, Bhat RS, Khedikar YP, Nadaf HL, Gautami B, Sarvamangala C, Lingaraju S, Radhakrishan T, Knapp SJ, Varshney RK (2012) QTL analysis and construction of consensus genetic map for foliar diseases resistance based on two RIL populations in cultivated groundnut (*Arachis hypogaea* L.). Mol Breed 32: 773–788.
Sumathi P, Muralidharan V (2007) Character association and path coefficient analysis in confectionery type groundnut (*Arachis hypogaea* L.). Madras Agri J 94: 105–109.
Sumathi P, Amalabalu P, Muralidharan V (2009) Genetic variability for pod characters in large seeded genotypes of groundnut (*Arachis hypogaea* L.). Adv Plant Sci 22: 281–283.
Swamy BPM, Upadhyaya HD, Goudar PVK, Kullaiswamy BY, Singh S (2003) Phenotypic variation for agronomic characteristics in a groundnut core collection for Asia. Field Crops Res 84: 359–371.
Tatum LA (1971) The southern corn leaf blight epidemic. Science 171: 1113–1116.
Upadhyaya HD (2003) Phenotypic diversity in groundnut (*Arachis hypogaea* L.) core collection assessed by morphological and agronomical evaluations. Genet Resour Crop Evol 50: 539–550.
Upadhyaya HD (2005) Variability for drought resistance related traits in the mini core collection of peanut. Crop Sci 45: 1432–1440.
Upadhyaya HD (2008) Crop germplasm and wild relatives: a source of novel variation for crop improvement. Kor J Crop Sci 53: 12–17.
Upadhyaya HD, Ortiz R (2001) A mini core collection for capturing diversity and promoting utilization of chickpea genetic resources in crop improvement. Theor Appl Genet 102: 1292–1298.
Upadhyaya HD, Ortiz R, Bramel PJ, Singh S (2001) Development of Asia region groundnut core collection. In: Diamond Jubilee Symposium on Hundred years of Post-Mendelian Genetics—Retrospect and Prospects. 6–9 November 2001, Indian Agricultural Research Institute, New Delhi, India, pp 43–44.

Upadhyaya HD, Bramel PJ, Ortiz R, Singh S (2002a) Geographical pattern of diversity for morphological and agronomic traits in the groundnut germplasm collection. Euphytica 128: 191–204.

Upadhyaya HD, Bramel PJ, Ortiz R, Singh S (2002b) Developing a mini core of peanut for utilization of genetic resources. Crop Sci. 42: 2150–2156.

Upadhyaya HD, Ortiz R, Bramel PJ, Singh S (2003) Development of a groundnut core collection using taxonomical, geographical and morphological descriptors. Genet Resour Crop Evol 50: 139–148.

Upadhyaya HD, Mallikarjuna Swamy BP, Goudar PVK, Kullaiswamy BY, Singh S (2005) Identification of diverse groundnut germplasm through multienvironment evaluation of a core collection for Asia. Field Crops Res 93: 293–299.

Upadhyaya HD, Reddy LJ, Gowda CLL, Singh S (2006) Identification of diverse groundnut germplasm: Sources of early-maturity in a core collection. Field Crops Res 97: 261–267.

Upadhyaya HD, Reddy LJ, Gowda CLL, Singh S (2009) Phenotypic diversity in cold-tolerant peanut (*Arachis hypogaea* L.) germplasm. Euphytica 165: 279–291.

Upadhyaya HD, Yadav D, Dronavalli N, Gowda CLL, Singh S (2010) Mini core germplasm collections for infusing genetic diversity in plant breeding programs. Elec J Plant Breed 1: 1294–1309.

Upadhyaya HD, Sharma S, Dwivedi SL (2011) *Arachis*. In: Kole C (ed) Wild Crop Relatives: Genomic and Breeding Resources. Vol: Legume Crops and Forages. Springer Berlin, Heidelberg, Germany, pp 1–19.

Upadhyaya HD, Dwivedi SL, Nadaf HL, Singh S (2011b) Phenotypic diversity and identification of wild *Arachis* accessions with useful agronomic and nutritional traits. Euphytica 182: 103–115.

Upadhyaya HD, Mukri G, Nadaf HL, Singh S (2012) Variability and stability analysis for nutritional traits in the mini core collection of peanut. Crop Sci 52, doi: 10.2135/cropsci2011.05.0248.

Vaddoria MA, Patel VJ (1990) Genetic variability in Virginia runner groundnut (*Arachis hypogaea* Linn.). J Oilseeds Res 7: 130–132.

Varma TSN, Dwivedi SL, Pande S, Gowda MVC (2005) SSR markers associated with resistance to rust (*Puccinia arachidis* Speg.) in groundnut (*Arachis hypogaea* L.). SABRAO J Breed Genet 37: 107–119.

Varshney RK, Bertioli DJ, Moretzsohn MC, Vadez V, Krishnamurthy L, Aruna R, Nigam SN, Moss BJ, Seetha K, Ravi K, He G, Knapp SJ, Hoisington DA (2009) The first SSR-based genetic linkage map for cultivated groundnut (*Arachis hypogaea* L.). Theor Appl Genet 118: 729–739.

Varshney RK, Gowda MVC, Radhakrishnan T, Pandey MK, Gautami B, Sujay V, Koppolu R, Senthilvel S, Vadez V, Nigam SN, Upadhyaya HD, Isobe S, He G, Bertioli DJ, Knapp SJ, Cook DR (2010) Development and application of genomic resources for molecular breeding in groundnut (*Arachis hypogaea* L). In: Proceedings of the 3rd International Conference on Plant Molecular Breeding (ICPMB), 5–9th September 2010a, Beijing, China.

Vekariya HB, Khanpara MD, Vachhani JH, Jivani LL, Vagadiya KJ, Revar HJ (2011) Correlation and path analysis in bunch groundnut (*Arachis hypogaea* L.). J Plant Sci (Muzzaffarnagar) 6: 11–15.

Wang ML, Sukumaran S, Barkley NA, Chen Z, Chen CY, Guo B, Pittman RN, Stalker HT, Holbrook CC, Pederson GA, Yu J (2011) Population structure and marker-trait association analysis of the US peanut (*Arachis hypogaea* L.) mini-core collection. Theor Appl Genet 123: 1307–1317.

3

Classical and Molecular Cytogenetics in *Arachis*

J Guillermo Seijo, Germán Robledo, Alejandra Ortíz and *Graciela I Lavia**

ABSTRACT

The progress that has been achieved in understanding the chromosome complements of peanut and its related wild species is reviewed here. Chromosome markers developed by using fluorescent *in situ* hybridization (FISH) with rDNA probes and fluorescent banding allowed the construction of the first chromosome map in peanut and in some diploid taxa. They also revealed the existence of different karyotype structures within the section *Arachis*. Each group of species having different karyotype structure was defined as having different genomes. FISH analysis together with genomic *in situ* hybridization allowed the differentiation of chromosomes from *A. duranensis* and *A. ipaënsis*, the two species involved in the origin of cultivated peanut. The results obtained by these techniques also suggest that the wild tetraploid *A. monticola* is the direct ancestor of peanut and that the Northwest of Argentina and South of Bolivia are the regions from which the AABB tetraploids of section *Arachis* have arisen.

Keywords: Chromosome numbers, karyotypes, FISH markers, genetic and geographic origin, genomes

Instituto de Botánica del Nordeste (CONICET-UNNE, Fac. de Ciencias Agrarias) and Facultad de Ciencias Exactas y Naturales y Agrimensura, Universidad Nacional del Nordeste, Corrientes, Argentina.
*Corresponding author: *lavia@agr.unne.edu.ar; seijo@agr.unne.edu.ar*

3.1 Introduction

Arachis hypogaea (v.n. "peanut", "groundnut") is a cultigen that has become the third most important grain legume crop of the world (Duke 1981) because of its multiple use as human food, vegetable oil, feedstock and ground cover (Wynne and Halward 1989). This species is adapted to a wide ecological range of tropical and subtropical regions and is cultivated under diverse agricultural production systems in Asia, Africa and the Americas (Holbrook and Isleib 2001).

On the bases of morphological features, crossing experiments and seed protein electrophoretic profiles, Krapovickas and Gregory (1994) recognized two subspecies within the cultigen, *hypogaea* and *fastigiata*. Additionally, six botanical varieties have been described, two of them within subsp. *hypogaea* (*hypogaea* and *hirsuta*) and four within subsp. *fastigiata* (*fastigiata*, *aequatoriana*, *peruviana*, and *vulgaris*). Moreover, there are numerous landraces within varieties that are very diverse in their vegetative and reproductive traits (cf. Krapovickas and Rigoni 1960; Grosso et al. 1994; Krapovickas and Gregory 1994; Krapovickas et al. 2009).

Peanut genetic resources are available in the form of naturally evolved landraces of cultivated peanut in the various centers of diversity, breeding lines or materials developed in different peanut producing areas, and also the numerous wild *Arachis* species. In spite of the morphological variability, the major constraint to the genetic improvement of peanut is the narrow genetic base of the extant crop. Wild *Arachis* species, by contrast, are diverse and have the genetic variability and agronomically useful characters needed to improve the peanut (Holbrook and Stalker 2003) and constitute valuable resources for the genetic upgrading of peanut. In this sense, information on the cytogenetics and phylogenetic relationships among wild species and between these species and the peanut is critical for the rational development of breeding programs and complete utilization of the wild materials.

The genus *Arachis* is a native of South America and naturally distributed over a large area that extends from the eastern foothills of the Andes Mountains in Bolivia and northern Argentina to the Atlantic coast in Brazil and from the southern limit of the Amazonian rainforest towards the northern coast of La Plata River in Uruguay. The species grow spontaneously from sea level on the Atlantic coast in Brazil and Uruguay to around 1,450 m a.s.l. in the Andes Mountains of Northwest Argentina (Krapovickas and Gregory 1994; Valls and Simpson 2005). Along this area wild *Arachis* species are adapted to a wide variety of habitats. Their ecological preferences range from the xerophytic forests to temporarily flooded areas, and from temperate grasslands to open patches of the subtropical rainforest. Soil preferences are also diverse, ranging from rock outcrops, layers of laterite pebble, heavy soils, poorly drained areas to well drained sandy soils. In spite of the ample range of ecological preferences displayed by the wild species, the genus as a whole is mainly associated with the savannah-like Cerrado biogeographical region as defined by Cabrera and Willink (1973).

The genus consists of a diverse group of 80 autogamous (some probably with variable percentages of allogamy) and geocarpic taxa (Krapovickas and Gregory 1994; Valls and Simpson 2005). According to morphology, cross-compatibility, viability of the hybrids, geographic distribution and cytogenetics, they have been taxonomically arranged in nine sections —Trierectoides, Erectoides, Procumbentes, Rhizomatosae, Heteranthae, Caulorrhizae, Extranervosae, Triseminatae and Arachis (Krapovickas and Gregory 1994). Trierectoides is considered to be the most ancestral section since the species included in it have tuberous hypocotyls or roots, trifoliated leaves and vaginated stipules, resembling those characters present in the genus *Stylosanthes*. On the other hand, since section Arachis has species with annual and perennial life cycles, with different basic chromosome numbers, ploidy levels and karyotype structures it is considered as the most diverse and derived. Between these two sections, species that belong to sections Erectoides and Procumbentes seem to be the most related to those of section Arachis. Some of the members of sections Rhizomatosae, Heteranthae and Caulorrhizae may produce hybrids with the most derived sections, but others show a strong genetic isolation. Sections Extranervosae and Triseminatae are the most isolated, and their evolutionary position still has to be determined (Krapovickas and Gregory 1994).

Among all the sections, Arachis has received particular attention because it contains the cultivated peanut and its putative wild progenitors. In accordance with its status as the most evolutionarily derived section, geographically it is the most widely distributed. It extends in an east–west direction between the Chapada dos Parecis in the central west of Mato Grosso State (Brazil) and the northern edge of the Chacoan region. From this latitudinal central axis, in the east, the species extend towards the northeast along the Tocantins River (central Brazil) and southward along the Paraguay–Paraná and Uruguay River Basins (Paraguay, Argentina and Uruguay) reaching the northern shore of La Plata River. In the west, they are found towards the northwest along the Mamoré and Guaporé Rivers in north Bolivia and towards the southwest along the Parapetí, Pilcomayo, Bermejo, San Francisco and Juramento River Basins in southern Bolivia and northern Argentina (Krapovickas and Gregory 1994).

Cytogenetics has played a main role among the biosystematic studies carried out in the genus. Many classical cytological studies have been performed in *Arachis*, which delivered important information about the complexity of the peanut genome. The chromosome numbers and the karyotype features of the species as well as the study of the meiotic divisions of interspecific hybrids have provided irreplaceable information to establish the relationships among species and the taxonomic sections. More recently, chromosome studies using fluorescent banding and *in situ* hybridization of DNA have been important to reveal the genetic origin of the cultigen, to provide a more realistic genome arrangement of the section Arachis and to understand the chromosome structure and genome evolution within the genus.

3.2 Chromosome Numbers

Classical cytogenetic studies began as early as the 1930s decade when the chromosome number of $2n = 40$ (Fig. 3-1a) for the cultigen *A. hypogaea* was determined (Husted 1933, 1936). After these pioneer data, several authors have reported the chromosome numbers for different species, although most of them included few entities and generally those that were considered most related to peanut (Gregory 1946; Mendes 1947; Smartt et al. 1978; Singh and Moss 1982; Smartt and Stalker 1982; Stalker 1991). High-quality chromosome plates from mitotic divisions, however, were obtained only

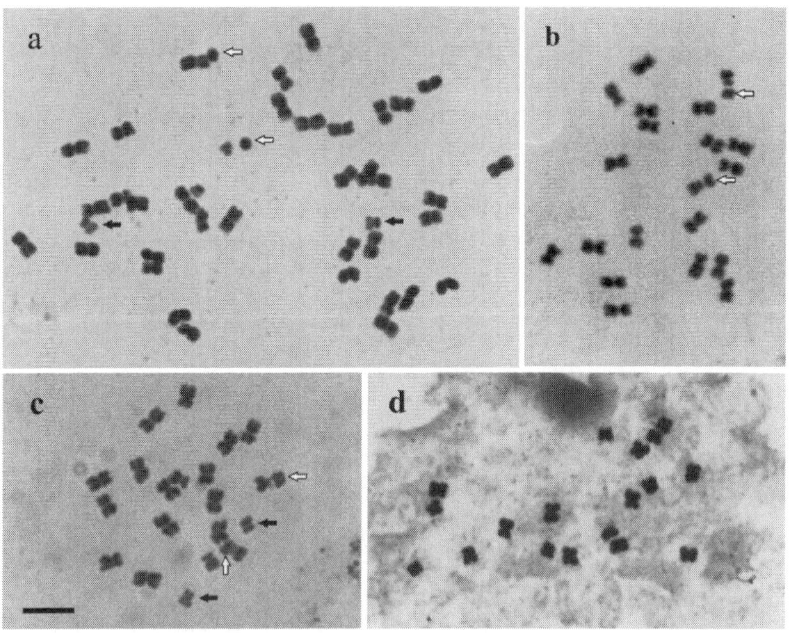

Figure 3-1 Mitotic metaphases of *Arachis* species stained with the Feulgen's technique. **(a)** *Arachis hypogaea* ($2n = 40$) showing A chromosomes and SAT chromosomes type 3. **(b)** *A. ipaënsis* ($2n = 20$) without "A" chromosomes. **(c)** *A. duranensis* ($2n = 20$) showing A chromosomes and SAT chromosomes type 6. **(d)** *A. praecox* with $2n = 18$ chromosomes. Black arrows indicate A chromosomes and white arrows the satellites of SAT chromosomes. Scale bar = 5 μm for all the pictures.

after Fernández and Krapovickas (1994). These authors carried out the most comprehensive work on cytogenetics by classical techniques in *Arachis* analyzing 41 species belonging to eight out of the nine recognized sections. Further contributions made in the last two decades elevated the knowledge of chromosome numbers to 95% of the species of the genus (Lavia 1996, 1998, 2000, 2001; Peñaloza et al. 1997; Lavia and Fernández 2004; Custodio et al. 2005; Peñaloza and Valls 2005). All these studies have revealed that the genus is dibasic, with a great predominance of the species with $x = 10$ (Fig. 3-1b and c) and only four species with $x = 9$ (Fig. 3-1d) (Lavia 1996, 1998; Peñaloza and Valls 1997; Peñaloza and Valls 2005). Three of the latter belong to section Arachis while the remaining species belongs to section Erectoides. They have also revealed that most of the species are diploid and few are polyploids with $2n = 4x = 40$. The latter are restricted to section Arachis, with *A. hypogaea* and *A. monticola*, and to sect. Rhizomatosae with *A. glabrata*, *A. pseudovillosa* and *A. nitida* (Gregory et al. 1973; Fernández and Krapovickas 1994; Peñaloza and Valls 2005). *Arachis pintoi* (section Caulorrhizae) is the only known species of the genus with diploid and triploid cytotypes (Lavia et al. 2011).

3.3 Karyotypes

Chromosome identification in peanut began with Husted (1933), who distinguished two pairs of chromosomes: one of them that borne a secondary constriction (SAT chromosome) and the other that has around half the size of the other chromosomes of the complement (A chromosome) (Fig. 3-1a). Based on these observations and in the meiotic behavior it was established that *A. hypogaea* is an allotetraploid species with two A and two B sets of chromosomes (Smartt et al. 1978). Several investigations carried out on different peanut varieties (D'Cruz and Tankasale 1961; Singh and Moss 1982; Stalker and Dalmacio 1986; Fernández and Krapovickas 1994; Lavia and Fernández 2004) have shown that the karyotypes are highly symmetric with a predominance of metacentric chromosomes. The most common karyotype formula found is $38m + 2sm$, although in some landraces of the *fastigiata* and *hypogaea* botanical varieties the formula $36m + 4sm$ has been reported (Fernández and Krapovickas 1994; Lavia and Fenández 2004). The average size of chromosomes is 1.88 µm ranging from 0.92 to 2.80 µm and it is impossible to distinguish the A genome chromosomes from those of the B genome. Only one pair of chromosomes with satellites is usually distinguished in all the varieties, although the morphology of these pairs varies among the landraces (Lavia and Fernández 2004). In spite of these minor differences, the karyotypes of the varieties and subspecies are very similar. Even the karyotype described for *A. monticola*, which is a wild tetraploid closely related to peanut, is indistinguishable from that of the cultigen (Fernández and Krapovickas 2004).

The chromosomes of diploid wild *Arachis* species are also relatively small and their size ranges from 1.11 to 4 μm. The karyotypes are completely or mainly composed of metacentric chromosomes. One or two pairs (up to six only in *A. glandulifera*) of submetacentric or subtelocentric chromosomes can be also present in the karyotypes of some species. In general, the karyotypes are moderately symmetric and fall into the 1A category of the Stebbins's asymmetry classification (Stebbins 1971). The chromosomes within a particular karyotype are very similar to each other and only two types of them are clearly distinguished. One of these types corresponds to the SAT chromosomes, which bear a conspicuous secondary constriction and a satellite (Fig. 3-1b). The other type corresponds to small A chromosomes, which are present only in a group of species (Fig. 3-1c). Almost all species have only one pair of secondary constrictions localized on the long arms of metacentric or submetacentric chromosomes, which delimits satellites of different sizes. Eleven types of SAT chromosomes have been identified in *Arachis* on the basis of the relative size of the satellites and the proximal segment of the chromosome arm. The variability found in the morphology of these chromosomes has been used to develop a scheme of species arrangement (Fernández and Krapovickas 1994) and was useful to delimit the taxonomic sections.

All the species show allocyclic condensation of the chromosomes. Chromatin condensation begins in the centromeric region and extends toward the telomeres. This phenomenon is particularly observed in A chromosomes, which usually has its distal region uncondensed until late metaphase (Fernández and Krapovickas 1994). In general, secondary constrictions are extended at prophase and early metaphase, and the satellites remain far from the corresponding proximal segments of the chromosome arms; a fact that has frequently conducted to errors in chromosome counts. In late metaphase the chromosomes are uniformly stained along their length and secondary constrictions are very short.

3.4 Chromosome Banding

The first attempt to provide additional chromosome markers involved the application of Giemsa C-banding in a few set of species (Cai et al. 1987). This analysis has revealed large centromeric bands in all the chromosomes of diploid species of the section Arachis, of the peanut, and of *A. rigonii* of section Procumbentes. Several interstitial and terminal bands have been also identified in different chromosomes of the species included in that analysis. Giemsa C-banding was later applied to a few other species belonging to different sections (Pierozzi et al. 2001) also revealing a high predominance of pericentromeric bands. In general, C-banding was helpful to identify unequivocally 3–4 chromosome pairs in addition to the A and the SAT chromosomes.

Counterstaining with DAPI after FISH treatment produced a banding pattern that reveal differences in chromosome structure among different groups of species (Raina and Mukai 1999; Seijo et al. 2004). As opposed to the C-banding, DAPI staining after FISH has revealed that in the tetraploid species, *A. hypogaea* (Fig. 3-2f) and *A. monticola*, half of the chromosomes—those belonging to the A genome—have centromeric C-DAPI$^+$ bands, while the remainder (corresponding to the B genome) have a uniform staining. All *A. hypogaea* varieties and *A. monticola* have a similar distribution and amount of C-DAPI$^+$ heterochromatin, which accounts for about 7% of the total karyotype length (Seijo et al. 2004).

Among the wild species, those included in section Arachis and Heteranthae have been the most comprehensively studied by fluorochrome banding. The heterochromatin patterns revealed the existence of different karyotypes among the species included in section Arachis. All the species with A chromosomes have a similar pattern of heterochromatic bands (Fig. 3-2d) with a total amount of heterochromatin that range from 10.28 to 14.67% of the karyotype length. Within this group, nine species have conspicuous C-DAPI$^+$ centromeric bands in all the chromosome pairs, while the others have conspicuous bands in all the chromosomes except in one or two pairs (A7 and A4) with small and faint bands or without them (Robledo et al. 2009).

By contrast, the species without A chromosomes showed three different patterns of heterochromatic bands (Robledo and Seijo 2010). The group composed of *A. batizocoi* (Fig. 3-2c), *A. cruziana*, and *A. krapovickasii* is characterized by having karyotypes with conspicuous centromeric bands in nine chromosome pairs and a total amount of centromeric heterochromatin that varies from 11.36 to 12.55% of the karyotype length. The karyotypes of *A. trinitensis* (Fig. 3-2a) and *A. benensis* characteristically have small and faint centromeric bands in only seven or eight chromosome pairs and the total amount of heterochromatin per complement varies from 5.89 to 7.52%. The remaining non-A genome species have karyotypes completely devoid of centromeric C-DAPI$^+$ bands. However, while *A. ipaënsis* (Fig. 3-2e) and *A. magna* are completely devoid of detectable heterochromatin, *A. gregoryi*, *A. valida*, and *A. williamsii* have one small interstitial or distal band in the short arms of pair 3. *Arachis glandulifera* is very particular in the banding pattern since it shows large centromeric blocks in the four subtelocentric chromosomes, small centromeric blocks in the other three chromosomes, and two interstitial blocks of medium size (Fig. 3-2b). This pattern revealed that its karyotype structure differs from any other *Arachis* species (Robledo and Seijo 2008).

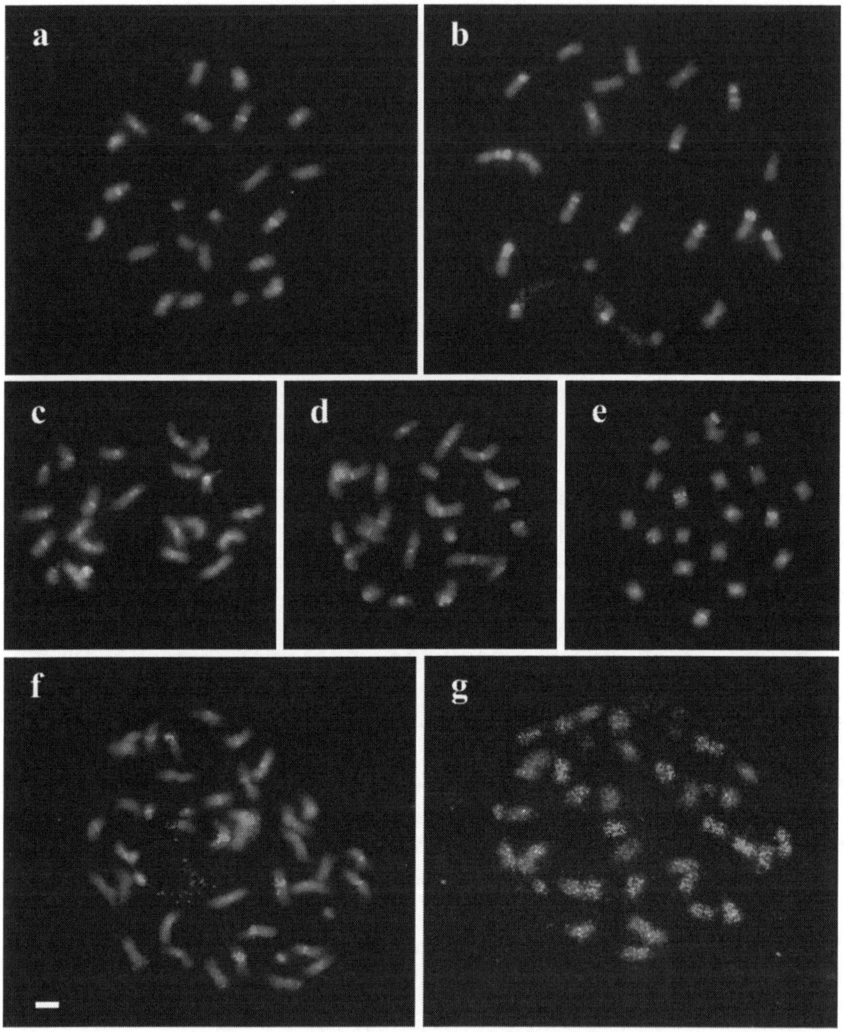

Figure 3-2 Somatic metaphases of *Arachis* species following double fluorescent *in situ* hybridization (a–f), showing yellow-green FITC signals from the 5S rDNA probe and red TRITC signals from the 18S–26S rDNA probe. DAPI counterstaining (light blue) subsequent to the FISH procedure was used to highlight the heterochromatin bands and to stain euchromatin. **(a)** *A. trinitensis* (F genome). **(b)** *A. glanduifera* (D genome). **(c)** *A. batizocoi* (K genome). **(d)** *A. duranensis* (A genome). **(e)** *A. ipaënsis* (B s.s. genome). **(f)** *A. hypogaea* ($2n$ = 40). **(g)** Somatic metaphase of *Arachis hypogaea* after double genomic *in situ* hybridization (GISH) using total DNA probes of *A. ipaënsis* (red) and of *A. duranensis* (green). Scale bar = 3 μm for all the pictures.

Color image of this figure appears in the color plate section at the end of the book.

Concerning the base composition of the heterochromatic bands found in the species of section Arachis, direct CMA/DAPI staining has demonstrated that all the centromeric ones are composed of AT-rich sequences. By contrast, bands associated to the rDNA clusters, mainly those located at the secondary constrictions, are CMA_3^+ bands and thus composed of GC rich sequences (Seijo et al. unpubli.).

As opposed to section Arachis, karyomorphological evaluation of five species of section Heteranthae (Silva et al. 2010) revealed that the pattern of GC rich heterochromatin is diverse and different from those described for section Arachis. There are species with CMA_3^+ centromeric bands in all the chromosomes and others completely deprived of such bands. $DAPI^+$ bands were only found in *A. pusilla* flanking some of the centromeric CMA_3^+ bands.

According to the available data, the analysis of the heterochromatin amount, composition and distribution by fluorochromes constitute one of techniques that provided many useful chromosome markers for the construction of karyotypes and the identification of chromosome homeologies among species. In addition, the variation in the patterns of heterochromatin that has been reported for species within and among sections reflects that it constitutes one of the most dynamic genomic fractions in the evolution of karyotypes within *Arachis*.

3.5 DNA Content

The nuclear DNA content has an important function in the evolution and adaptation of plants (Bennett 1982; Price 1976, 1988) and its comparison among related taxa contributed to clarify phylogenetic relationships and to establish evolutionary trends in several groups of organism. The 2C nuclear DNA amounts have been reported for around 40 *Arachis* species belonging to eight out of the nine taxonomic sections (Resslar et al. 1981; Singh et al. 1996; Temsch and Greilhuber 2000, 2001; Lavia and Fernández 2008). However, there are incongruences between the absolute DNA content values obtained by different authors. Most of the microdensitometric studies found that mean 2C DNA amounts varied from 10.26 to 11.82 pg among accessions of *A. hypogaea* ($2n = 4x = 40$) and from approximately 3 to 7 pg in diploid ($2n = 2x = 20$) species of the genus (Resslar et al. 1981; Singh et al. 1996; Lavia and Fernández 2008). By contrast, flow cytometric measurements, corroborated with microdensitometric analysis, indicated that *A. hypogaea* has a mean 2C = 5.914 pg and *A. monticola* a mean 2C = 5.979 pg (Temsch and Greilhuber 2000). Similar analyzes determined that *A. duranensis* has a mean 2C value of 2.63 pg that represents about half of the values obtained in other reports (Temsch and Greilhuber 2001). Discrepancies in the DNA content have been attributed by Temsch and Greilhuber (2001) to technical measurement problems that may have remained unrecognized in the other studies; thus more research is needed to precisely determine the absolute DNA content of *Arachis* species.

Another point of debate concerns the existence of variation in DNA content among populations of any particular species. Intraspecific variation among different landraces of *A. hypogaea* has been cited by Singh et al. (1996) and by Lavia and Fernández (2008). According to these authors, the accessions belonging to *A. hypogaea* ssp. *hypogaea* (mean value 11.27 pg) with a longer life cycle have significantly larger mean DNA content than the accessions of *A. hypogaea* ssp. *fastigiata* (mean value 10.97 pg) with a shorter life cycle. However, Temsh and Greilhuber (2000) were not able to find that variation in DNA content.

By contrast, significant variation in the 2C amount of DNA has been reported among populations of *A. duranensis*, in spite of the differences in absolute values observed by different authors (Singh et al. 1996; Temsh and Greilhuber 2001). There is a negative correlation of genome size with latitude and altitude above sea level of the collection sites, and it has been postulated that this variation may be due to selection for favorable genome sizes in particular environmental conditions (Singh et al. 1996).

An overview of the published data showed that, in general, the diploid perennial species of section Arachis have about 12% more DNA than the annual species (Singh et al. 1996). Concerning the sections, the species within Procumbentes, Caulorrhizae, Rhizomatosae and Arachis sections have higher values of DNA content than those included in Erectoides, Extranervosae, Triseminatae and Heteranthae sections (Singh et al. 1996; Lavia and Fernández 2008). There is no available data for section Trierectoides. Comparisons of DNA content within section Arachis revealed a great variation in DNA content among the species, in part due to the presence of species with different basic chromosome numbers and levels of ploidy. Considering the Cx values, the genome size per haploid complement is smaller in polyploids than in the parental diploids, suggesting that a postpolyploidization reduction of DNA content has occurred (Lavia and Fernández 2008). Taking in account the whole data for the genus, it has been suggested that there is an apparent tendency to increase the genome size during the evolution of *Arachis* (Lavia and Fernández 2008), as it is the general tendency in angiosperms (Leitch et al. 1998; Soltis et al. 2004). However, the values published for section Arachis suggest that the evolution of the genome size may have been dynamics and complex, involving different cycles of expansions and reductions within particular lineages of species.

3.6 Molecular Cytogenetics

Modern cytogenetic tools, such as fluorescence *in situ* hybridization (FISH) or genomic *in situ* hybridization (GISH) techniques have helped to resolve the chromosome evolution history and enhanced the knowledge on the mechanism that conducted to the genome differentiation in *Arachis*. Localization of the 5S and 18S–26S rDNA on the chromosomes by FISH was initially applied to a set of species from different sections of *Arachis* (Raina and Mukai 1999) revealing their usefulness for the characterization of the group. Chromosome mapping of rDNA families by FISH was later used to analyze in detail the karyotypes of all the species included in section Arachis (Seijo et al. 2004; Robledo and Seijo 2008; Robledo et al. 2009; Robledo and Seijo 2010), to propose a new genome arrangement of them and to bring light on the genetic and geographic origin of peanut.

Physical mapping by FISH of the rDNA loci in the six botanical varieties of *A. hypogaea* (Fig. 3-2f) and in *A. monticola* revealed two pairs of 5S and five pairs of 18S–26S rDNA sites. In both species, the 5S loci are proximally located in short arms (pairs A3 and B3), while the 18S–26S rDNA clusters are proximally (pairs A2, A10, B3 and B10) or subterminally placed (B7). One 5S site is syntenic to an 18S–26S site (B3). The high degree of homeology detected between *A. monticola* and *A. hypogaea* strongly shows evidence that they are very closely related taxa. The mapping of the rDNA loci, together with the heterochromatin analysis, provided the first chromosome map for peanut, which by anchoring other different sequence base markers, will be useful to generate an integrated map for the AABB tetraploids of *Arachis*.

Concerning diploid species, those with A chromosomes have only one pair of interstitial (or rarely proximal) 5S rDNA loci located on the A3 pair (Fig. 3-2d). However, the number, size and chromosomal localization of the 18S–26S rDNA loci are variable among the species (Robledo et al. 2009). The number of these gene clusters ranges from two to four pairs (Fig. 3-2d). In general, the largest loci are located in A10 pair, those of intermediate size are borne by the A2 pair, while the smallest and faintest signals (in the cases that the species have more than two loci) were detected in the A7 and A4 chromosomes.

In spite of the general homogeneous structure of the karyotypes, according to the pattern of rDNA loci and heterochromatic bands, the species with A chromosomes have been arranged into three subgroups (Robledo et al. 2009). The first one is defined by having one or two chromosome pairs without or with small heterochromatic bands, and by three to four pairs of 18S–26S rDNA sites. This subgroup includes *A. cardenasii*, *A. herzogii* and *A. kempff-mercadoi*. The remaining species, which characteristically have heterochromatic bands in all the chromosomes, have been arranged in two different subgroups according to the number of 18S–26S rDNA sites.

One of them, includes the species with five or six sites (*A. diogoi*, *A. helodes*, *A. kuhlmannii*, *A. simpsonii* and *A. stenosperma*), and the other, is composed of the taxa with two pairs of sites (*A. correntina*, *A. duranensis*, *A. schininii* and *A. villosa*).

Arachis glandulifera has only one pair of 5S rDNA loci subterminally located and five pairs of 18–26S rDNA loci (Fig. 3-2b). These markers, together with the pattern of heterochromatic bands, were sufficient to identify precisely all the chromosome pairs of the karyotype and to the construction of the first wholly-resolved idiogram for an *Arachis* species (Robledo and Seijo 2008).

FISH analysis within the species without A chromosomes revealed that all of them have one pair of 5S rDNA loci localized in proximal (or rarely interstitial) position on the metacentric pair 3. *Arachis cruziana*, *A. batizocoi* (Fig. 3-2c), and *A. krapovickasii* have two additional pairs of loci. The number of 18S–26S rDNA loci ranges from one in *A. gregoryi* and *A. trinitensis* (Fig. 3-2a) to four pairs in *A. magna* and *A. valida*. Most of them are located in proximal or interstitial position, but few are distally or subterminally located.

3.7 Genome Organization of Section Arachis

The first genome constitution established within the genus *Arachis* was the AABB for *A. hypogaea* (Smartt et al. 1978). On the basis of the crossing experiments, chromosome features of the karyotypes and chromosome pairing in interspecific hybrids, diploid species within the section *Arachis* have been traditionally arranged into three different genome groups. The species characterized by symmetric karyotypes and the small A pair of chromosomes were included within the A genome (Smartt et al. 1978; Fernández and Krapovickas 1994). The species with symmetric karyotypes but without A chromosomes have been considered members of the non-A or B genome (Smartt et al. 1978; Smartt and Stalker 1982; Fernández and Krapovickas 1994), while the only species with an asymmetric karyotype (*A. glandulifera*) has been considered to have the D genome (Stalker 1991).

Molecular cytogenetics revealed a high degree of homogeneity among the karyotypes of the species classically included within the A genome. However, considering the statement that closeness of taxa is usually correlated with the similarity of their heterochromatin and rDNA FISH patterns (Hizume et al. 2002; Liu et al. 2003), the variation in number and positions of C-DAPI$^+$ bands and 18S–26S rDNA sites among species was used to establish three subgroups of karyotype homeologies. Interestingly, the species within each subgroup tend to be more closely distributed geographically than those belonging to different subgroups; therefore each subgroup was named using a geographical reference (Robledo et al. 2009). The first one, named Chiquitano, comprised the species (*A. cardenasii*, *A. chiquitana*, *A. herzogii*, and *A. kempff-mercadoi*) that grow in the southern and western portion of the Chiquitanía biogeographic region in Santa Cruz Department of Bolivia. The second subgroup, called Pantanal, includes the species (*A. diogoi*, *A. kuhlmannii*, *A. helodes*, *A. simpsonii* and *A. stenospema*) distributed in the Pantanal biogeographic region, in western Brazil, northern Paraguay and eastern Bolivia. The third subgroup, called La Plata River Basin, corresponds to the species (*A. duranensis*, *A. schininii*, *A. correntina* and *A. villosa*) distributed along the La Plata River Basin (except for the region that comprises the upper stream of the Paraguay River in the Pantanal). Within this scheme, the A genome of *A. hypogaea* falls into the subgroup of La Plata River Basin.

The particular karyotype features of *A. glandulifera*, i.e., assymetric karyotype, the unique pattern of heterochromatin distribution and the 5 pairs of 18–26 rDNA loci, justify the maintenance of this species as having the D genome (Robledo and Seijo 2008).

The other species with $2n = 20$ included within the non-A genome have been arranged in three different karyotype groups according to the number, size and distribution of the rDNA sites and the features of the heterochromatic bands. Moreover, each group of species having different karyotype structure has been further segregated into different genomes (Robledo and Seijo 2010). One of the karyotype groups includes *A. batizocoi*, *A. cruziana*, and *A. krapovickasii* which have conspicuous heterochromatic bands in nine chromosome pairs and three pairs of 5S rDNA loci. This group has been named K genome. The second one, named F genome, is integrated by *A. benensis* and *A. trinitensis* that have karyotypes with small and faint heterochromatic bands in seven or eight chromosome pairs and only two 5S rDNA loci. The last group is composed of all the species without pericentromeric heterochromatic bands in their karyotypes, namely, *A. ipaënsis* (Fig. 3-2e), *A. gregoryi*, *A. magna*, *A. valida* and *A. williamsii*. It has been proposed that this group should retain the B genome *sensu stricto* designation (Robledo and Seijo 2010). This proposal was based on the fact that the B genome has been originally assigned to one of the chromosome complements of the cultigen (Smartt et al. 1978). Thus, the wild donor of this genome, *A. ipaënsis* (see below), and all the relatives that share the same karyotype structure should belong to the same genome.

In accordance with the particular geographic distribution described for each karyotype subgroup of the A genome species (Robledo et al. 2009), the species included within each of these new genome groups also showed a strong tendency to be co-distributed, mainly those of the F and K genomes. The K genome species are distributed in the northern and northwestern edges of the Chaco Boreal region, whereas the species having the F genome are restricted to the savannas around Trinidad city, Beni department in Bolivia. The B genome species are distributed over a large area of semi-deciduous forests and savannas of the Cerrado associated with the chiquitano highlands and peripheral mountain ranges of the western portion of the Brazilian Precambrian Shield. The only exception is *A. ipaënsis*, from which the only known population grows at the top of the sand banks of streams in an ecotone between the Tucumano-Oranense deciduous forest and the chacoan xerophytic forest (Robledo and Seijo 2010).

Diploid species with $2n = 18$ of section Arachis are less characterized, but their heterochromatin patterns and rDNA loci features revealed that they have a different genome constitution from those observed in the species with $2n = 20$. Therefore, it was proposed that they should be included in a new genome, named G genome (Silvestri et al. unpubli.).

The new genome arrangement is supported by a set of different kind of data. For *Arachis*, records of pollen stainability lower than 25% in F_1 hybrids are widely accepted as indicative of intergenomic crosses. In all the hybrids obtained from crosses between species that have different karyotype structures (different genomes after Robledo and Seijo 2010) the pollen viability of the F_1 are below 25% (Smartt et al. 1978; Stalker 1991; Krapovickas and Gregory 1994; Tallury et al. 2005; Burrow et al. 2009), thus supporting the arrangements of the groups with different karyotypes as different genomes. Bivalent formation at meiotic metaphases of hybrids obtained by crosses between species with the same genome is usually higher than 9.5 out of the 10 expected. However, in intergenomic hybrids of *Arachis* species, the number of bivalents formed is lower than 7.5 (Stalker et al. 1991; Tallury et al. 2005). The number of bivalents formed in F_1 obtained by crosses between species with large heterochromatic bands (*A. batizocoi*) and those without them (*A. ipaënsis* and *A. williamsii*) is in the range expected for intergenomic hybrids, also supporting the new genome arrangement.

Molecular analyzes performed in section Arachis have usually included incomplete sets of non-A genome species (Milla et al. 2005; Tallury et al. 2005; Burow et al. 2009). However, in all of them, the group of species without bands (i.e., the B genome *s.s.* after Robledo and Seijo 2010) always formed a cluster separated from that formed by *A. batizocoi*, *A. cruziana*, and *A. krapovickasii* (K genome) (Bechara et al. 2010). Similarly, whenever *A. benensis* and *A. trinitensis* (F genome) were included in those analyzes, they were always grouped together and in a separated cluster (Milla et al. 2005). Therefore, the clustering of species observed in those dendrograms highly supports the arrangement of non-A genome species in three different genomes (B, F and K genome) as proposed by Robledo and Seijo (2010).

3.8 Genetic and Geographical Origin of Peanut

The origin of peanut has long interested to plant taxonomists, geneticists and breeders. However, the knowledge about its origin is still limited compared with other major crops. Concerning the genetic origin of peanut, more than eight wild diploid species having either the A or non-A genomes have been considered as involved in the origin of peanut. Since the early 50s, when the first hybrid between *A. hypogaea* and the diploid *A. correntina* (A genome) was obtained (Krapovickas and Rigoni 1954), several other diploid species, have produced hybrids with *A. hypogaea* and, thus, been proposed as putative progenitors of the tetraploids (Krapovickas and Rigoni 1957; Raman 1960; Smartt and Gregory 1967; Krapovickas 1973; Stalker and Wynne 1979; Singh and Moss 1984). Different authors, using morphology and cross compatibility data, have further proposed *A. correntina*, *A. duranensis*, *A. cardenasii* (all with A genome), and *A. batizocoi* (K genome) as probable parents of *A. hypogaea* (Krapovickas 1973; Gregory and Gregory 1976; Singh and Smartt 1998). Classical chromosome analyzes suggested that *A. duranensis* and *A. ipaënsis* (B genome) (Fernández and Krapovickas 1994), or *A. trinitensis* (F genome) and *A. williamsii* (B genome), could also be the genome donors of the cultigen (Lavia 1996). Assays on molecular markers have also revealed different species as probable ancestors of peanut. For instance, from Restriction Fragment Length Polymorphism (RFLP) analysis, *A. duranensis* and *A. ipaënsis* were proposed as the most likely progenitors of *A. hypogaea* (Kochert et al. 1991, 1996; Burrow et al. 2009), while from Randomly Amplified Polymorphic DNA (RAPD) and Inter-Simple Sequence Repeats (ISSR) analyzes, *A. villosa* (A genome) and *A. ipaënsis* were considered as the best candidates (Raina et al. 2001). On the other hand, Amplified Fragment Length Polymorphism (AFLP) data showed that at least three diploid species with the A genome and three with the non-A genome display small genetic distance when compared with the cultigen (Milla et al. 2005), while microsatellite markers demonstrated that, although *A. duranensis* and *A. ipaënsis* are closely related to the cultigen, a group of other species having the A or non-A genome have small genetic distance with peanut (Moretzsohn et al. 2004; Koppolu et al. 2010). Similarly, ITS analysis revealed that a set of species having the A or B genome *s.s.* could have participated in the origin of the tetraploids (Bechara et al. 2010).

The molecular cytogenetic approach involving the analysis of the number (Raina and Mukai 1999) and the mapping (Seijo et al. 2004; Robledo et al. 2009) of ribosomal gene clusters by FISH and heterochromatin distribution showed that the species included in the La Plata River Basin group of the A genome are the most related to the A genome of the tetraploids. Concerning the donors of the B genome of peanut, detailed karyotype analysis supports *A. ipaënsis* from the B genome *s.s.* group as the most probable B genome donor because (1) its chromosomes are completely deprived of heterochromatic bands (Raina and Mukai 1999; Seijo et al. 2004) and (2) its pattern of rDNA loci is the most similar (among the wild diploids) to that observed in the complement without bands of the cultigen (Seijo et al. 2004; Robledo and Seijo 2010).

Double GISH experiments on metaphase spreads of the wild *A. monticola* and the cultigen *A. hypogaea* (Fig. 3-2g), showed that from all the possible combinations among the diploid *Arachis* species ($2n = 2x = 20$) that have been proposed as parents of the cultigen, the DNA of *A. duranensis* and *A. ipaënsis* yielded the most intense and uniform hybridization pattern onto the respective *A. hypogaea* chromosome subsets (Seijo et al. 2007). A similar pattern was observed in all the varieties of the cultigen and also in the wild tetraploid *A. monticola*, which was interpreted as an evidence of a single origin of these taxa.

Long-standing ideas about the center of origin of the cultivated peanut, which were based on the morphological variability of the landraces and biogeography (Gregory et al. 1980; Krapovickas and Gregory 1994), place the origin of the domesticated peanut in northern Argentina and southern Bolivia, in a transition area between the Tucumano-Boliviano forest and the Chaco lowlands. The distributions of the most accepted putative A and B genome donors for *A. hypogaea* and the location of *A. monticola* in this area provided additional support for this hypothesis. However, archeological studies indicate the presence of *A. hypogaea* in the Huarmey Valley, near the Peruvian coast, around 5000 yr BP (Bonavia 1982). Also pod samples that strongly resemble those of wild species have been found in the Casma Valley on the Pacific coast of Perú that date between 3500 and 3800 yr BP and suggests that ancient people from northwest Perú may have used and even cultivated some wild *Arachis* species (Simpson and Faries 2001). More recently, radiocarbon-dated macrobotanical remains dating around 7840 yr BP, that appear to correspond morphologically to a wild *Arachis* species or to peanut fruits in early stages of domestication, were recovered from sealed house floors and hearths in buried preceramic sites in a tropical dry forest of the Ñanchoc Valley, on the lower western slopes of the Andes in northern Perú (Dillehay et al. 2007). However, since there is no evidence to indicate that the Ñanchoc Valley was a domestication center for any major economic plants, the early adoption of peanut, and other crops in these valley, suggests that *Arachis* species must have been cultivated elsewhere earlier than 8000 yr BP. In spite of these findings, the region of spontaneous origin of peanut is thought to have been far from this valley because no wild species of *Arachis* are presently found in the area. Wild *Arachis* species would not be expected to have occurred in those valleys spontaneously because the genus originated in the Amambay ranges and nowadays it is mostly distributed in Central-East Brazil, East Bolivia, Paraguay, and North Argentina below 1,200–1,500 m a.s.l. Due to the geocarpic fruits, the Andes mountains would have been an insurmountable barrier, preventing the species from reaching the Pacific coast by natural dispersion. An alternative hypothesis contemplates that peanut may have originated in the gardens of ancient people from South America, who may have used and cultivated wild *Arachis* species (Simpson and Faries 2001). In this scenario, the gardens of those people may have also served as a possible site for the origin of *A. hypogaea*.

The fact that both subspecies and all the botanical varieties of the cultigen had identical patterns of genomic hybridization (GISH) suggests that the same wild species participated in their origin. This statement implies that all presently known varieties and subspecies of peanut arose from a single, unique allotetraploid plant population, i.e., they have a common origin. The common ancestry of all infraspecific taxa of *A. hypogaea* is supported by the low genetic variability so far detected with most molecular markers in the cultivated peanut (Halward et al. 1991; Kochert et al. 1996; Herselman 2003). Although possible introgression from other diploid species of *Arachis* cannot be fully discarded, the fact that fragments of alien chromatin could not be identified in the karyotype of *A. hypogaea* by GISH suggests that if introgressions occurred during the history of peanut culture, the mechanism of gene transference may have not involved large chromosome segments or entire chromosomes (Seijo et al. 2007).

It has been postulated that diploid ancestors of *A. hypogaea* could first have given origin to a wild allotetraploid plant (Krapovickas and Gregory 1994). The unique extant wild tetraploid species so far known within section Arachis is *A. monticola*, which has several morphological traits similar to *A. hypogaea*. It should be noted that the wild condition of *A. monticola* is based mainly on its fruit structure (wherein each seed has its own shell separated by an isthmus) and on its ability to persist as natural populations, unlike the cultivated peanut. FISH and GISH results (Seijo et al. 2004, 2007), as well as molecular marker data (Gimenes et al. 2002; Milla et al. 2005), revealed a very close genomic relationship between both tetraploid species and strongly supports the hypothesis that *A. monticola* is the immediate wild antecessor of *A. hypogaea*. Accordingly, it was proposed that the origin of *A. hypogaea* occurred by an initial hybridization event followed by chromosome duplication or fusion of unreduced gametes that have given rise to a wild tetraploid with two sets of the A genome chromosomes and two sets of the B genome. It has been recently published that sexual polyploidization may have been the mechanism by which the tetraploid *A. glabrata* and the triploid cytotype of *A. pintoi* have arisen (Ortíz et al. 2011; Lavia et al. 2011). The production of unreduced gametes was reported for few diploid species in section Arachis, but they are common in hybrids (Chatuverdi et al. 1990). These data supports the hypothesis advanced by Seijo et al. (2007) which considers that the AABB tetraploid would have arisen by means of bilateral sexual polyploidization of AB diploid hybrids.

After the origin of the wild allotetraploid, probably *A. monticola* (which possibly had larger seeds than any of the progenitors as a result of the *gigas* effect in polyploids), *A. hypogaea* arose by domestication. High selective pressure in different agroecological environments may have led to the origin of the different subspecies and varieties of the cultigen. Therefore, morphological variability would mainly result from artificial selection, as it is the case in most domesticated crops, rather than from the participation of several species in the origin of *A. hypogaea*.

The artificial resynthesis of an amphidiploid from *A. ipaënsis* and *A. duranensis* (Fávero et al. 2006), that is morphologically very similar to *A. monticola* and that can hybridize with all varieties of the cultigen producing fertile offspring, supports the model advanced.

3.9 Chromosome Evolution in Section Arachis

The analyzes of chromosome numbers have revealed the existence of two basic chromosome numbers and two ploidy levels within the section. Since $x = 10$ is the most widespread basic chromosome number within the genus and since it is present in all the sections considered as ancestral, the $x = 9$ should be considered as a derived character. Aneuploidy and disploidy have been proposed as the mechanisms responsible for the reduction of the chromosome number in *Arachis* (Lavia et al. 2008). The former implies the loss of chromosomes, and consequently of all the genetic information contained by them. On the contrary, the latter involves the distribution of the chromatin of one or more chromosomes in some or all the other chromosomes of the complement by successive unidirectional translocations. In spite of the many cytogenetic studies performed in the genus, evidences to support any of these mechanism as the responsible for the origin of the $x = 9$ are still needed.

The two spontaneous polyploids (*A. monticola* and *A. hypogaea*) of section Arachis are allopolyploids. This type of polyploids combines and maintains diploid sets of chromosomes from two or more parental species (Leitch and Bennett 1997). However, genome restructuring usually occurs during their establishment in order to stabilize the newly built complex genome (Soltis and Soltis 1999; Kenton et al. 1993; Lim et al. 2000). In opposition to this general rule, the GISH patterns observed in the AABB tetraploids of *Arachis* did not reveal large structural rearrangements of chromosomes between the A and B chromosome sets of *A. monticola*/*A. hypogaea* (Seijo et al. 2007).

These findings are in agreement with the conservative localizations of the 5S and 18S–26S rDNA loci in the A and B complements of the tetraploids compared to those localizations observed in their wild diploid progenitors, *A. duranensis* and *A. ipaënsis*, respectively (Seijo et al. 2004). Moreover, the analysis of few dispersed (Nielen et al. 2010, 2011) and tandemly repetitive (Robledo et al. unpubli.) sequences showed that in the AABB tetraploids those genomic elements are in almost the same amount than that expected from the addition of the amounts present in their diploid progenitors.

Concerning the structural changes of chromosomes at diploid level, since a karyotype formula is conserved among most of the species, it has been postulated that these types of rearrangements may have not played an important role during the evolution of the section Arachis (Lavia et al. 2009). The exception to this general statement is *A. glandulifera* because it has been reported that its asymmetrical karyotype formula arose by the occurrence of several translocations (Stalker 1991). In spite of the stasis found in the karyotype formula of this group of species, chromosome markers revealed by banding and FISH showed that changes in the amount and distribution of AT rich heterochromatin are one of the most relevant factors that determined the differentiation of the A, B, D, F and K genomes (Raina and Mukai 1999; Seijo et al. 2004; Robledo and Seijo 2010).

GISH on AABB species demonstrated a clear differential hybridization pattern between the chromosomes of the A genome and those of the B genome. Since this technique relies largely on the hybridization of genome-specific repetitive sequences, it was advanced that divergence in the content of the genomic repetitive elements accompanied speciation and has driven genome differentiation in *Arachis* (Seijo et al. 2007). A recent analysis on one of the members of the Ty3-*gypsy* elements present in the *Arachis* genomes, named FIDEL, demonstrated that it is dispersely distributed and that it is more frequent in the A- than in the B-genome, with copy numbers of about 3,000 in *A. duranensis* and 820 in *A. ipaënsis* per haploid genome (Nielen et al. 2010). This was the first evidence of uneven distribution of a dispersed repetitive element among different genomes of *Arachis*. In conclusion, the investigations carried out on dispersed elements and tandemly repetitive sequences of the centromeric heterochomatin strongly support that the changes in the repetitive fraction have played a key role in the chromosomal and genome evolution of *Arachis*.

Acknowledgements

The authors are grateful to Ing. Agr. A. Krapovickas and Ing. Agr. A. Fernández for their valuable comments and financial support extended by: CGIAR, Challenge Program; ANPCYT—National Agency for the

Promotion of Science and Technology, CONICET—National Council of Scientific and Technological Researches, and National University of the Northeast, Argentina.

References

Bechara MD, Moretzsohn MC, Palmieri DA, Monteiro JP, Bacci M Jr, Martins J Jr, Valls JFM, Lopes CR, Gimenes MA (2010) Phylogenetic relationships in genus *Arachis* based on ITS and 5.8S rDNA sequences. BMC Plant Biol 10: 255.

Bennett MD (1982) Nucleotypic basis of the spatial ordering of chromosomes in eukaryotes and the implications of the order for genome evolution and phenotypic variation. In: Dover GA, Flavel RB (eds) Genome Evolution. Academic Press, London, UK, pp 239–261.

Bonavia D (1982) Precerámico peruano, Los Gavilanes, mar, desierto y oasis en la historia del hombre. Corporación Financiera de Desarrollo e Instituto Arqueológico Alemán. Lima, Perú.

Burow MD, Simpson CE, Faries MW, Starr JL, Paterson A (2009) Molecular biogeographic study of recently described B- and A-genome *Arachis* species, also providing new insights into the origins of cultivated peanut. Genome 52: 107–119.

Cabrera AL, Willink A (1973) Biogeografía de América Latina. Monografía 13, Serie de Biología. OEA, Washington DC, USA.

Cai Q, Lu S, Chinnappa CC (1987) Analysis of karyotypes and Giemsa C-banding patterns in eight species of *Arachis*. Genome 29: 187–194.

Chaturvedi M, Yunus D, Nair PKK (1990) Cytopalynological studies of *Arachis* (Leguminosae). Grana 29: 109–117.

Custodio AR, Peñaloza APS, Valls JFM (2005) Further cytogenetic information on *Arachis stenosperma* (Leguminosae). Cytologia 70: 331–335.

D'Cruz R, Tankasale MP (1961) A note on chromosome complement of four groundnut varieties. Indian Oilseeds J 5: 58–59.

Dillehay TD, Rossen J, Andres TC, Williams DE (2007) Preceramic adoption of peanut, squash, and cotton in Northern Peru. Science 29: 1890–1893.

Duke JA (1981) Handbook of Legumes of World Economic Importance. Plenum Press, New York, USA.

Fávero AP, Simpson CE, Valls JFM, Vello NA (2006) Study of the evolution of cultivated peanut through crossability studies among *Arachis ipaënsis*, *A. duranensis*, and *A. hypogaea*. Crop Sci 46: 1546–1552.

Fernández A, Krapovickas A (1994) Cromosomas y evolución en *Arachis* (Leguminosae). Bonplandia 8: 187–220.

Gimenes MA, Lopez CR, Valls JFM (2002) Genetic relationships among *Arachis* species based on AFLP. Genet Mol Biol 25: 349–353.

Gregory WC (1946) Peanuts breeding program underway. In: Research and Farming. 69th Annual Report, North Carolina Agricultural Experiment Station, North Carolina State University, Raleigh, NC, USA, pp 42–44.

Gregory WC, Gregory MP (1976) Groundnut. *Arachis hypogaea* (Leguminosae-Papilionatae). In: Simmonds NW (ed) Evolution of Crop Plants. Longman Group, London, UK, pp 151–154.

Gregory WC, Krapovickas A, Gregory MP (1980) Structure, variation, evolution and classification in *Arachis*. In: Summerfield RJ, Bunting AH (eds) Advances in Legume Science. Kew, London, UK, pp 469–481.

Gregory WC, Gregory MP, Krapovickas A, Smith BW, Yarbrough JA (1973) Structures and genetic resources of peanuts. In: Wilson CT (ed) Peanuts—Culture and Uses. American Peanut Research Education Society, Stillwater, USA, pp 47–134.

Grosso NR, Krapovickas A, Pietrarelli JR, Guzmán CA (1994) Las proteínas seminales del maní (*Arachis hypogaea*, Leguminosae) y su relación con las categorías infraespecíficas. Bonplandia 8: 221–233.

Halward TM, Stalker HT, Larue EA, Kochert G (1991) Genetic variation detectable with molecular markers among unadapted germplasm resources of cultivated peanut and related species. Genome 34: 1013–1020.

Herselman L (2003) Genetic variation among southern African cultivated peanut (*A. hypogaea* L.) genotypes as revealed by AFLP analysis. Euphytica 133: 319–327.

Hizume M, Shibata F, Matsusaki Y, Garajova Z (2002) Chromosome identification and comparative karyotypic analyses of four *Pinus* species. Theor Appl Genet 105: 491–497.

Holbrook CC, Isleib TG (2001) Geographical distribution and genetic diversity in *Arachis hypogaea*. Peanut Sci 28: 80–84.

Holbrook CC, Stalker HT (2003) Peanut breeding and genetic resources. In: Janick J (ed) Plant Breeding Reviews. John Wiley & Sons, Oxford, UK, pp 297–356.

Husted L (1933) Cytological studies on the peanut, *Arachis*. I. Chromosome number and morphology. Cytologia 5: 109–117.

Husted L (1936) Cytological studies on the peanut, *Arachis*. II. Chromosome number, morphology and behaviour, and their application to the problem of the origin of the cultivated forms. Cytologia 7: 396–423.

Kenton A, Parokonny AS, Gleba YY, Bennett MD (1993) Characterization of the *Nicotiana tabacum* L. genome by molecular cytogenetics. Molec Gen Genet 240: 159–169.

Kochert G, Halward T, Branch WD, Simpson CE (1991) RFLP variability in peanut (*Arachis hypogaea* L.) cultivars and wild species. Theor Appl Genet 81: 565–570.

Kochert G, Stalker HT, Gimenes M, Galgaro L, Romero Lopes C, Moore K (1996) RFLP and cytogenetic evidence on the origin and evolution of allotetraploid domesticated peanut, *Arachis hypogaea* (Leguminosae). Am J Bot 83: 1282–1291.

Koppolu R, Upadhyaya HD, Dwivedi SL, Hoisington DA, Varshney RK (2010) Genetic relationships among seven sections of genus *Arachis* studied by using SSR markers. BMC Plant Biol 10: 15.

Krapovickas A, Rigoni VA (1954) Cruzamientos interespecíficos en *Arachis* y consideraciones sobre el origen del maní cultivado. Proccedings of 28th Congreso Panamericano de Agronomía. Piracicaba, Sao Paulo, Brazil.

Krapovickas A, Rigoni VA (1957) Nuevas especies de *Arachis* vinculadas al origen del maní. Darwiniana 11: 431–455.

Krapovickas A, Rigoni VA (1960) La nomenclatura de las subespecies y variedades de *Arachis hypogaea* L. Rev Investig Agrícolas 14: 197–228.

Krapovickas A (1973) Evolution of the genus *Arachis*. In: Moav R (ed) Agricultural Genetics. Selected Topics. National Council for Research and Development, Jerusalem, Israel, pp 135–151.

Krapovickas A, Gregory WC (1994) Taxonomía del género *Arachis* (Leguminosae). Bonplandia 8: 1–186.

Krapovickas A, Vanni RO, Pietrarelli JR, Williams DE, Simpson CE (2009) Las razas de maní de Bolivia. Bonplandia 18: 95–189.

Lavia GI (1996) Estudios cromosómicos en *Arachis* (Leguminosae). Bonplandia 9: 111–120.

Lavia GI (1998) Karyotypes of *Arachis palustris* and *A. praecox* (Section Arachis), two species with basic chromosome number $x = 9$. Cytologia 63: 177–181.

Lavia GI (2000) Chromosomes studies of wild *Arachis* (Leguminosae). Caryologia 53: 177–181.

Lavia GI (2001) Chromosomal characterization of germplasm of wild species of *Arachis* L. belonging to sections Trierectoides, Erectoides and Procumbentes. Caryologia 54: 115–119.

Lavia GI, Fernández A (2004) Karyotypic studies in *Arachis hypogaea* L. varieties. Caryologia 57: 353–359.

Lavia GI, Fernández A (2008) Genome size in wild and cultivated peanut germplasm. Plant Syst Evol 272: 1–10.

Lavia GI, Fernández A, Seijo G (2008) Cytogenetic and molecular evidences on the evolutionary relationships among *Arachis* species. In: Sharma AK, Sharma A (eds) Plant Genome: Biodiversity and Evolution. Volume 1E: Phanerogam-Angiosperm. Science Publishers, Enfield, New Hampshire, USA, pp 101–134.

Lavia GI, Ortiz AM, Fernández A (2009) Karyotypic studies in wild germplasm of *Arachis* (Leguminosae). Genet Resour Crop Evol 56: 755–764.

Lavia GI, Ortiz AM, Robledo G, Fernández A, Seijo G (2011) Origin of triploid *Arachis pintoi* (Leguminosae) by autopolyploidy evidenced by FISH and meiotic behaviour. Ann Bot 108: 103–111.

Leitch IJ, Bennett MD (1997) Polyploidy in angiosperms. Trends Plant Sci 2: 470–476.

Leitch IJ, Chase MW, Bennett MD (1998) Phylogenetic analysis of DNA C-values provides evidence for a small ancestral genome size in flowering plants. Ann Bot 82: 85–94.

Liu ZL, Zhang D, Hong DY, Wang XR (2003) Chromosomal localization of 5S and 18S-5.8S-25S ribosomal DNA sites in five Asian pines using Fluorescence *in situ* hybridization. Theor Appl Genet 106: 198–204.

Mendes AJT (1947) Estudios citológicos no gênero *Arachis.* Bragantia 7: 257–268.

Milla SR, Isleib TG, Stalker HT (2005) Taxonomic relationships among *Arachis* sect. Arachis species as revealed by AFLP markers. Genome 48: 1–11.

Moretzshon MC, Hopkins MS, Mitchell SE, Kresovich S, Valls JFM, Ferreira ME (2004) Genetic diversity of peanut (*Arachis hypogaea* L.) and its wild relatives based on the analysis of hypervariable regions of the genome. BMC Plant Biol 4: 11.

Nielen S, Campos-Fonseca F, Leal-Bertioli SCM, Guimaraes PM, Seijo G, Town C, Arrial R, Bertioli D (2010) FIDEL—a retrovirus-like retrotransposon and its distinct evolutionary histories in the A- and B-genome components of cultivated peanut. Chrom Res 18: 227–246.

Nielen S, Vidigal BS, Leal-Bertioli SCM, Ratnaparkhe M, Paterson AH, Garsmeur O, D'Hont A, Guimaraes PM, Bertioli DJ (2011) Matita, a new retroelement from peanut: characterization and evolutionary context in the light of the *Arachis* A-B genome divergence. Mol Genet Genomics 287: 21–38.

Ortíz AM, Seijo JG, Fernández A, Lavia GI (2011) Meiotic behavior and pollen viability of tetraploid *Arachis glabrata* and *A. nitida* species (Section *Rhizomatosae*, Leguminosae): implications concerning their polyploid nature and seed set production. Plant Syst Evol 292: 73–83.

Peñaloza APS, Valls JFM (1997) Contagem do número cromossômico em acessos de *Arachis decora* (Leguminosae) In: Simpósio Latino-Americano de Recursos Genéticos Vegetais 1. Programa e Resumos. Instituto Agronômico de Campinas, Campinas, Brazil, p 39.

Peñaloza APS, Valls JFM (2005) Chromosome number and satellited chromosome morphology of eleven species of *Arachis* (Leguminosae). Bonplandia 15: 65–72.

Pierozzi NI, Galgaro ML, Lopes CR (2001) Application of C-banding in two *Arachis* wild species, *Arachis pintoi* Krapov. and W.C. Gregory and *A. villosulicarpa* Hoehne to mitotic chromosome analyses. Caryologia 54: 377–384.

Price HJ (1976) Evolution of DNA content in higher plants. Bot Rev 42: 27–52.

Price HJ (1988) DNA content variation among higher plants. Ann Mo Bot Gard 75: 1248–1257.

Raina SN, Rani V, Kojima T, Ogihara Y, Singh KP, Devarumath RM (2001) RAPD and ISSR fingerprints as usefull genetic markers for analysis of genetic diversity, varietal identification and phylogenetics relationships in peanut (*Arachis hypogaea*) cultivars and wild species. Genome 34: 1013–1020.

Raina SN, Mukai Y (1999) Detection of a variable number of 18S-5.8S-26S and 5S ribosomal DNA loci by fluorescent *in situ* hybridization in diploid and tetraploid *Arachis* species. Genome 42: 52–59.

Raman VS (1960) Studies in the genus *Arachis*. IX. A fertile synthetic tetraploid groundnut from the interspecific backcross *A. hypogaea* and *A. villosa*. Indian Oilseeds J 4: 90–92.

Resslar PM, Stucky JM, Miksche JP (1981) Cytophotometric determination of the amount of DNA in *Arachis* L. sect. *Arachis* (Leguminosae). Am J Bot 68: 149–153.

Robledo G, Seijo JG (2008) Characterization of *Arachis* D genome by FISH chromosome markers and total genome DNA hybridization. Genet Mol Biol 31: 717–724.

Robledo G, Seijo JG (2010) Species relationships among the wild B genome of *Arachis* species (section Arachis) based on FISH mapping of rDNA loci and heterochromatin detection: A new proposal for genome arrangement. Theor Appl Genet 121: 1033–1046.

Robledo G, Lavia GI, Seijo JG (2009) Species relations among wild *Arachis* species with the A genome as revealed by FISH mapping of rDNA loci and heterochromatin detection. Theor Appl Genet 118: 1295–1307.

Robledo G, Lavia GI, Seijo JG (2010) Genome re-assignation of *Arachis trinitensis* (Sect. Arachis, Leguminosae) and considerations on its implication in the genetic origin of peanut. Genet Mol Biol 33: 714–718.

Seijo JG, Lavia GI, Fernández A, Krapovickas A, Ducasse DA, Moscone EA (2004) Physical mapping of 5S and 18S-25S rRNA genes evidences that *Arachis duranensis* and *A. ipaënsis* are the wild diploid species involved in the origin of *A. hypogaea* (Leguminosae). Am J Bot 91: 2293–2303.

Seijo JG, Lavia GI, Fernández A, Krapovickas A, Ducasse DA, Bertioli DJ, Moscone EA (2007) Genomic relationships between the cultivated peanut (*Arachis hypogaea*—Leguminosae) and its close relatives revealed by double GISH. Am J Bot 94: 1963–1971.

Silva SC, Martins MIG, Santos RC, Peñaloza APS, Melo Filho PA, Benko-Iseppon AM, Valls JFM, Carvalho R (2010) Karyological features and banding patterns in *Arachis* species belonging to the Heteranthae section. Plant Syst Evol 285: 201–207.

Simpson CE, Faries MJ (2001) Advances in the characterization of diversity in section Arachis: archeological evidence, crossing results and their relationship in understanding the origin of *Arachis hypogaea* L. In: Proc 3rd Simposio de recursos genéticos para a America Latina e Caribe. Londrina, Brazil, pp 103–104.

Singh AK, Moss JP (1982) Utilization of wild relatives in genetic improvement of *Arachis hypogaea* L. Part 2. Chromosome complements of species in section Arachis. Theor Appl Genet 61: 305–314.

Singh AK, Moss JP (1984) Utilization of wild relatives in genetic improvement of *Arachis hypogaea* L. Part 5. Genome analysis in section Arachis and its implications in gene transfer. Theor Appl Genet 68: 355–364.

Singh AK, Smartt JP (1998) The genome donors of the groundnut/peanut (*Arachis hypogaea* L.) revisited. Genet Resour Crop Evol 45: 113–118.

Singh KP, Raina SN, Singh AK (1996) Variation in chromosomal DNA associated with the evolution of *Arachis* species. Genome 39: 890–897.

Smartt J, Stalker HT (1982) Speciation and cytogenetics in *Arachis*. In: Pattee HE, Young CT (eds) Peanut Science and Technology. American Peanut Research and Education Society, Yoakum, Texas, USA, pp 21–49.

Smartt J, Gregory WC (1967) Interspecific cross-compatibility between the cultivated peanut *Arachis hypogaea* L. and other species of the genus *Arachis*. Oleagineux 22: 455–459.

Smartt J, Gregory WC, Gregory MP (1978) The genomes of *Arachis hypogaea*. 1. Cytogenetic studies of putative genome donors. Euphytica 27: 665–675.

Soltis DE, Soltis PS (1999) Polyploidy: recurrent formation and genome evolution. TREE 14: 348–352.

Soltis DE, Soltis PS, Tate JA (2004) Advances in the study of polyploidy since Plant Speciation. New Phytol 161: 173–191.

Stalker HT (1991) A new species in section Arachis of peanuts with a D genome. Am J Bot 78: 630–637.

Stalker HT, Wynne JC (1979) Cytology of interspecific hybrids in section Arachis of peanuts. Peanut Sci 6: 110–114.

Stalker HT, Dalmacio RD (1986) Karyotype analysis and relationships among varieties of *Arachis hypogaea* L. Cytologia 51: 617–629.

Stalker HT, Dhesi JS, Parry DC, Hahn JH (1991) Cytological and interfertility relationships of *Arachis* section Arachis. Am J Bot 78: 238–246.

Stebbins GL (1971) Chromosomal evolution in higher plants. Addison-Wesley, London, UK, 216.

Tallury SP, Hilu KW, Milla SR, Friend SA, Alsaghir M, Stalker, HT, Guandt D (2005) Genomic affinities in *Arachis* section Arachis (Fabaceae): molecular and cytogenetic evidence. Theor Appl Genet 111: 1229–1237.

Temsch EM, Greilhuber J (2000) Genome size variation in *Arachis hypogaea* and *A. monticola* re-evaluated. Genome 43: 449–451.

Temsch EM, Greilhuber J (2001) Genome size in *Arachis duranensis*: a critical study. Genome 44: 826–830.

Valls JFM, Simpson CE (2005) New species of *Arachis* (Leguminosae) from Brazil, Paraguay and Bolivia. Bonplandia 14: 35–64.

Wynne JC, Halward T (1989) Cytogenetics and genetics of *Arachis*. Crit Rev Plant Sci 8: 189–220.

4

Peanut Breeding

Boshou Liao

ABSTRACT

Peanut is an important oilseed and cash crop in many countries, with the global harvest area as 24.01 million hectares and production as 37.95 million tons in 2010. Breeding for improved cultivars is crucial for increasing yield, enhancing quality and broadening adaptation of the crop. Even high yield was the key objective in most peanut breeding programs, more breeding efforts have been made for improving quality and resistance to biotic and abiotic stresses in the recent two decades. For enhancing efficiency in germplasm research and utilization in breeding, core and mini core collections of peanut have been selected and evaluated in several institutions. Breeding methods including selection of variants from landraces or improved varieties, hybridization breeding, interspecific hybridization for transferring elite traits from wild *Arachis* accessions, and Marker Assisted Selection (MAS) have been popularly applied in many breeding programs. Remarkable progress has been achieved in peanut breeding for high yield, high oil content, high oleic acid, resistance to leafspot, rust, bacterial wilt, Tomato Spotted Wilt Virus (TSWV), Groundnut Rosette Virus (GRV), aflatoxin contamination and drought stress. Application of improved cultivars has largely contributed to the global peanut production increase during the past two decades. Further research emphasis will be given to enhancing and stabilizing peanut productivity in different production systems, improving quality for diversified utilization, and reducing aflatoxin and other harmful contamination in peanuts for ensuring food safety.

Keywords: Landraces, Improved varieties, Aflatoxin contamination, Edible oil, Hybridization breeding, Recurrent selection, Marker assisted slection (MAS)

Oil Crops Research Institute of Chinese Academy of Agricultural Sciences, Wuhan, 430062, China.
Email: *lboshou@hotmail.com*

4.1 Introduction

Peanut or groundnut is an important source of plant oil and protein worldwide. The oil content in peanut seeds ranges 45–60% and protein ranges 22–35% in different varieties (Sun 1998). Globally, peanut is planted in various enviornments with different soils, moisture, cropping patterns, biotic and abiotic stresses and economic and social status in more than 100 countries between latitudes 40ºN and 40ºS (Liao and Holbrook 2007). In 2010, the global sowing area under peanut was 24.01 million hectares with a total production of 37.95 million tons thus reaching a productivity of 1,581 kg/ha (FAO 2010). During the first decade (2001–2010) of the 21st century, the global peanut harvested area and production increased by 4.05 and 5.78%, respectively while India had the largest peanut sowing area (6 million ha in average) and China had the largest production (14.24 million tons in average) (FAO 2010). Other important peanut producing countries are Nigeria, the USA, Myanmar, Senegal, Indonesia, Sudan, Argentina and Vietnam. Approximately 53% of the total global peanut production is crushed for edible oil, 32% is used for confectionery consumption and the remaining 15% is used for animal feed and seed (Dwivedi et al. 2003). The current utilization ratio of peanut in the world would not change much in the coming decade and the crop will continue to be a major vegetable oil source in certain developing countries especially in the largest peanut producing countries such as India and China where edible oil supply is likely still in shortage even though most of the peanut is used as direct food consumption in a few developed countries where peanut is grown.

Breeding for improved cultivars is fundamental to the peanut industry development. The history of peanut germplasm and breeding research can be traced back to the early 20th century. Improved cultivars have predominated in production for more than 50 years in most peanut producing regions in the world. Several researchers have estimated the impact of breeding on promoting peanut industry development. Isleib et al. (2001) estimated that resistant peanut cultivars especially those with resistance to *Sclerotinia* blight, root-knot nematode, and tomato spotted wilt virus (TSWV) have had an economic impact of more than US$200 million annually for US peanut producers. Wan (2003) also estimated that application of improved cultivars has contributed at least 30% of the peanut yield increase in China during the past two decades. It is common that fewer peanut cultivars have been used in the developed countries than those in the developing countries. The number of released cultivars may not necessarily reflect the breeding effort or its effectiveness, but the diversified production conditions and lack of effective seed system for quick seed extension in developing countries have been the reasons for a large number of peanut cultivars used in production. Marked progress has been made in collecting peanut germplasm, evaluating germplasm by selecting core and mini core collections (Holbrook et al. 1993; Jiang and Ren 2006; Liao and Holbrook 2007; Upadhyaya et al. 2001, 2003) and broadening genetic diversity of peanut cultivars. Combining traditional and advanced technologies including introgression of genes from wild *Arachis* species, marker assisted selection and transformation of alien genes has been the most important strategic approaches in peanut genetic enhancement (Nigam et al. 2000; Dwivedi et al. 2003; Liao and Holbrook 2007; Varshney et al. 2009). This chapter reviews the recent progress in peanut breeding worldwide.

4.2 Breeding Objectives

The breeding objectives for peanut are determined by ecological characteristics in the producing zones and utilization purpose of the produce. Breeding for high yield is, of course, a common objective in most peanut producing regions worldwide. Besides high yield, breeding for high oil or protein contents and improved fatty acid profiles have also become important breeding objectives. For edible peanuts, breeding objectives for quality are more diversified and relatively limited research has been conducted in some crucial traits such as flavor, texture and lowed saturated fatty acids.

Resistance or tolerance to various biotic and abiotic stresses has received increased breeding efforts during the past two decades. There are many constraints to peanut production in different ecological environments. The important widespread biotic constraints in peanut are foliar diseases including late leaf spot [*Phaeoisariopsis personata* (Berk. & Curtis) Deighton], early leafspot [*Cercospora arachidicola* Hori], rust [*Puccinia arachidis* Sperg.] and web blotch [*Didymella arachidicola* (Chock.) Taber, Pettit & Philley], stem or root rot caused by soil borne fungi such as *Sclerotium rolfsii* Sacc., bacterial wilt [*Ralstonia solanacearum* (E.F. Smith)], groundnut rosette virus (GRV), Peanut Bud Necrosis (PBND), Peanut Stunt Virus (PSV), Peanut Strip Virus (PStV), Tomato Spotted Wilt Virus (TSWV) and nematodes. The abiotic constraints included drought, acid soil, low soil fertility and low temperature. Late leaf spot, drought and rust are the most important constraints in terms of economic losses globally (Dwivedi et al. 2003). In Southeast USA, TSWV has been an important constraint to peanut production. In certain regions of Africa, GRV is the most serious peanut disease. In Central and South China and several Southeast Asian countries, bacterial wilt can cause serious yield loss to peanut. Several fungi blight diseases are becoming more serious in many peanut producing areas (Liao and Holbrook 2007). Peanuts are susceptible to *Aspergillus* infection, which can result in aflatoxin contamination during production, processing, storage and transportation. In the USA, Aflatoxin contamination costs peanut industry over US$20 million annually (Holbrook et al. 2009). Allergins in peanuts is also a food safety issue (Holbrook and Stalker 2003). Residual of heavy metals such as Cadmium (Cd) is also a food safety concern in certain regions, and varaiation in heavy metal accumulation has been observed among different genotypes (Yu 2011) hinting that there is a potential for breeding cultivars with reduced contamination of heavy metals. Under this circumstance, much of the plant breeding effort in the world is being redirected from only developing cultivars with high yields to ones that are also incorporating genes conferring resistance or tolerance to the important constraints to genotypes with adaptability to certain locations. Breeding for resistance or tolerance to acid soil, iron deficiency, heat and cold stress in peanut has not been properly addressed up to now.

4.3 Methods of Peanut Breeding

The general strategy for plant breeding is to create, identify and select useful genetic variation from breeding materials. Selection of desirable variants and cultivars from landraces of peanut in the originating centers in South America and consequently in other continents has a long history. Introduction and exchange of landraces and improved germplasm lines among different regions have greatly contributed to breeding, production and utilization of peanut. Modern standard breeding approaches for self-pollinated plants have been extensively applied to develop peanut cultivars (Isleib and Wynne 1992; Isleib et al. 1994; Knauft and Ozias-Akins 1995; Knauft and Wynne 1995; Sun 1998; Nigam et al. 2000; Liao and Holbrook 2007; Yu 2011). Cultivars that have been widely distributed are commonly used as parents in hybridization programs, and thus the genetic base of peanut cultivars has historically been quite narrow (Hammons 1973). In China, two peanut landraces, Fuhuasheng and Shitouqi, have been extensively used as crossing parents in breeding programs from 1950s to 1970s and their pedigree could be found in more than 60% of the peanut cultivars released (Jiang et al. 1998; Sun 1998; Yu 2011). However, since the late 1980s, a large number of diverse cultivars have been released in major peanut growing countries and consequently the genetic base of the commercially produced germplasm is much broader (Liao and Holbrook 2007). Parental selection is the key consideration in plant breeding, and with yield potential, resistance to pests, quality and uniformity requirements imposed by the peanut growers and processing industries, the genetic base of peanut might continue to be relatively narrow in the future, which remains a great concern in future breeding programs (Nigam et al. 2000; Nigam and Aruna 2008; Yu 2011).

Pedigree breeding is commonly used in peanut as in many other self-pollinated crops. Backcross method is becoming more frequent with the identification of useful qualitatively inherited traits. Mass selection is used infrequently in peanut because of negative correlations between most disease resistance and yield potential (Knauft and Wynne 1995). Use of single-seed descent method has greatly increased in recent years because this method could save space and other resources in breeding (Isleib et al. 1994; Liao et al. 2010). Recurrent selection has received little attention in peanut breeding because of the efforts needed to make a large number of crosses. It is clear that most improved peanut cultivars were developed through conventional hybridization and pedigree selection. For example, about 300 peanut cultivars were developed by hybridization among the 400 improved cultivars released in China since 1950s (Yu 2011). Besides selection for productivity, more emphasis is being given to resistance to diseases, pests and abiotic stresses and quality traits that could be effectively selected by pedigree breeding.

Transferring genes conferring important resistance or quality traits from wild relative species of the genus *Arachis* has been a key strategy in modern breeding programs. High levels of resistance to many diseases and insect pests have generated much interest in creating interspecific hybrids. Introgressing useful genes from wild species into *A. hypogaea* are normally influenced by sterility barriers due to different ploidy levels, genomic incompatibilities and cryptic genetic differences. Constraints to obtaining hybrids may occur at the time of fertilization, during early cell division of the embryo or during later embryo development. Even when hybrids are obtained, genetic recombination is often restricted, and desired genes may not be incorporated into the *A. hypogaea* genome properly. The progress of research and utilization of wild *Arachis* species have been well reviewed by several authors (Isleib et al. 2001a; Holbrook and Stalker 2003; Liao and Holbrook 2007; Yu 2011). In China, extensive efforts have been made for introgressing useful genes from wild *Arachis* species to the cultivated peanut. A bacterial wilt resistant cultivar, Yuanza 9102, a selected progeny of *A. hypogaea* cv. "Baisha 1016" with *A. diogoi*, has been released in several provinces in China (Yu 2011). Several germplasm lines derived from interspecific crossing with high oil content have been obtained (Sun 1998; Yu 2011). In India, development for foliar disease resistant peanut cultivars by using interspecific hybrid derivatives has progressed well, and GPBD 4, a Spanish bunch cultivar with desirable resistance both to late leafspot and rust was released to production (Gowda et al. 2002).

Mutation breeding was used extensively in the late 1950s to early 1970s in the USA, but the materials produced were not widely utilized in peanut improvement or production. In China, mutation breeding of peanut has been conducted in several institutions since 1960s, and several cultivars have been developed with mutants as crossing parents (Qiu 1992; Yu 2011), among which, Yueyou 551 and 8130 have been extensively cultivated in production.

Peanut mutants with variation in content of protein, oil, amino acid and fatty acid components have been obtained by treating a Spanish type cultivar, Baisha 1016 (Qiu and Feng 1998). Mutants with improved resistance or tolerance to salt, leaf spot and drought stress have been reported (Qiu and Feng 1998). The effectiveness of mutation breeding can be enhanced with other approaches such as hybridization. Wang (2002) reported production of extra large and small pod mutants by chemical mutagenesis. Several peanut cultivars have been developed by mutation breeding and cultivated in India (Nadaf et al. 2009; Mothilal et al. 2010).

More recently, advances in molecular genetic technology have allowed peanut researchers to more precisely measure genetic polymorphism and enabled the identification of molecular marker or Quantitative Trait Loci (QTLs) for economically significant traits (Holbrook et al. 2011). Marker Assisted Selection (MAS) approaches would be available for high olerate (Chu and Ozias-Akins 2009), resistance to rust (Khedikar et al. 2010), late leaf spot (Xia 2007), bacterial wilt (Jiang et al. 2007), infection of *A. flavus* (Lei et al. 2005) and drought tolerance (Ravi et al. 2011). The research progress in molecular approaches and genetic transformation that could be used in peanut breeding is reviewed in concerned chapters in this book.

4.4 Breeding Progress for Important Traits

4.4.1 High Yield and Early Maturity

Breeding peanut for high yield has been successfully conducted in many countries. Breeding for other objectives such as quality traits and resistance to biotic and abiotic stresses is also based on high yield. In the north producing regions in China, some improved peanut cultivars could harvest a yield over 9 t/ha under favorable cultivation conditions (Yu 2011). Traditional breeding strategies have been widely used in high yield breeding programs. Most of the cultivars with highest yield potential have large pods and are developed through hybridization. Suitable growth period of new cultivars is an important character for their adaptation to special cropping systems. Early maturity (not more than 120 days from sowing to harvest) is necessary for the summer-sown cropping pattern in which the peanut is planted in late May to early June after wheat harvest in North China. A similar growth period is also needed in South China where peanut is rotated with paddy rice. In most hilly areas where drought stress is frequent in autum, early maturing cultivars are also needed to avoid serious yield losses from drought stress in late growth stage. In some tropical areas including India and African countries, early maturity (100 days or a little shorter) is necessary to avoid end-season drought stress (Nigam and Aruna 2008).

4.4.2 High Oil Content and Improved Oil Quality

High oil content of peanut cultivars is a crucial trait for the oil processing industry, especially in developing countries where most peanuts are for a major source of cooking oil. Limited attention has been paid to breeding for oil content in the developed countries where peanut is rarely used for oil. It is estimated that each 1% increase of oil content would increase the processor's benefit by 7% (Liao 2003). Around 55% of the peanut production in China is crushed for oil with an annual oil production of 2.3 million tonnes (Liao and Holbrook 2007) which is just second to rapeseed oil among the domestic production of plant oil. As peanut oil is relatively less competitive with other plant oils such as rapeseed and soybean oils because of its relatively higher market price, breeding for cultivars with high oil content could not only increase oil production but also enhance the market competitiveness of peanut oil.

Extensive peanut germplasm screening has been conducted for high oil content. Among the 5,947 peanut germplasm accessions tested in China, the oil content ranged from 32.3 to 60.2% with an average of 50.6% (Jiang et al. 2006). Most peanut accessions with relatively high oil content (over 55%) belong to Spanish type with early maturity. Jiang et al. (2010) also reported that the oil contents of 87 wild *Arachis* accessions ranged from 51.4 to 62.9% with an average of 55.8%, and 12 of the 87 accessions possess oil content higher than 58%, indicating the potential of enhancing oil content in *A. hypogaea* by introgressing concerned genes from the wild *Arachis* species. Liao et al. (2010) reported several high oil content lines from a Recombinant Inbred Line (RIL) population derived from the cross of "Yuanza 9102 × Zhonghua 5". Currently, the average oil content of the most popular 30 peanut cultivars extensively grown in China is just 51.4% (Liao and Holbrook 2007) even more than 20 cultivars with oil content over 55% have been developed and released during the past decade (Yu 2011). Low speed of peanut seed reproduction and extension system has limited the coverage of high oil cultivars in production. Meanwhile, there is variation among the high oil genotypes in terms of stability of oil content across seasons and environments (Dwivedi et al. 1993).

The oil content in peanut has been reported as independently inherited with most other agronomic traits (Yu 2011), but it is controlled by multiple genes. The yield level of cultivars for oil extraction could be enhanced through crossing high oil content parents with large-pod and high yielding ones. Currently, it is relatively easy to screen oil content in a large scale with the improvement of breeding facilities. However, it is relatively more difficult to develop high oil cultivars with resistance to certain foliar diseases such as late leafspot. Unfortunately, most of the available high oil cultivars are found to be more supportive to seed infection of *Aspergillus flavus* and consequent aflatoxin contamination (Liao and Holbrook 2007).

Quality of peanut oil is determined by its fatty acid components. Oleic (O) and linoleic (L) acids comprise over 80% of the oil content in peanut while linoleic acid is less saturated and less stable than oleic acid. The oxidative stability and shelf-life of peanut and peanut products can be enhanced by increasing the O/L ratio. Among the 5,947 germplasm accessions tested in China (Jiang and Ren 2006), there are 22 lines with oleic acid content higher than 67% and three lines with palmatic acid less than 8%. The dragon type (var. *hirsuta*) has highest oleic acid content at 53.6% and Valencia type has the lowest one at 43.4% (Jiang and Ren 2006). Germplasm evaluation for quality traits in peanut was also conducted in the USA (Isleib et al. 2001; Holbrook and Stalker 2003) and ICRISAT (Upadhyaya et al. 2001, 2003). Norden et al. (1987) tested fatty acid composition of 500 peanut genotypes and fortunately obtained two lines with 80% oleic acid and very low linoleic acid (2%). Since then, breeding for high oleic acid has been attracting more research efforts in many countries. Several high olerate peanut cultivars have been developed in the USA, Australia and other countires (Liao and Holbrook 2007; Yu 2011).

4.4.3 Resistance to Foliar Diseases

Worldwide, three foliar diseases in peanut including early leaf spot, late leaf spot and rust are most widespread (Dwivedi et al. 2003). In terms of early leaf spot, the geographic distribution is relatively limited therefore breeding for this resistance has been limited. More than 30 germplasm lines have been identified as resistant (Singh et al. 1997). Most of these resistant genotypes originated from secondary centers of diversity in South America. Resistant lines have been identified in var. *hypogaea*, var. *fastigiata* and var. *peruviana* but none in var. *vulgaris* (Spanish type). However, some of the resistant lines showed less stable in resistance acrossing locations (Singh et al. 1997). ICG 13917 is an interspecific derivative and showed stable resistance to early leaf spot in Malawi and three locations in Asia. In the USA, Wilson is a peanut cultivar with improved resistance to early leaf spot and the resistance is reported as transferred from a plant introduction PI 476823 that is a genotype named "Xuxi 4" introduced from China (Porter et al. 1992).

Late leaf spot in peanut is more widespread and destructive compared to early leaf spot. Scientists at ICRISAT, India have tested over 13,000 germplasm accessions for their reaction to late leaf spot. In total 69 peanut lines have been identified as resistant with disease scores ranging between three and five based on a one to nine scale (Singh et al. 1997). In China, 53 genotypes have been identified as resistant to late leaf spot from 5,700 accessions tested, and nearly 60% of these resistant materials belong to var. *fastigiata* introduced from foreign countries (Liao 2003). Undesirable genetic linkage between the resistance and low yield, late maturity, low shelling outturn and heavy pod reticulation is believed to be the reason that most of the resistant germplasm lines are difficult to be efficiently used in breeding. A number of interspecific populations have shown high levels of resistance to late leaf spot and better breeding efficiency and promising breeding lines have been selected from them. Agronomically, the pod traits such as shelling percentage and pod reticulation in the progenies generated from the resistant interspecific populations are more acceptable compared to those from the resistant ssp. *fastigiata* lines. Some interspecific germplasm lines with resistance to leaf spot have been released in the USA (Stalker et al. 2002). Until the release of "Southern Runner" with moderate resistance to late leaf spot in 1984, no commercial cultivar was resistant to this disease in the United States. Moderate to high levels of resistance are also available in the cultivars Florida MDR 98 (Gorbet and Shokes 2002b), C-99R (Gorbet and Shokes 2002a) and "Georgia-01R" (Branch 2002).

Peanut rust has been reported in most tropical and subtropical countries. ICRISAT has also tested over 13,000 peanut germplasm accessions, from which, 169 lines with disease scores of five or less on a one—nine scale (Subrahmanyam et al. 1995) have been reported as resistant to rust (Upadhyaya et al. 2001). In China, 92 accessions have been identified as resistant to rust from 5,700 accessions screened (Liao 2003). Similar to the resistance sources to early and late leaf spot, most of the rust resistant landraces belonging to var. *fastigiata* and var. *peruviana* have poor agronomic traits including low shelling outturns, thick pod shells, heavy pod reticulation, and even unacceptable seed coat colors. Undesirable genetic linkage between rust resistance and the poor pod characters has been observed in most breeding programs and impeded the progress of breeding. Several interspecific derivatives with rust resistance transferred from *A. batizocoi* and *A. duranensis* might be more valuable and their use may lead to release of cultivars with better agronomic characteristics.

Resistance to rust and late leaf spot is reported to be somehow correlated (Anderson et al. 1990). Forty two genotypes resistant to late leaf spot are also resistant to rust (Singh et al. 1997). ICG 13917, an interspecific derivative, has shown high levels of resistance to late leaf spot, early leaf spot and rust (Singh et al. 1997). Reddy et al. (2000) reported that ICGV 87853 was released with high level resistance to rust and moderate resistance to late leaf spot. Dwivedi et al. (2002) reported that the remaining green leaf area is an important criterion for selecting resistance to late leaf spot and rust. Several interspecific derivatives, ICGVs 99005, 99003, 99012, and 99015 for rust and ICGVs 99006, 99013, 99004, 99003, and 99001 for late leaf spot, would be desirable parents for use in resistance breeding programs (Dwivedi et al.

2002). Yueyou 114 developed at Crops Research Institute of Guangdong Academy of Agricultural Sciences (China) was resistant to late leaf spot and bacterial wilt (Li et al. 2007). Zhonghua 4 and Shanyou 27 with moderate resistance to rust have been extensively cultivated in central and southern regions in China where rust is a problem (Liao 2003).

4.4.4 Resistance to Bacterial Wilt

In certain regions in China, Indonesia, Vietnam and Uganda, Bacterial Wilt (BW) caused by *Ralstonia solanacearum* is a serious problem in peanut production. Planting resistant cultivar is the most useful approach to control the disease and breeding for resistance started in the early 20th century in Indonesia (Sun 1998). Peanut germplasm screening for resistance has been conducted during the past century in China and Indonesia (Singh et al. 1997). High survival ratio under heavy disease pressure is a key criteria for resistance assessment. Latent infection by the bacteria in peanut plants has been reported (Liao et al. 1998). Resistant germplasm lines have also been identified in Vietnam (Hong et al. 1999). Worldwide, more than 120 peanut genotypes with high level resistance to bacterial wilt have been identified including landraces and improved cultivars. Bacterial wilt resistance has also been reported in wild *Arachis* species (Liao et al. 1998; Tang and Zhou 2000). Evidence in transferring of bacterial wilt resistance from wild species into peanut without obvious undesirable genetic linkage verified that the resistance in some wild species is controlled by major genes. Interestingly, many dragon type landraces collected from South China where bacterial wilt has been prevalent have been identified as highly resistant, but no landrace collected from the northern regions (without serious BW incidence) in this botanical type has been identified as resistant.

Even though many sources of resistance to bacterial wilt have been identified, only a few have been successfully used in peanut breeding (Sun 1998). In China, most released resistant peanut cultivars have Xiekangqing or Taishan Sanlirou as resistant parents. In Indonesia, only Schwarz 21 and its derivatives have been used as resistant parents. All released resistant cultivars are derived from resistant parents belonging to subspecies *fastigiata*, even though many resistant lines in subspecies *hypogaea* have been used in breeding programs. Several peanut cultivars including Zhonghua 4, Zhonghua 6, Tianfu 11, Yuanza 9102, Yueyou 202-35, Yueyou 79, Yuhua 14 and Zhonghua 21 with desirable resistance to bacterial wilt and improved agronomic traits have been developed and released for production in China (Yu 2011). Several cultivars with bacterial wilt resistance developed at Guangdong Academy of Agricultural Scienecs in China are also resistant to rust. Besides being highly resistant to bacterial wilt, Zhonghua 6 developed at the Oil Crops Research Institute of Chinese Academy of Agricultural

Sciences has also been identified as relatively more resistant to aflatoxin production compared to other cultivars (Liao et al. 2003). Application of the resistant cultivars has greatly reduced wilting ratio in production to less than 5% in hotspot areas.

4.4.5 Resistance to Viral Diseases

Several viral diseases are causing more yield loss in peanut production, breeding for resistance has only been effectively conducted for Tomato Spot Wilt Virus (TSWV) in the USA and Groundnut Rosette Virus (GRV) in Africa. For many other viruses affecting peanut, no meaningful resistance source has been identified even in the wild species. In the USA, Southern Runner is the first commercial peanut cultivar with modertate level of resistance to TSWV. Like Southern Runner, other TSWV resistant cultivars such as Georgia Green, C-99R and Florida MDR98 all have PI 203396 (a line with resistance to late leaf sopt and TSWV) as their direct or indirect parent are also relatively resistant late leaf spot. More recently, Tifguard has been developed in the USA with resistance to both root knot nematodes and tomato spotted wilt virus by hybridizing a TSWV-resistant cultivar with a nematode-resistant cultivar. Resistance to GRV has been successfully transferred from wild *Arachis* species to the cultivated peanut at ICRISAT. Several GRV resistant peanut cultivars have been developed by using the resistant interspecific derivatives and they can be grown in Africa where GRV has been a serious problem in production.

4.4.6 Resistance to Aflatoxin Contamination

With the increased food safety concern caused by aflatoxin contamination in agricultural products including peanuts, improvement of genetic resistance to aflatoxin contamination in the host plant has received many research efforts which may lead to effective reduction of contamination and food safety risk (Nigam et al. 2009). Three types of resistance to aflatoxin in peanut have been reported including resistance to *Aspergillus* invasion, resistance to aflatoxin production and resistance to Preharvest Aflatoxin Contamination (PAC). In most cases, resistance to seed infection by the fungi and resistance to aflatoxin formation are assessed by *in vitro* inoculation of healthy peanut seeds under controlled conditions (Mehan et al. 1991; Cole et al. 1995; Xiao et al. 1999). Research efforts have also been made for techniques of screening field ressistance. Several authors have investigated the effect of temperature and moisture stress on colonization of kernels and aflatoxin production (Hill et al. 1983; Sanders et al. 1985) and described techniques for accurately measuring aflatoxins in breeding (Sanders et al. 1993; Holbrook et al. 1994; Anderson et al. 1996; Holbrook et al. 1998). In general, evaluation of resistance in peanut to aflatoxin contamination is much more difficult compared to other traits, and only a small portion of germplasm accessions have been tested.

Inconsistency between *in vitro* resistance screening and field resistance testing has been reported (Blankenship et al. 1985; Anderson et al. 1995), correlations between preharvest aflatoxin contamination or field resistance and *in vitro* resistance have been reported in Africa (Zambettakis et al. 1981; Waliyar et al. 1994), India (Mehan et al. 1986, 1987), and China (Wang et al. 2004). *In vitro* resistance to aflatoxin formation in peanut seeds that had been stored for two to four years with poor emergence capacity has been observed (Wang et al. 2003), indicating that certain storage proteins could be involved in resistance in these genotypes. Twenty one peanut genotypes have been identified as resistant to seed invasion of *Aspergillus flavus* with a seed infection frequency less than 2% in a sick plot under imposed drought stress at ICRISAT (Singh et al. 1997). Although significant genotype × environment interactions for seed infection have been reported, some accessions including ICGs 1326, 3263, 3700, 4749 and 7633 have shown consistent resistance to seed infection in India and Senegal. Most of these lines also possess resistance to seed colonization in laboratory inoculation (Mehan et al. 1991). It is interesting to note that most peanut genotypes resistant to seed infection and colonization belong to *A. hypogaea* subsp. *fastigiata* var. *vulgaris* with small seeds. Latha et al. (2007) analyzed the relationship between total phenols and aflatoxin production of peanut genotypes under end-of-season drought conditions. Milla-Lewis et al. (2006) reported reduced aflatoxin production in interspecific hybrid derivatives.

Considerable variation in the ability to support aflatoxin production has been observed among peanut genotypes, the resistance level is not high and less stable. Xiao et al. (1999) reported two lines with lowest aflatoxin production under artificial inoculation conditions among 1,517 peanut accessions screened. U 4-7-5 and VRR 245 were indentified as resistant to aflatoxin production even they were susceptible to seed colonization by *A. flavus* (Singh et al. 1997). There is no significant correlation between resistance to aflatoxin contamination, oil content, pod mass and BW resistance and several breeding lines with high oil content, large seeds, bacterial wilt resistance and reduced aflatoxin contamination have been selected (Liao et al. 2010). Lei et al. (2004) reported that Taishan Zhengzhu and Zhonghua 6 with resistance to bacterial wilt were relatively resistant to aflatoxin formation *in vitro*. Some peanut accessions with field resistance to PAC have been identified from the US core collection, and these accessions exhibited a 70 to 90% reduction in aflatoxin contamination in comparison to susceptible accessions (Holbrook et al. 1998). Asis et al. (2000) identified some peanut breeding lines including Colorado Irradiado moderately resistant to aflatoxin production. In Thailand, some peanut lines including

ICGVs 98305, 98330, 98348, 98353 and Tifton-8 were reported to have low aflatoxin contamination (Girdtha et al. 2009).

4.4.7 Drought Tolerance

Improved drought tolerance of peanut can help in sustaining pod yield and reducing aflatoxin contamination risk (Wright et al. 1994; Reddy et al. 2003). Drought tolerance has been reported to be associated with field aflatoxin contamination of peanut (Arunyanark et al. 2009; Girdthai et al. 2010). Drought tolerance in peanut may be improved through enhancing capability of extracting water from deep soil (Wright and Nageswara Rao 1994) or enhancing water use efficiency of the plant, or both (Hebbar et al. 1994). Drought tolerance is associated with root weight, length and size. Selecting large root volume is effective in breeding for drought tolerance (Nigam et al. 2005; Sun 1998). Dwarfed plants with more nodes, thick leaflets and more hair on leaf surface are also meaningful traits (Sun 1998). Other physiological traits are also associated with drought tolerance including stomatal conductance, surface wax, water potential, membrane stability, accumulation of cytokinins and evapotranspiration. Normally, selection of drought tolerance in peanut can be based on performace of morphological and physiological traits. In conventional breeding programs, biomass and pod yield under drought stress have been used for drought tolerance assessment. For physiological selection approaches, some are more expensive than others. Specific leaf area can be inexpensively and easily measured, thus it has been used in breeding for drought resistance in Australia (Nageswara Rao et al. 2000). Arunyanark et al. (2008) reported a promising selection index for drought tolerance based on chlorophyll stability. Drought tolerance traits including Special Leaf Area (SLA) and Root Length Density (RLD) could be contributing to resistance to aflatoxin contamination (Arunyanark et al. 2009). In the US, C76-16 is identified with improved resistance to drought as well as to aflatoxin contamination (Holbrook et al. 2009). Considerable progress in breeding drought tolerant peanut has been made at ICRISAT, several Indian institutions and Australia (Nigam et al. 2005).

4.5 Prospectives of Peanut Breeding

The global peanut production is expected to further increase with population growth and market demand expansion. However, it is commonly recognized that further production increase in peanut, as in many other crops, will mostly come from yield increase rather than expansion of planting area. Genetic enhancement for improved cultivars will continue to play a key role in peanut industry development. Generally, research emphasis will be given to enhancing and stabilizing peanut productivity in different production systems especially in low yielding areas, improving quality for diversified utilization, and ensuring food safety by reducing aflatoxin and other harmful contamination. Peanut breeding objectives will still be diverse among locations. For the developed countries, breeding efforts have shifted emphasis during recent years from mostly selecting for increased yields to selecting high-yielding cultivars with better resistance to biotic stresses and enhanced quality traits. For the developing countries, most efforts will still be made for enhancing productivity, but more efforts will also be made for quality improvement. High oil content will attract more breeding efforts in regions where more peanuts are crushed for oil. Breeding for tolerance to drought stress and aflatoxin contamination is of high priority in both developed and developing nations. Further utilization of wild *Arachis* species would contribute more to enhancing resistance to foliar diseases, nematodes and viruses. Hopefully, interspecific hybridization will also contribute more to improving quality in particular for high oil content. More molecular techniques are being developed in peanut and more efforts will be made in identifying DNA markers for breeding application (Varshney et al. 2009). Even though no transgenic peanut cultivars has been released for production, the concerned research would serve as a long-term strategy for traits that are difficult to be addressed through conventional breeding technology (Dwivedi et al. 2003).

References

Anderson WF, Beute MK, Wynne JC, Wongkaew SE (1990) Statistical procedure for assessment of resistance in a multiple foliar disease complex of peanut. Phytopathology 80: 1451–1459.

Anderson WF, Holbrook CC, Wilson DM, Matheron ME (1995) Evaluation of preharvest aflatoxin contamination in some potentially resistant peanut genotypes. Peanut Sci 22: 29–32.

Anderson WF, Holbrook CC, Wilson DM (1996) Development of greenhouse screening for resistance to *Aspergillus parasiticus* infection and preharvest aflatoxin contamination in peanut. Mycopathologia 135: 115–118.

Arunyanark A, Jogloy S, Akkasaeng C, Vorasoot N, Kesmala T, Nageswara Rao RC, Wright GC, Patanothai A (2008) Chlorophyll stability is an indicator of drought tolerance in peanut. J Agron Crop Sci 194: 113–125.

Arunyanark A, Jogloy S, Wongkaew S, Akkasaeng C, Vorasoot N, Wright GC, Nageswara Rao RC, Patanothai A (2009) Association between aflatoxin contamination and drought tolerance traits in peanut. Field Crops Res 114: 14–22.

Asis R, Barrionuevo DL, Giorda LM, Nores ML, Aldao MA (2005) Aflatoxin production in six peanut (*Arachis hypogaea* L.) genotypes infected with *Aspergillus flavus* and *Aspergillus parasiticus*, isolated from peanut production areas of cordoba. Arg J Agri Food Chem 53: 9274–9280.

Blankenship PD, Cole RJ, Sanders TH (1985) Comparative susceptibility of four experimental peanut lines and the cultivar Florunner to preharvest aflatoxin contamination. Peanut Sci 12: 70–72.

Branch WD (2002) Registration of 'Georgia-01R' peanut. Crop Sci 42: 1750–1751.
Chu Y, Ozias-Akins P (2009) Marker assisted selection (MAS) for breeding high oleic Tifguard. In: Proceedings of the American Peanut Research Education Society, (abstract) Annual meeting held at Raleigh, North Carolina on 14–17 July, 41: 61.
Cole RJ, Dorner JW, Holbrook CC (1995) Advances in mycotoxin elimination and resistance. In: Pattee HE, Stalker HT (eds) Advances in Peanut Science. American Peanut Research and Education Society, Stillwater, OK, USA, pp 456–474.
Dwivedi SL, Nigam SN, Jambunathan R, Sahrawat KL, Nagabhushanam GVS, Raghunath K (1993) Effect of genotypes and environments on oil content and oil quality parameters and their association in peanut (*Arachis hypogaea* L.). Peanut Sci 20: 84–89.
Dwivedi SL, Pande S, Narayan Rao J, Nigam SN (2002) Components of resistance to late leaf spot and rust among interspecific derivatives and their significance in a foliar disease resistance breeding in groundnut (*Arachis hypogaea* L.). Euphytica 125: 81–88.
Dwivedi SL, Crouch JH, Nigam SN, Ferguson ME, Paterson AH (2003) Molecular breeding of groundnut for enhanced productivity and food security in the semi-arid tropics: opportunities and challenges. Adv Agron 80: 153–221.
FAO web, http://www.fao.org/.
Gorbet DW, Shokes FM (2002a) Registration of 'C-99R' peanut. Crop Sci 42: 2207.
Gorbet, DW, Shokes FM (2002b) Registration of 'Florida MDR 98' peanut. Crop Sci 42: 2207–2208.
Girdthai T, Jogloy S, Vorasoot N, Akkasaeng C, Wongkaew S, Holbrook CC, Patanothai A (2010) Associations between physiological traits for drought tolerance and aflatoxin contamination in peanut genotypes under terminal drought. Plant Breed 129: 693–699.
Gowda MVC, Motagi BN, Naidu GK, Diddimani SB, Sheshagiri R (2002) GPBD 4: A Spanish bunch groundnut genotype resistant to rust and late leaf spot. Int *Arachis* Newsl 22: 29–32.
Hammons RO (1973) Genetics of *Arachis hypogaea*. In: Peanuts—culture and uses. American Peanut Research and Education Association, Stillwater, OK, USA, pp 135–173.
Hebbar KB, Sashidhar VR, Udayakumar M, Devendra R, Nageswara Rao RC (1994) A comparative assessment of water use efficiency in groundnut (*Arachis hypogaea*) grown in containers and in the field under water-limited conditions. J Agri Sci 122: 429–434.
Hill RA, Blankenship PD, Cole RJ, Sanders TH (1983) Effects of soil moisture and temperature on preharvest invasion of peanuts by the *Aspergillus flavus* group and subsequent aflatoxin development. Appl Environ Microbiol 45: 628–633.
Holbrook CC, Anderson WF, Pittman PN (1993) Selection of a core collection from the US germplasm collections of peanut. Crop Sci 33: 859–861.
Holbrook CC, Matheron ME, Wilson DM, Anderson WF, Will ME, Norden AJ (1994) Development of a large-scale field system for screening peanut for resistance to preharvest aflatoxin contamination. Peanut Sci 21: 20–22.
Holbrook CC, Wilson DM, Matheron ME (1998) Sources of Resistance to preharvest aflatoxin contamination in peanut. Proc Am Peanut Res Educ Soc 30: 54 (abstr).
Holbrook CC, Isleib TG (2001) Geographical distribution of genetic diversity in *Arachis hypogaea*. Peanut Sci 28: 80–84.
Holbrook CC, Stalker HT (2003) Peanut breeding and genetic resources. Plant Breed Rev 22: 297–456.
Holbrook CC, Guo BZ, Wilson DM, Timper P (2009) The US breeding program to develop peanut with drought tolerance and reduced aflatoxin contamination. Peanut Sci 36: 50–53.
Hong NX, Mehan VK, Lieu NV, Yen NT (1999) Identification of groundnut genotypes resistant to bacterial wilt in Vietnam. Int J Pest Manag 45(4): 239–243.
Isleib TG, Wynne JC (1992) Use of plant introductions in peanut improvement. In: Shands HL, Weisner LE (eds) Use of Plant Introductions in Cultivar Development, Part 2, CSSA Special Publ, Madison, WI, USA, pp 75–116.

Isleib TG, WynneJC, Nigam SN (1994) Groundnut breeding. In: Smartt J (ed) The Groundnut Crop: A Scientific Basis for Improvement. Chapman and Hall, London, UK, pp 552–623.

Isleib TG, Holbrook CC, Gorbet DW (2001a) Use of *Arachis* spp. plant introductions in peanut cultivar development. Peanut Sci 28: 96–113.

Isleib TG, Pattee HE, Gorbet DW, Giesbrecht FG (2001b) Genotypic variation in roasted peanut flavor quality across 60 years of breeding. Peanut Sci 29: 92–98.

Jiang H, Duan N (1998) Utilization of groundnut germplasm resources in breeding programs. Crop Genet Resour 2: 24–25.

Jiang H, Ren XP (2006) Genetic diversity of peanut resource on morphological characters and seed chemical components in China. Chin J Oil Crop Sci 28: 421–426.

Jiang HF, Ren XP, Wang SY, Liao BS (2006) Durability of resistance to *Aspergillus flavus* infection and effect of intact testa without injury on aflatoxin production in peanut. Acta Agron Sin 32: 851–855.

Jiang HF, Chen BY, Ren XP, Liao BS, Lei Y, Fu TD, Ma CZ, Mace E, Crouch JH (2007) Identification of SSR markers linked to bacterial wilt resistance of peanut with RILs. Chin J Oil Crop Sci 29: 26–30.

Jiang HF, Ren XP, Wang SY, Huang JQ, Lei Y, Liao BS (2010) Identification and evaluation of high oil content in wild *Arachis* species. Chin J Oil Crop Sci 32: 30–34.

Khedikar YP,Gowda MVC, Sarvamangala C, Patgar KV, Upadhyaya HD, Varshney RK (2010) A QTL study on late leaf spot and rust revealed one major QTL for molecular breeding for rust resistance in groundnut (*Arachis hypogaea* L.). Theor Appl Genet 121: 971–984.

Knauft DA, Ozias-Akins P (1995) Recent methodologies for germplasm enhancement and breeding. In: Pattee HE, Stalker HT (eds). Advances in Peanut Science. American Peanut Research and Education Society, Stillwater, OK, USA, pp 54–94.

Knauft DA, Wynne JC (1995) Peanut breeding and genetics. Adv Agron 55: 393–445.

Latha P, Sudhakar P, Sreenivasulu Y, Naidu PH, Reddy PV (2007) Relationship between total phenols and aflatoxin production of peanut genotypes under end-of-season drought conditions. Acta Physiol Plant 29: 563–566.

Lei Y, Wang SY, Li D, Jiang HF, Liao BS (2004) Evaluation of resistance to aflatoxin contamination among peanut germplasm with resistance to bacterial wilt. Chin J Oil Crop Sci 26: 69–71.

Lei Y, Liao BS, Wang SY, Li D, Jiang HF (2005) Identification of AFLP markers for resistance to seed infection by *Aspergillus flavus* in peanut (*Arachis hypogae*a L.), Acta Agron Sin 31: 1349–1353.

Li S X, Zhou GY, Liang XQ, Li SL, Lin YK, Hong YB, Chen XP, Liu HY (2007) Breeding of a new peanut variety Yueyou 114 and its cultivation technologies. J Peanut Sci 36: 34–37.

Liao BS (ed) (2003) The Groundnut. Hubei Press for Science and Technology, Wuhan, China (in Chinese) ISBN 7-5352-2919-0/S.327.

Liao BS, Duan NX, Jiang HF, Liang XQ, Gao GQ (1998) Germplasm screening and breeding for resistance to bacterial wilt in China. In: Groundnut Bacterial Wilt: Proceedings of the Fourth Working Group Meeting, 11–13 May 1998, Vietnam Agricultural Science Institute, Hanoi, Vietnam. S Pande, B Liao, NX Hong, Johansen C, Gowda CLL (eds) International Crops Research Institute for the Semi-Arid Tropics, India, pp 75–81.

Liao BS, Lei Y, Wang SY, Li D, Jiang HF, Ren XP (2003) Aflatoxin resistance in bacterial wilt resistant groundnut germplasm. Int *Arachis* Newsl 23: 23.

Liao BS, Holbrook CC (2007) Groundnut. In: Singh RJ (ed) Genetic Resources, Chromosome Engineering and Crop Improvement. CRC Press, Boca Raton, FL, USA, pp 51–87

Liao BS, Lei Y, Li D, Wang SY, Huang JQ, Ren XP, Jiang HF, Yan LY (2010) Novel high oil germplasm with resistance to *Aspergillus flavus* and bacterial wilt developed from recombinant inbred lines. Acta Agron Sin 36: 1296–1301.

Mehan VK, McDonald D, Haravu LJ, Jayanthi S (1991) The Groundnut Aflatoxin Problem: Review and Literature Database. International Crops Research Institute for Semi-Arid Tropics, Patancheru, Andhra Pradesh, India.

Mehan VK, McDonald D, Rajagopalan K (1987) Resistance of peanut genotypes to seed infection by *Aspergillus flavus* in field trials in India. Peanut Sci 14: 17–21.

Mehan VK, McDonald D, Ranakrishna N, Williams JH (1986) Effect of genotype and date of harvest on infection of peanut seed by *Aspergillus flavus* and subsequent contamination with aflatoxin. Peanut Sci 13: 46–50.

Milla-Lewis SR, Swift JE, Isleib TG, Tallury SP, Stalker HT (2006) AFLP markers associated with reduced aflatoxin accumulation in interspecific peanut lines. In: Proceedings of American Peanut Research and Education Society, (abstr) Annual meeting held at Savannah, Georgia on 11–14 July, 38: 57.

Mothilal A, Karunanithi K, Babu C (2010) Gamma ray induced Spanish bunch mutant with foliar disease resistance in groundnut (*Arachis hypogaea* L.). Elec J Plant Breed 1(3): 342–345

Nageswara Rao RC, Wright GC, Cruickshank AL (2000) Genetic enhancement of drought resistance in Australian peanuts. Proceedings American Peanut Research and Education Society 32: 71 (abstr).

Nadaf HL, Kaver SB, Madhusudan K, Motagi BN (2009) Induced genetic variability for yield and yield components in peanut (*Arachis hypogaea* L.). In: Shu QY (ed). Induced Plant Mutations in the Genomics Era. Food and Agriculture Organization of the United Nations, Rome, Italy, pp 346–348.

Nigam SN (2000) Some strategic issues in breeding for high and stable yield in groundnut in India. J Oilseeds Res 17: 1–10.

Nigam SN, Aruna R (2008) Improving breeding efficiency for early maturity in peanut. In: Janick J (ed) Plant Breeding Reviews, vol 30. Wiley, Hoboken, UK, pp 295–323.

Nigam SN, Chandra S, Rupa Sridevi K, Bhukta M, Reddy AGS, Rahaputi NR, Wright GC, Reddy PV, Deshmukh MP, Mathur RK, Basu MS, Vasundhara S, Vindhiyavarman P, Nagda AK (2005) Efficiency of physiological trait-based and empirical selection approaches for drought tolerance in groundnut. Ann Appl Biol 146: 433–439.

Nigam SN, Waliyar F, Aruna R, Reddy SV, Lava Kumar P, Craufurd PQ, Diallo AT, Ntare NR, Upadhyaya HD (2009) Breeding peanut for resistance to aflatoxin contamination at ICRISAT. Peanut Sci 36: 42–49.

Norden AJ, Gorbet DW, Knauft DA, Young CT (1987) Variability in oil quality among peanut genotypes in the Florida breeding program. Peanut Sci 14: 7–11.

Porter DM, Coffelt TA, Wright FS, Mozingo RW (1992) Resistance to sclerotinia blight and early leafspot in Chinese germplasm. Peanut Sci 19: 41–43.

Qiu QS (1992) The major progress on groundnut breeding by radiation mutation and prospectives. Oil Crops of China 1: 6–7.

Qiu QS, Feng HS (1998) Breeding groundnut by radiation. In: Sun D (ed) Breeding of Groundnut. China Agricultural Press, Beijing (in Chinese), pp 190–221.

Ravi K, Vadez V, Isobe S, Mir RR, Guo Y, Nigam SN, Gowda MVC, Radhakrishnan T, Bertioli DJ, Knapp SJ, Varshney RK (2011) Identification of several small main-effect QTLs and a large number of epistatic QTLs for drought tolerance related traits in groundnut (*Arachis hypogaea* L.). Theor Appl Genet 122: 1119–1132.

Reddy LJ, Nigam SN, Subrahmanyam P, Ismael FM, Govinden N, van der Merwe PJA (2000) Registration of groundnut cultivar Venus (ICGV 87853). Int *Arachis* Newsl 20: 29–31.

Reddy TY, Reddy VR, Anbumozhi V (2003) Physiological responses of groundnut (*Arachis hypogaea* L.) to drought stress and its amelioration: a critical review. Plant Growth Regul 41: 75–88.

Sanders TH, Cole RJ, Blankenship PD, Dorner JW (1993) Aflatoxin contamination of peanut from plants drought stressed in pod or root zones. Peanut Sci 20: 5–8.

Sanders TH, Cole RJ, Blankenship PD, Hill RA (1985) Relation of environmental stress duration to *Aspergillus flavus* invasion and aflatoxin production in preharvest peanuts. Peanut Sci 12: 90–93.

Singh AK, Mehan VK, Nigam SN (1997) Sources of resistance to groundnut fungi and bacterial diseases: an update and appraisal. Information Bulletin No 50. International Crops Rsearch Institute for the Semi-Arid Tropcis, Patancheru, Andhra Pradesh, India.

Stalker HT, Beute MK, Shew BB, Isleib TG (2002) Registration of five leaf spot-resistant peanut germplasm lines. Crop Sci 42: 314–316.

Subrahmanyam P, McDonald D, Waliyar F, Reddy LJ, Nigam SN, Gibbons RW, Rao VR, Singh AK, Pande S, Reddy PM, Subba Rao PV (1995) Screening methods and sources of resistance to rust and late leafspot of groundnut. Information Bulletin no 47. International Crops Research Institute for the Semi-Arid Tropics, Patancheru, Andhra Pradesh, India.

Sun DR (1998) Breeding of Groundnut. China Agricultural Press, Beijing (in Chinese). ISBN 7-109-05154-4/S.3261.

Tang RH, Zhou HQ (2000) Resistance to bacterial wilt in some interspecific derivatives of hybrids between cultivated peanut and wild species. Chin J Oil Crop Sci 22(3): 61–65.

Upadhyaya HD, Ferguson ME, Bramel PJ (2001) Status of the *Arachis* germplasm collection at ICRISAT. Peanut Sci 28: 89–96.

Upadhyaya HD, Oritz R, Bramel PJ, Singh S (2003) Development of a groundnut core collection using taxonomical, geographical and morphological descriptors. Genet Resour Crop Evol 50: 139–148.

Varshney RK, Bertioli DJ, Moretzsohn MC, Vadez V, Krishnamurthy L, Aruna R, Nigam SN, Moss BJ, Kannan S, Ravi K, He G, Knapp SJ, Hoisington DA (2009) The first SSR-based genetic linkage map for cultivated groundnut (*Arachis hypogaea* L.). Theor Appl Genet 118: 729–739.

Waliyar F, Hassan H, Bonkoungou S, Bosc JP (1994) Sources of resistance to *Aspergillus flavus* and aflatoxin contamination in groundnut genotypes in West Africa. Plant Dis 78: 704–708.

Wan SB (2003) Science of Peanut Cultivation in China. Shanghai Press for Science and Technology, Shanghai, China (in Chinese). ISBN 7-5323-7172-7.

Wang SY, Lei Y, Li D, Xiao DR, Liao BS (2003) Effect of storage period on infection of *Aspergillus flavus* and aflatoxin formation in groundnut. J Peanut Sci 32(Suppl): 390–393.

Wang SY, Liao BS (2004) Aflatoxin contamination under drought stress in groundnut. Int *Arachis* Newsl No 24.

Wang CT (2002) Production of extra large poded and small poded peanut mutants following chemical mutagen treatment. J Peanut Sci 31: 5–8.

Wright GC, Nageswara Rao RC (1994) Groundnut water relations. In: Smartt J (ed) The Groundnut Crop: A Scientific Basis for Improvement. Chapman and Hall, London, UK, pp 281–335.

Wright GC, Nageswara Rao RC, Farquhar GD (1994) Water-use efficiency and carbon isotope discrimination in peanut under water deficit conditions. Crop Sci 34: 92–97.

Xia YL, Liao BS, Li JN, Lei Y, Jiang HF, Cui FH, Zeng XP (2007) Identification of AFLP markers linked to resistance to late leaf spot in peanut (*Arachis hypogaea* L.). Chin J Oil Crop Sci 29: 318–321.

Xiao DR, Wang SY, Zhang HL (1999) Progress of research on resistance to aflatoxin contamination in groundnut. Peanut Sci Technol (Suppl): 124–129.

Yu SL (2011). Peanut Genetics and Breeing in China. Shanghai Scientific and Technology Press, Shanghai, China.

Zambettakis C, Wliyar F, Bockelee Murvan A, de Pins O (1981) Results of four years of research on resistance of groundnut varieties to *Aspergillus flavus*. Oleágineux 36: 377–385.

5

Molecular Markers, Genetic Maps and QTLs for Molecular Breeding in Peanut

Manish K Pandey,[1,2] *Baozhu Guo,*[2] *C Corley Holbrook,*[3] *P Janila,*[1] *Xinyou Zhang,*[4] *David J Bertioli,*[5] *Sachiko Isobe,*[6] *Xuanqiang Liang*[7] and *Rajeev K Varshney*[1,7,8,9,*]

ABSTRACT

Integration of plant breeding, genetics and genomics promises to foster genetic enhancement leading to increased productivity, oil quality and resistance/tolerance to biotic and abiotic stresses. Recent advances in peanut have resulted in the development of genomic resources such as SSR markers, and genetic maps for diploids and tetraploids. Even though the tetraploid species have both the genomes, the genetic diversity observed in cultivated peanut maps has been low. Therefore, only partial (<100 loci) to low-moderate (<300 loci) genetic maps could be constructed. Consensus genetic maps were, therefore, constructed with thousands of marker loci using mapping information of multiple mapping populations in order to integrate as many markers as possible

[1]International Crops Research Institute for the Semi-Arid Tropics (ICRISAT), Hyderabad, India.
[2]USDA-ARS, Crop Protection and Management Research Unit, Tifton, USA.
[3]USDA-ARS, Crop Genetics and Breeding Research Unit, Tifton, USA.
[4]Henan Academy of Agricultural Sciences, Zhengzhou, China.
[5]Universidade de Brasília, Brasília, Brazil.
[6]Kazusa DNA Research Institute, Kisarazu, Japan.
[7]Crops Research Institute, Guandong Academy of Agricultural Sciences, Guangzhou, China.
[8]CGIAR Generation Challenge Programme, c/o CIMMYT, Mexico DF, Mexico.
[9]The University of Western Australia, Crawley, Australia.
*Corresponding author: *r.k.varshney@cgiar.org*

List of abbreviations after the text.

on a single genetic map. Development of SNP markers should lead to even more dense genetic maps and use of these markers in routine breeding and genetic applications. Efforts with the available limited genomic resources led to the identification of linked markers for drought tolerance, oil quality and disease resistance in peanut through trait mapping. These developments also led to deployment of linked markers to improve disease resistance and oil quality. Ongoing efforts should lead to the availability of the whole-genome sequence in the near future, providing huge genomic resources, which will hasten the much needed linking of phenotype with markers/genome sequences. However, this can only be achieved with precise and high-throughput phenotyping for complex traits. Recent advances in peanut genomics and molecular breeding efforts provide hope for efficient genetic enhancement of peanut for production as well as quality constraints.

Keywords: Groundnut, Genetic maps, QTL mapping, Molecular markers, Molecular breeding, Genomic resources, Genetic improvement

5.1 Introduction

Peanut or groundnut (*Arachis hypogaea* L.), with current annual production of 38.0 million tons from an area of 24.0 m ha (http://faostat.fao.org), is the fourth-largest oilseed crop in the world and is mostly grown in semi-arid regions with relatively low inputs of chemical fertilizers. The crop is cultivated in more than 100 countries of Asia, Africa and the Americas with the largest (more than two-third) contributions coming from China and India. Peanut plays important roles in food and nutritional security along with improving the livelihood of resource-poor farmers. Peanut seeds contain edible oil (40–60%), protein (20–40%), carbohydrate (10–20%) and several nutritional components such as vitamin E, niacin, calcium, phosphorus, magnesium, zinc, iron, riboflavin, thiamine and potassium. Several uses of peanut make it an excellent cash crop for domestic as well as international trade. The major share goes towards extraction of vegetable oil for use in cooking apart from its use in the confectionary industry and fodder, a major source for protein feed for animals.

Since peanut is generally grown in marginal environments in Asia and Africa, the crop is challenged by several stress factors including biotic and abiotic stresses. The lack of genetic and genomic resources has significantly hampered peanut improvement programs. The major constraints for low genetic enhancement of cultivated peanut is attributed to: (i) very recent origin and highly conserved genome (Young et al. 1996), (ii) availability of only one related tetraploid wild species (*A. monticola*) (Krapovickas and Gregory 1994), (iii) the species in other sections are mostly diploid and hence limited sexual compatibility with cultivated peanut, (iv) lack of information on genetic architecture of economically important traits of peanut, and (v) limited availability of molecular markers, genetic maps and Quantitative Trait Loci (QTLs). Genomics tools offer great promise to overcome the complex genetic makeup of peanut but lack of minimum genomic and genetic resources has hampered such efforts. Major biotic stresses include Early Leaf Spot (ELS), Late Leaf Spot (LLS), leaf rust, mottle virus, rosette virus, aphids, jassids and thrips/Tomato Spotted Wilt Virus (TSWV). Drought is the major abiotic stress as 70% of the crop is grown in the semi-arid tropics, which are characterized by low and erratic rainfall. In spite of the genetical obstacles listed above, some efforts were made towards crop improvement through stress management using conventional approaches. Furthermore, restricted gene flow due to differences in ploidy level has severely hampered transfer of desired alleles from diploid wild relatives and hence, the much needed broadening of the genetic base of the species could not be achieved so far. Thus, the increasing population pressure seems not to be managed alone with conventional approaches and needs integration of genomics tools with the peanut improvement programs.

Due to the increased availability of genomic tools in recent years, Genomics-Assisted Breeding (GAB) offers hope for accelerated peanut improvement. Additionally, integration of genomics tools should aid in diversifying the existing narrow genetic base of the peanut gene pool with useful alleles and in understanding the complexity of the large tetraploid genome for genetic enhancement of cultivated peanut. Recent years have witnessed much progress in better understanding of crop genomics and its integration with conventional breeding, referred to as genomics-assisted breeding (GAB) to practice precision breeding for target traits (Varshney et al. 2005, 2010a). This advancement has not been achieved uniformly for all important crops and most importantly, could handle only simpler traits. Nevertheless, recent results showed significant advantages over conventional breeding in handling traits which are difficult to manage through conventional phenotypic selection and GAB has been successfully demonstrated in several temperate cereal crops (Varshney et al. 2006) and some legume crops (Varshney et al. 2010b, 2012a, 2013). In addition, introgression/pyramiding of multiple recessive alleles can be achieved very efficiently in less time and with more accuracy along with pyramiding of several monogenic traits or QTLs for a single trait (Ribaut and Hoisington 1998; Xu and Crouch 2008; Varshney et al. 2009a,b) such as in the case of marker-assisted improvement to develop a high oleic version of the nematode resistant cultivar, Tifguard, less than three years (Chu et al. 2011; Holbrook et al. 2011). However, to advance GAB in peanut, information on available genetic variation in germplasm, availability of appropriate molecular markers and genotyping platforms, suitable genetic maps, precise phenotyping platforms and QTLs with high phenotypic variance are required.

In spite of the potential of molecular markers in crop improvement, peanut experienced slow progress in the area of developing genomic resources such as molecular markers and genetic maps until 2005. Since then significant progress has been achieved as a result of concerted efforts of the international peanut community resulting in the development of several thousands of markers, several genetic maps, dense consensus genetic maps, QTL mapping and molecular breeding for resistance/tolerance to biotic stresses for peanut improvement (Guo et al. 2011; Holbrook et al. 2011; Pandey et al. 2012a). The progress made in genomic resources such as molecular markers, genetic maps, QTL identification and marker-assisted breeding in peanut has started to make progress with the help of genomic resources and should help to overcome genetic bottlenecks, and result in accelerated breeding progress.

5.2 Marker Development

Among all the genomic resources, molecular markers have proved to have the most direct applications towards characterizing and harnessing available genetic variation. These markers have been used in several genetic studies such as germplasm characterization, trait mapping and most importantly molecular marker-assisted breeding (Guo et al. 2011; Holbrook et al. 2011; Pandey et al. 2012a). Although several marker systems such as Restriction Fragment Length Polymorphisms (RFLPs), Random Amplified Polymorphic DNAs (RAPDs), Amplified Fragment Length Polymorphisms (AFLPs) and Diversity Arrays Technology (DArT) markers became available and proved their utility from time to time (Varshney et al. 2006; Gupta et al. 2010), Simple Sequence Repeats (SSRs) or microsatellites and Single Nucleotide Polymorphism (SNP) markers are currently the most preferred marker systems for genetic studies and breeding applications. Although SSR markers are very much in use in current plant breeding applications, due to high-throughput genotyping amenability, SNPs seem to have more potential for future marker systems.

Early generation marker systems (RFLPs, RAPDs and AFLPs) were used primarily for studying genetic diversity of peanut (Hilu and Stalker 1995; Kochert et al. 1996; Subramaniyan et al. 2000, Dwivedi et al. 2001; He and Prakash 2001; Herselman 2003; Bravo et al. 2006). In some cases, these markers were also used for construction of genetic maps (Halward et al. 1993; Burow et al. 2001; Milla 2003; Herselman et al. 2004; Garcia et al. 2005; Leal-Bertioli et al. 2009) and identification of associated QTLs (Herselman et al. 2004). However, the insufficient number of these markers and other discouraging reasons associated with them motivated researchers towards development and use of better marker systems. As a result, several hundred SSR markers were generated (Pandey et al. 2012a). Low diversity detected with SSR markers in the cultivated gene pool, however, demanded development of large-scale SSR markers for effective use in routine genetic and breeding applications. Therefore, aggressive efforts made worldwide during the last few years resulted in the development of >13,000 SSR markers from SSR-enriched libraries, Bacterial Artificial Chromosome (BAC)-end sequences, Expressed Sequence Tag (EST) sequences and transcript sequences generated by using 454/FLX sequencing technology (Table 5-1). After screening 4,485 SSR markers on a set of parental genotypes of several mapping populations, a set of highly informative SSR markers (199 SSRs with >0.50 PIC) along with polymorphism features of 946 novel SSR markers have been identified and these SSRs have been used for several genetic and breeding studies in peanut (Pandey et al. 2012b). Similarly, Zhao et al. (2012) and Macedo et al. (2012) have reported 143 and 66 highly informative (≥0.50 PIC) SSR markers of the 1,343 and 78 polymorphic markers detected after screening 9,274 and 146 markers, respectively.

In addition, a DArT platform (ca. 15,000 features) has been developed at DArT Pty Ltd (Australia) in collaboration with ICRISAT (India), CIRAD (France) and Catholic University of Brasília and EMBRAPA (Brazil). However, the use of DArT arrays showed a very low level of polymorphism in tetraploid (AABB) genotypes as compared to moderate level of diversity among diploid (AA and BB) genotypes (Kilian 2008; Varshney et al. 2013a). The results indicated potential use of DArT markers in monitoring genome introgression from wild relatives into peanut lines but limited use in genetics and breeding applications in cultivated peanut.

Recently, SNP markers have also been developed but mainly in diploid *Arachis* species. In the case of cultivated species, these SNPs have not been very polymorphic. For instance, The University of Georgia (USA) identified 8,486 SNPs after comparing the 454/FLX transcript sequences of 17 genotypes (over 350 Mb transcriptome data) with reference transcriptome of "Tifrunner" with moderately stringent filtering. An Illumina GoldenGate SNP array with 1,536-SNPs with high confidence was designed and used for genotyping on a diverse panel of *Arachis* genotypes. The newly designed array worked successfully (>95%) but very low polymorphism was detected for cultivated tetraploid genotypes (http://nespal.org/oziasakinslab/projects/plant-biotechnology-peanut-grasses/peanut-snp-discovery/). Another parallel effort resulted in identification of SNPs between diploid genotypes for Tentative Orthologous Genes (TOGs) at the University of California-Davis (Douglas Cook, pers. comm.) and development of 768-SNP Illumina GoldenGate array. Despite these arrays being very informative for diploid species, the study showed that homoeology between AA- and BB-genomes posed a major constraint in proper use of these arrays for

Table 5-1 *Arachis* markers available in public domain for genetic and breeding applications.

Marker series	Markers	References	Research Institute/University
Development of novel simple sequence repeat (SSR) markers			
Ah, Lec	26	Hopkins et al. 1999	USDA-ARS, USA
pPGPseq, pPGSseq	226	Ferguson et al. 2004	University of Georgia, USA/Cornell University, USA
Ah, Lec, Ap	32	Palmieri et al. 2002, 2005; Gimenes et al. 2007	Universidade Estadual Paulista (UNESP), Brazil
PM	103	He et al. 2003; Luo et al. 2005	USDA-ARS/Tuskegee University, USA
AC, Ah, gi, RN, TC, Seq	338	Moretzsohn et al. 2004, 2005	EMBRAPA, Brazil/USDA-ARS, USA
S	103	Nelson et al. 2006	University of Western Australia, Australia
LG, Lup	188	Proite et al. 2007	University of Brasília/EMBRAPA, Brazil
RN, RM	123	Wang et al. 2007	Shandong Peanut Research Institute, China
Lup, Dal, Stylo, Ades, Amor, Chaet, IPAHM, ICGM	178	Mace et al. 2007; Cuc et al. 2008; Gautami et al. 2009	International Crops Research Institute for the Semi-Arid Tropics (ICRISAT), India
EM	290	Liang et al. 2009b	Guangdong Academy of Agricultural Sciences, China/USDA-ARS, USA
ES	685	Hong et al. 2009b	Guangdong Academy of Agricultural Sciences, China
PM	138	Yuan et al. 2010	Shandong Peanut Research Institute, China/USDA-ARS/Tuskegee University, USA
F, H, PD	94	Song et al. 2010	Shandong Academy of Agricultural Sciences, China
AHGS	6680	Shirasawa et al. 2012b	Kazusa DNA Research Institute (KDRI), Japan

AhM	63	Naito et al. 2008	Ibaraki University/Mitsubishi Chemical Medience Co. , Japan
GM710-GM2847	2138	Nagy et al. 2010a	University of Georgia, USA
Fl, Ahl	1152	Douglas R Cook, unpublished	University of California, USA
Transposon markers	1039	Shirasawa et al. 2012b, 2013	Kazusa DNA Research Institute (KDRI), Japan
Development of highly informative SSR markers (PIC >0.50)			
PGPSeq, TC, gi, IPAHM, PM, S, GM, GNB	199	Pandey et al. 2012b	International Crops Research Institute for the Semi-Arid Tropics (ICRISAT), India
Ah, S, GNB, IPAHM, gi, PGPseq, AS, PGSseq	143	Zhao et al. 2012	Tuskegee University, USA
TC	66	Macedo et al. 2012	University of Brasilia, Brazil
Diversity arrays technology (DArT) markers			
DArT features	15,000	Killian 2008; Varshney 2012	DArT Pty Ltd (Australia), ICRISAT (India), CIRAD (France), and Catholic University of Brasília & EMBRAPA (Brazil)
Single nucleotide polymorphism (SNP) markers			
Illumina GoldenGate SNP array	1,536	Ozias-Akins, Peggy, pers. comm.	University of Georgia, USA
Illumina GoldenGate SNP array	768	Cook, Douglas, pers. comm.	University of California-Davis, USA

cultivated peanut. Hence, SSR markers remain the best choice for genetic and breeding studies in cultivated peanut until the whole genome sequence project is completed (www.PeanutBioscience.com). Also, in a collaborative effort with Peggy Ozias-Akins (University of Georgia, USA), ICRISAT has used a set of 96 highly informative SNPs in cultivated germplasm for conversion into KASPar assays. This assay was validated successfully for 91 SNPs (Khera et al. 2013).

Thus, thousands of molecular markers such as SSRs (13,596), DArTs (15,000) and SNPs (2304) are available (Table 5-1) for use in different genetical and breeding applications in peanut.

5.3 Construction of Individual and Integrated Genetic Maps

Although initial efforts for construction of genetic maps with 1st generation markers were reported in the last two decades of the 20th century, the majority of genetic maps were constructed between 2005–2012. Most of the initial genetic maps were developed based on mapping populations derived using diverse diploid parental genotypes in order to put the maximum number of markers on the maps. However, tetraploid populations have recently been used for construction of genetic maps as well as identification of QTLs for agronomically important traits.

5.3.1 Genetic Maps for AA-Genome

Genetic mapping in peanut was first started for AA-genome and marker systems such as RFLP (Halward et al. 1993), AFLP (Milla 2003), RAPD (Garcia et al. 2005), SSR (Moretzsohn et al. 2005; Leal-Bertioli et al. 2009) and SNP (Leal-Bertioli et al. 2009; Nagy et al. 2010a) were deployed for construction of several genetic maps (Table 5-2). The first genetic map of *Arachis* species was constructed with RFLP markers using F_2 population (*A. stenosperma* × *A. cardenasii*) mapping a total of 117 RFLP loci (Halward et al. 1993). This map was followed by construction of three more genetic maps, all with different marker systems such as AFLP (*A. kuhlmanni* × *A. diagoi*, 102 AFLP loci; Milla 2003), RAPD (*A. stenosperma* × [*A. stenosperma* × *A. cardenassi*], 167 RAPDs; Garcia et al. 2005) and SSR markers (*A. duranensis* × *A. stenosperma*, 170 SSRs; Moretzsohn et al. 2005). Since, dense maps could not be constructed using one particular marker system, efforts were then made to use a range of marker systems for genetic mapping. These efforts resulted in the development of comparatively more saturated maps. For example, one of the above-described maps (Moretzsohn et al. 2005) with 170 SSR marker loci was then saturated with an additional 199 markers including AFLP, RFLP, SCAR and SNP markers and a consolidated map with 369 marker loci was prepared (Leal-Bertioli et al. 2009). Recently, the use of

Table 5-2 Details of genetic maps constructed in *Arachis* species.

Population	Population size	Marker loci mapped	Linkage groups	Total map distance (cM)	References
A-genome genetic maps					
A. stenosperma × *A. cardenassi*	F_2	117 RFLPs	11	1,063	Halward et al. 1993
A. kuhlmanni × *A. diogoi*	179 F_2	102 AFLPs		1,068	Milla 2003
A. stenosperma × (*A. stenosperma* × *A. cardenassi*)	44 BC_1F_1	167 RAPDs, 39 RFLPs	11	800	Garcia et al. 2005
A. duranensis × *A. stenosperma*	93 F_2	170 SSRs	11	1,231	Moretzsohn et al. 2005
A. duranensis × *A. stenosperma*	93 F_2	369 markers (SSR, AFLP, SNP, RFLP, SCAR)	10	-	Leal-Bertioli et al. 2009
A duranensis × *A. duranensis*	94 F_2	2,319 markers (1,127 SNPs, 971 SSRs, 221 SSCPs)	10	-	Nagy et al. 2010a
B-genome genetic maps					
A. ipäensis × *A. magna*	93 F_2	149 SSRs	10	1,294	Gobbi et al. 2006; Moretzsohn et al. 2009
A. batizocoi PI 298639 × *A. batizocoi* PI 468327	94 F_2	449 SSRs	-	-	Guo et al. 2010
AB genome genetic maps					
A. hypogaea × *A. cardenasii*	46 $F_{10}C_9$	167 RAPDs, 39 RFLPs	11	800	Garcia et al. 1995
A. hypogaea × (*A. batizocoi* × (*A. cardenasii* × *A. diogoi*))	78 BC_1F_1	370 RFLPs	23	2,210	Burow et al. 2001
ICG 12991 × ICGV-SM 93541	200 F_2	12 AFLPs	5	139	Herselman et al. 2004
TAG 24 × ICGV 86031	318 RILs	191 SSRs	20	1,785.4	Varshney et al. 2009c; Ravi et al. 2011

Table 5-2 contd....

Table 5-2 contd.

Population	Population size	Marker loci mapped	Linkage groups	Total map distance (cM)	References
A. duranensis × (*A. ipäensis* × *A. duranensis*)	88 BC_1F_1	298 SSRs	21	1,843.7	Foncéka et al. 2009
Yueyou 13 × Zhen Zhuhei	142 RILs	133 SSRs	19	793.1	Hong et al. 2010a
Yueyou 13 × FU 95-5	84 RILs	109 SSRs	21	503.1	Hong et al. 2010a
Yueyou 13 × J 11	136 RILs	46 SSRs	13	357.4	Hong et al. 2010a
TAG 24 × GPBD 4	266 RILs	188 SSRs	20	1,922.4	Khedikar et al. 2010; Sujay et al. 2012
ICGS 44 × ICGS 76	188 RILs	82 SSRs	15	831.4	Gautami et al. 2012a
ICGS 76 × CSMG 84-1	176 RIL	119 SSRs	20	2,208.2	Gautami et al. 2012a
TG 26 × GPBD 4	146 RILs	181 SSRs	21	1,963	Sarvamangala et al. 2011; Sujay et al. 2012
SunOleic 97R × NC94022	190 RILs	170 SSR, 2 CAPS	26	1,304.9	Qin et al. 2012
Tifrunner × GT-C20	158 RILs	238 SSR, 1 CAPS	22	917.45	Qin et al. 2012
Tifrunner × GT-C20	94 F_2	381 marker loci	21	1,674.4	Wang et al. 2012, 2013
Satonoka × Kintoki	94 F_2	351 SSR, 165transposon	21	2166.4	Shirasawa et al. 2012a, 2013
Nakateyutaka × YI-0311	186 F_2	186 SSR, 107 transposon	19	1332.9	Shirasawa et al. 2012a, 2013

newly developed markers resulted in the development of an even denser genetic map using the F_2 population derived from the cross (*A. duranenis* × *A. duranensis*) with 2,319 markers (971 SSRs, 221 single stranded DNA conformation polymorphism (SSCP) markers and 1,127 SNPs) mapped on 10 linkage groups (Nagy et al. 2010a). This map has the distinction of being the densest genetic map among all peanut diploid and tetraploid genetic maps. The latter two maps combined different marker systems such as AFLP, RFLP, SSR, SCAR, SSCPs and SNP markers.

5.3.2 Genetic Maps for BB-Genome

Only two maps have been reported for the BB-genome. One genetic map with 149 SSR loci on 11 linkage groups covering 1,294 cM genome, which was developed based on an F_2 population (93 lines) derived from the cross between *A. ipäensis* (KG30076) and *A. magna* (KG30097) (Gobbi et al. 2006; Moretzsohn et al. 2009). The other genetic map was constructed with 449 SSR loci using again a F_2 population derived from the cross *A. batizocoi* (PI298639) × *A. batizocoi* (PI468327) (Guo et al. 2010) (Table 5-2). Less polymorphism has been observed in BB-genome genetic maps compared to AA-genome genetic maps.

5.3.3 Genetic Maps for Tetraploid (AABB) Genome

Realizing the difficulty of transforming full information from diploids to cultivated peanuts, intensive efforts have recently been made for development of good genetic maps for tetraploid peanut. The very first effort to construct a genetic map for AABB genome was with RAPD and RFLP markers using the cross *A. hypogaea* and *A. cardenassi*. A total of 167 RAPD and 39 RFLP loci were mapped on 11 linkage groups covering 800 cM genome distance (Garcia et al. 1995) (Table 5-2). The second tetraploid genetic map was developed six years later with 370 RFLP loci mapped on 23 linkage groups (2,210 cM genome coverage) using a backcross population (78 BC_1F_1 lines) generated from the cross of T*x*AG-6 {a synthetic amphidiploid line ([*A. batizocoi* × (*A. cardensii* × *A. diogoi*)]^{4x})} and Florunner (Burow et al. 2001). The next genetic map was constructed using AFLP markers, which resulted in development of a partial map with only 12 AFLP marker loci (Herselman et al. 2004). The comparison of diploid and tetraploid linkage maps revealed a high degree of colinearity between linkage groups (Burow et al. 2001; Jesubatham and Burow 2006) and identification of genome specific markers to assign A- and B-genome linkage groups in tetraploid genetic maps.

Low number of markers (RAPDs, RFLPs and AFLPs) and low genetic diversity among cultivated peanut seriously hampered the construction of dense genetic maps with 1st generation markers. Meanwhile, SSR markers have become more popular among geneticists and breeders due to their easy, reliable, cost-effective and robust genotyping nature. During the last decade we have witnessed the development of >13,000 SSR markers (Pandey et al. 2012a) and even identification of highly polymorphic genic and genomic SSR markers (Macedo et al. 2012; Pandey et al. 2012b; Zhao et al. 2012) that can be efficiently used in genetic diversity, mapping, QTL analysis and molecular breeding applications (Varshney et al. 2012). The first SSR-based genetic map using a Recombinant Inbred Line (RIL) population derived from TAG 24 × ICGV 86031 was constructed with 135 SSR loci after screening a total of 1,145 SSR markers (Varshney et al. 2009c) (Table 5-2). This genetic map was further saturated to 191 SSR loci mapped on 20 linkage groups with 1,785 cM genome coverage (Ravi et al. 2011). Later genetic maps were all constructed using RIL populations (Hong et al. 2010a; Khedikar et al. 2010; Sarvamangala et al. 2011; Gautami et al. 2012a; Sujay et al. 2012; Qin et al. 2012) in addition to four maps, which are based on backcross (Fonceka et al. 2009) and F_2 populations (Shirasawa et al. 2012a; Wang et al. 2012, 2013), respectively. As the mapping populations used for these maps also segregate for different traits, these maps have also been used for QTL analysis (see later).

The next genetic map based on SSRs was constructed with 298 marker loci on 21 linkage groups spanning a map distance of 1,843.7 cM using a backcross mapping population with 88 individuals from the cross between a cultivar (Fleur 11) and a synthetic amphidiploid (*A. duranensis* × *A. ipäensis*) (Fonceka et al. 2009). This map showed overall colinearity between homologous linkage groups of both the A and B genomes, and also shed light on chromosomal rearrangements events prior to tetraploidization of cultivated species. This effort was also significant towards diversification of narrow cultivated gene pool. Hong et al. (2010a) reported the next three genetic maps based on three RILs namely Yueyou 13 × Zhen Zhuhei, Yueyou 13 × FU 95-5 and Yueyou 13 × J 11 with 133 (793.1 cM), 109 (503.1 cM) and 46 (357.4 cM) marker loci, respectively. Using genotyping data from these three populations, a composite map containing 175 SSR markers in 22 linkage groups was developed (Table 5-2).

ICRISAT in collaboration with the University of Agricultural Sciences-Dharwad (UAS-D) initiated work on mapping QTLs for foliar diseases and in the process developed two new partial genetic maps using the RILs derived from the crosses TAG 24 × GPBD 4 (Khedikar et al. 2010, 462.24 cM genome coverage) and TG 26 × GPBD 4 (Sarvamangala et al. 2011, 657.9 cM genome coverage) with 56 and 45 marker loci, respectively. Upon availability of more markers, these two maps were then saturated to 188 (1,922.4 cM) and 181 (1,963 cM) marker loci, respectively (Sujay et al. 2012). In addition to the above three updated maps (TAG 24 × ICGV

86031, TAG 24 × GPBD 4 and TG 26 × GPBD 4), two more genetic maps based on RIL populations namely ICGS 76 × CSMG 84-1 and ICGS 44 × ICGS 76 were developed with 119 (2,208.2 cM genome coverage) and 82 (831.4 cM genome coverage) marker loci, respectively. In parallel, Qin et al. (2012) reported construction of two genetic maps based on the two RIL populations namely Tifrunner × GT-C20 (T population) and SunOleic 97R × NC94022 (S population). Individual genetic maps were constructed for T and S populations with 236 (1,213.4 cM) and 172 (920.7 cM) marker loci, respectively (Qin et al. 2012). The effort towards saturation of T and S population genetic maps based on RILs is ongoing (Pandey et al. 2012c; Wang et al. 2013). The genetic map based on T population has the distinction of being the densest genetic map for cultivated peanut using an RIL mapping population. A segregating population (94 F_2 individuals) of the T population was used to develop a denser map with 333 marker loci on 28 linkage groups covering a genome distance of 1,674.4 cM (Wang et al. 2012). Most recently, Shirasawa et al. (2012a) has reported construction of two genetic maps using the F_2 population derived from the crosses, i.e., Satonoka × Kintoki (516 loci includes 351 SSRs and 165 transposon) and Nakateyutaka × YI-0311 (293 loci includes 186 SSRs and 107 transposon) covering map distance of 2166.4 and 1332.9 cM, respectively. These two maps report mapping of transposon markers for the first time in peanut making this map (Satonoka × Kintoki) the most dense genetic map so far in tetraploid peanut.

As SNP markers have gained significant popularity during the past five years and have shown promising results in several crops, efforts are underway to integrate SNPs in the tetraploid maps of *Arachis*. For example, efforts at the University of California-Davis, USA (Richard Michelmore, pers. comm.) have recently started for generating ultra-high density genetic maps through low coverage, shotgun sequencing of diploid and tetraploid mapping populations and of reference sets of germplasm (Froenicke et al. 2011). These genetic materials represent populations from the AA genome (*A. duranensis* × *A. stenosperma),* BB genome *(A. ipäensis* × *A. magna)* and *AA*BB genome (*A. hypogaea* cv. IAC Runner × synthetic amphidiploid of the two progenitor species) along with reference sets of ICRISAT (Upadhyaya et al. 2003), the US mini-core collection (Holbrook and Dong 2005) and the Chinese mini-core collection (Jiang et al. 2010). The idea behind this study is to identify SNPs in the diversity panel and to use these for estimating Linkage Disequilibrium (LD) and improving the genetic bins of highly dense genetic/consensus maps. Finally, these results will help in assisting and complementing the assembly of the reference genome sequence for peanut, which will be soon available for the peanut research community (www.PeanutBioscience.com).

5.3.4 Integrated or Consensus Genetic Maps

Dense genetic linkage maps have several genetic and breeding applications such as trait mapping through linkage mapping or association analysis, marker-assisted breeding, map-based cloning and physical map alignment. Genome sequence information regarding marker order and location is very important for judicious application in breeding. Since it is almost impossible to map a large number of markers on a single map, the best option is to combine marker information of many individual genetic maps on to an integrated/consensus genetic map so that a maximum number of marker loci are mapped. Consensus maps have several advantages over individual genetic maps. The major advantages include ability: (1) to map several marker loci onto a single map, (2) to determine the relative position and stability of markers across populations and genome, (3) to provide evidence for chromosomal rearrangements and gene duplication, (4) to assign linkage groups to chromosome, and also (5) to provide the basic information for comparative genomic studies among related species and subspecies (Beavis and Grant 1991; Kianian and Quiros 1992; Hauge et al. 1993; Gentzbittel et al. 1995). Because of the above mentioned features, consensus genetic maps have been developed in many crop species like maize (Sharopova et al. 2002; Falque et al. 2005), wheat (Somers et al. 2004), barley (Varshney et al. 2007; Marcel et al. 2007), soybean (Song et al. 2004; Choi et al. 2007), pigeonpea (Bohra et al. 2012) and more recently for peanut (Gautami et al. 2012b; Shirasawa et al. 2013) (Table 5-3).

The initial integrated genetic maps were developed based on two or three mapping populations. The first integrated genetic map was based on three RIL populations ($F_{4:6}$) with 175 marker loci on 22 linkage groups with genome coverage of 885.4 cM (Hong et al. 2010a). The next integrated map was developed using two mapping populations with 225 SSR loci covering a total map distance of 1,152.9 cM (Sujay et al. 2012). Another integrated map was based on three populations with 293 marker loci onto 20 linkage groups covering genome distance of 2,840.8 cM (Gautami et al. 2012a). The latter two integrated maps were also used to show QTLs on the map, which were identified in individual populations for foliar disease resistance and drought related traits, respectively. The most recent integrated map was based on two mapping populations with 324 marker loci on 21 linkage groups covering a 1,352 cM genome distance (Qin et al. 2012).

Beside the effort towards development of integrated maps based on two or three individual maps, the marker density and number of markers has not been enhanced significantly. Therefore, a global effort was initiated to put maximum markers on the same genetic map through integrating markers from all published individual genetic maps. Marker information from one BackCross (BC) population (Fonceka et al. 2009) was also included

Table 5-3 Comparative features of different integrated/consensus genetic maps of tetraploid peanut.

S. No.	Populations used for construction of integrated/ consensus genetic maps	Number of markers integrated	Total genome distance (cM)	Linkage groups (LGs)	Marker density (marker/cM)	References
1.	Yueyou 13 × Zhen Zhuhei, Yueyou 13 × FU 95-5, Yueyou 13 × J 11	175	885.4	22	5.8	Hong et al. 2010a
2.	TAG 24 × GPBD4, TG 26 × GPBD 4	225	1152.9	20	5.15	Sujay et al. 2012
3.	TAG 24 × ICGV 86031, ICGS 44 × ICGS 76, ICGS 76 × CSMG 84-1	293	2840.8	20	9.69	Gautami et al. 2012a
4.	SunOleic 97R × NC94022, Tifrunner × GT-C20	324	1,352.1	21	4.5	Qin et al. 2012
5.	All the above 10 RILs and one BC population {*A. duranensis* x (*A. ipaensis* x *A. duranensis*)}	897	3,863.6	20	4.54	Gautami et al. 2012b
6.	Sixteen mapping populations	3,693	2,651	20	0.72	Shirasawa et al. 2013

in the development of a reference consensus map along with 10 individual genetic maps, which were all constructed using RIL populations. Finally, the reference consensus genetic map was constructed with 897 marker loci. These 897 marker loci (895 SSRs and 2 CAPS) could be mapped on 20 linkage groups spanning a total map distance of 3,607.97 cM with an average map density of 3.94 cM (Gautami et al. 2012b). More interestingly, this reference consensus genetic map was divided into 20 cM along with 203 BINs, which carry one to 20 loci with an average of four marker loci per BIN. Realizing the importance of dense consensus genetics maps, the above consensus genetic map has recently been improved further by international research partners. The mapping information from five new genetic maps (total 16 individual genetic maps) were utilized for improvement of an earlier consensus map from 897 to 3,693 markers spanning 2,651 cM of the genome and 20 linkage groups (Shirasawa et al. 2013). These dense consensus maps will have greater impact on peanut improvement because of their use in several applications such as aligning new genetic and physical maps, QTL analysis, genetic background effect on QTL expression and several other genetic and molecular breeding activities in peanut.

5.4 Trait Mapping

The ultimate goal of development of markers and genetic maps is to identify markers that are associated with traits of interest. Hence, denser genetic maps covering the full genome will enhance chances for identification of tightly-linked markers to agronomically important traits through linkage/association mapping. That is why almost all the genetic maps (except Wang et al. 2012) were constructed using immortal RIL populations segregating for important traits in cultivated peanut. Once tightly linked/perfect/functional markers are developed using these resources, these markers can be deployed in marker-assisted peanut improvement.

Initial mapping populations in peanut were developed in order to map the maximum number of loci on a single genetic map by selecting parents with diverse origins. Realizing the restricted use of these genetic maps in cultivated peanut improvement, later research focused on only development of mapping populations targeting mapping of economically important traits such as biotic stresses (TSWV, early leaf spot, late leaf spot, rust, aphid vector of groundnut rosette disease, Cylindrocladium black rot disease, Sclerotinia and nematode resistance), abiotic stress (drought tolerance), nutritional quality (aflatoxin contamination, oil content, oleic acid, linoleic acid, oleic/linoleic acid ratio) and several agronomic traits (Pandey et al. 2012a; Varshney et al. 2013a) (Table 5-4). The initial efforts towards mapping of economically important traits was through Bulked Segregant Analysis (BSA) for identifying the linked marker to nematode resistance (Burow et

Table 5-4 List of QTLs identified for some economically important traits in peanut.

Traits studied	QTLs identified	Phenotypic variance explained (%)	References
Resistance to disease resistance			
Late leaf spot (LLS)	39	1.70–67.98	Khedikar et al. 2010; Sujay et al. 2012; Wang et al. 2013
Leaf rust	27	1.70–82.96	Khedikar et al. 2010; Sujay et al. 2012
Resistance to *Aspergillus flavus* invasion	6	6.2–22.7	Liang et al. 2009a
Aphid vector of groundnut rosette disease	8	1.18–76.16	Herselman et al. 2004
Resistance to tomato spotted wilt virus (TSWV)	2	12.9–35.8	Qin et al. 2012
Drought tolerance related traits			
Transpiration (T)	15	4.36–18.17	Varshney et al. 2009c; Ravi et al. 2011; Gautami et al. 2012a
Transpiration efficiency (TE)	14	4.47–18.12	Varshney et al. 2009c; Ravi et al. 2011; Gautami et al. 2012a
Specific leaf area (SLA)	13	3.48–13.29	Varshney et al. 2009c; Ravi et al. 2011
Leaf area (LA)	4	7.24–11.51	Varshney et al. 2009c; Ravi et al. 2011
SPAD chlorophyll meter reading (SCMR)	29	5.72–19.53	Varshney et al. 2009c; Ravi et al. 2011
Biomass	7	4.25–20.32	Varshney et al. 2009c; Ravi et al. 2011
Canopy conductance (ISC)	7	3.28–22.24	Varshney et al. 2009c; Ravi et al. 2011
Total dry matter (TDM)	7	4.34–22.39	Varshney et al. 2009c; Ravi et al. 2011; Gautami et al. 2012a
Agronomic and yield component traits			
Shoot dry weight (ShDW)	11	5.03–22.09	Varshney et al. 2009c; Ravi et al. 2011; Gautami et al. 2012a

Table 5-4 contd....

Table 5-4 contd.

Traits studied	QTLs identified	Phenotypic variance explained (%)	References
Pod weight (PW)	7	4.17–8.73	Varshney et al. 2009c; Ravi et al. 2011
Seed weight (SW)	5	4.18–8.22	Varshney et al. 2009c; Ravi et al. 2011
Haulm weight (HW)	6	3.78–33.66	Varshney et al. 2009c; Ravi et al. 2011
Harvest index (HI)	3	6.39–40.10	Gautami et al. 2012a
Pod mass/plant	3	13.1–18.3	Liang et al. 2009a
Mature pods/plant	3	11.9–12.3	Liang et al. 2009a
Number of branches	7	8.1–17.3	Liang et al. 2009a
Number of fruit branches	1	17.5	Liang et al. 2009a
Height of main axis	7	8.2–12.8	Liang et al. 2009a
Stem diameter	4	7.8–24.1	Liang et al. 2009a
Leaf length, width and length/width ratio	7	12.4–18.9	Liang et al. 2009a
Length of main stem	3	4.8–15.7	Shirasawa et al. 2012a
Length of the longest branch	2	14.2–21.1	Shirasawa et al. 2012a
Number of branches	1	15.6	Shirasawa et al. 2012a
Weight of plant	1	11.8	Shirasawa et al. 2012a
Weight of mature pod per a plant	1	28.1	Shirasawa et al. 2012a
Length of pod	3	8.4–28.2	Shirasawa et al. 2012a
Thickness of pod	1	21.7	Shirasawa et al. 2012a

Width of pod	2	15.2–25.5	Shirasawa et al. 2012a
Shape of tip of pods	1	9.9	Shirasawa et al. 2012a
Weight of seeds	1	19.1	Shirasawa et al. 2012a
Number of seeds per a plant	1	6.8	Shirasawa et al. 2012a
Yield parameters	5	9.19–17.69	Selvaraj et al. 2009
Other morphological traits			
Flowering date	1	19.5	Shirasawa et al. 2012a
Angle of branch	2	11.9	Shirasawa et al. 2012a
Constriction of pod	2	6.9–18.1	Shirasawa et al. 2012a
Colour of seed coat	1	9.7	Shirasawa et al. 2012a
Seed and oil quality			
Oil content	14	1.5–20.7	Liang et al. 2009a; Selvaraj et al. 2009; Sarvamangala et al. 2011; Pandey et al. 2012c
Oil quality	38	1.4–74.03	Sarvamangala et al. 2011; Pandey et al. 2012c
Protein content	10	1.5–13.4	Liang et al. 2009a; Sarvamangala et al. 2011
High oleate trait	2	89.7	Shirasawa et al. 2012a; Pandey et al. 2012c

al. 1996; Garcia et al. 1996) and aphid vector of groundnut rosette disease (Herselman et al. 2004) using RAPD and AFLP markers, respectively. Similarly, the above strategy was also used for mapping the yield and yield components with SSR markers (Liang et al. 2009a; Selvaraj et al. 2009). The above mapping strategy is relatively simpler to use in crops lacking genomic resources and also for simply-inherited traits. Hence, with the availability of more SSR markers in public domains, a major shift was observed towards development of immortal populations in order to generate multiseason phenotyping data so that stable QTLs can be identified along with studying G x E interactions using advanced mapping tools. Such studies were conducted to identify the QTLs for drought tolerance related traits (Varshney et al. 2009c; Ravi et al. 2011; Gautami et al. 2012a), resistance to biotic resistance (Khedikar et al. 2010; Pandey et al. 2012c; Qin et al. 2012; Sujay et al. 2012; Wang et al. 2013), morphological and yield components (Varshney et al. 2009; Pandey et al. 2012c; Shirasawa et al. 2012a) and nutritional quality traits (Sarvamangala et al. 2011; Pandey et al. 2012c; Shirasawa et al. 2012a).

Three mapping populations (TAG 24 × ICGV 86031, ICGS 44 × ICGS 76, ICGS 76 × CSMG 84-1) were used for identification of QTLs controlling drought-tolerance related traits (Varshney et al. 2009c; Ravi et al. 2011; Gautami et al. 2012a) and mapping of all the identified QTLs onto an integrated genetic map (Gautami et al. 2012a). Multiseason phenotypic data were generated on these populations for drought-tolerance related traits such as transpiration, transpiration efficiency, biomass, specific leaf area, pod weight, total dry matter, SPAD chlorophyll meter reading, total dry weight, shoot dry weight and harvest index traits. Simultaneously, genotypic data were generated on these three mapping populations followed by construction of individual genetic maps with mapped loci ranging from 82 (ICGS 44 × ICGS 76) to 191 (TAG 24 × ICGV 86031) marker loci. Different QTL mapping programs such as QTL Cartographer, QTL Network and Genotype Matrix Mapping (GMM) were used for detailed QTL analysis using genotyping and multiseason phenotyping data. This analysis resulted in identification of a total of 153 main effects and 25 epistatic QTLs for drought-tolerance related traits (Varshney et al. 2009c; Ravi et al. 2011; Gautami et al. 2012a). In addition, 16 important genomic regions on the integrated maps were identified realizing their potential role towards drought tolerance (Table 5-4). The above study revealed that the majority of the identified QTLs contributed low phenotypic variation, and hence, molecular breeding approaches such as Marker-Assisted Back Crossing (MABC) will not be useful for introgressing drought tolerance. In order to handle such QTLs, other modern breeding approaches (marker-assisted recurrent selection and genomic selection) may be more appropriate.

Another notable QTL study was conducted for mapping QTLs for resistance to foliar diseases such as late leaf spot (LLS) and rust (Khedikar et al. 2010; Sujay et al. 2012). Two RIL populations, namely TAG 24 × GPBD 4 and TG 26 × GPBD 4, were extensively phenotyped for rust and LLS resistance for seven to eight seasons. Genotyping data were generated for 209 polymorphic markers for each of the two populations. Two individual genetic maps with 188 (TAG 24 × GPBD 4) and 181 (TG 26 × GPBD 4) marker loci were constructed along with development of an integrated map with 225 marker loci. Using the multiseason phenotyping data and genotyping information, a comprehensive QTL analysis identified a total of 28 QTLs for resistance against late leaf spot (LLS) and 13 QTLs for resistance against rust explaining 10.07 to 67.8% and 2.54 to 82.96% of phenotypic variation, respectively (Khedikar et al. 2010; Sujay et al. 2012). This study led to the identification of tightly linked markers and one major QTL each for leaf rust (55.2% PVE, Khedikar et al. 2010; 82.96% PVE, Sujay et al. 2012) and LLS (67.98% PVE, Sujay et al. 2012) resistance (Table 5-4). The tightly linked markers for rust resistance (IPAHM103, GM2079, GM2301 and GM1536) were identified in both the populations and were then validated among a set of resistant/susceptible breeding lines. Furthermore, phenotypic data on oil content and quality were also generated on one of these RIL populations (TG 26 × GPBD 4) to identify linked markers for important nutritional traits. QTL analysis using phenotypic data and partial genetic map information detected seven QTLs for protein content (2.54–9.78%), eight QTLs for oil content (1.5–10.2%) and six common QTLs for oleic and linoleic acid contents (3.3–9.7%) (Sarvamangala et al. 2011).

The next effort towards trait mapping was to identify linked markers for tomato spotted wilt virus (TSWV) resistance using two RIL populations, namely T (Tifrunner × GT-C20) and S (SunOleic 97R × NC 94022), populations. Genotyping data of both the maps were used for construction of an integrated map and identification of QTLs for TSWV resistance. QTL analysis using QTL Cartographer detected one QTL in each of the two populations with PVE ranging from 12.9 (qTSWV1) to 35.5% (qTSWV2) (Qin et al. 2012). The linked markers (IPAHM287 and Seq12F7) need validation before applying in routine MAS programs. Most recently, Shirasawa et al. (2012a) reported identification of QTLs for several agronomic traits for which PVE ranged from 11.8% (plant weight and angle of branch) to 28.2% (pod length). The other traits (PVE%) for which QTLs have been reported include flowering date (19.5%), length of main stem (15.7–19.2%), length of longest branch (14.2–21.1%), number of branches (15.6%), mature pod weight/plant (28.1%), pod thickness (21.7%), pod width (15.2–25.5%), pod constriction (18.1%), seed weight (19.1%) and seed diameter (24.1%).

Attempts were also made to identify linked markers from wide crosses for nematode resistance and as a result, two SCAR markers (Garcia et al.

1996) and three RAPD markers (Burow et al. 1996) were identified using the populations (*A. hypogaea* × *A. cardenasii*) and (*A. hypogaea* × TxAG-6), respectively. Since these markers produced inconsistent results and were complicated to use in routine molecular breeding programs, the RAPD marker (RKN440, Garcia et al. 1996) was converted into a new PCR-based dominant marker (S197) (Chu et al. 2007a). Nagy et al. (2010b) also identified a total of 13 markers (including S197 reported by Chu et al. 2007a) in two tetraploid crosses. A total of three markers namely S197 (PCR-based), 1169/1170 (CAPS) and GM565 (SSR) were used to select resistant, susceptible and heterozygous allele, respectively during development of the second marker-assisted product in peanut, i.e., Tifguard High O/L (Chu et al. 2011). Another study with diploids resulted in the identification of five QTLs for resistance to LLS from the cross *A. duranensis* × *A. stenosperma* (Leal-Bertioli et al. 2009). Initially, CAPS markers were developed for mutant *FAD* alleles in both genomes (Chu et al. 2009), but later PCR-based allele-specific markers were reported by the same research group (Chu et al. 2011). These allele-specific markers are now successfully mapped on the peanut genome along with identification of a total of 155 QTLs for oil quality and several agronomically important traits. QTL analysis also revealed that the *FAD2B* gene contributes more than the *FAD2A* gene for high oleic/linoleic (O/L) ratio (Pandey et al. 2012c). Further, very high PVE (65.20–89.7%) has been reported for high oleate traits (Pandey et al. 2012c; Shirasawa et al. 2012a).

Although linked markers to a few disease resistance traits such as nematode (Nagy et al. 2010b), leaf rust (Khedikar et al. 2010; Sujay et al. 2012), LLS (Sujay et al. 2012) and TSWV (Qin et al. 2012) and one oil quality trait, i.e., high-oleate trait (Chu et al. 2009; Chen et al. 2010; Pandey et al. 2012c; Shirasawa et al. 2012a) are currently available to use in molecular breeding, more research is needed for identifying tightly-linked molecular markers to several other important traits. It is anticipated that the availability of more genomic resources, such as SNPs, and the genome sequence will accelerate trait mapping efforts in the near future and will make available linked markers for many other traits (Varshney et al. 2012).

5.5 Genomics-assisted Breeding

Genomics-assisted breeding (GAB) offers a breeding platform where genomics tools are integrated with conventional breeding methods to develop improved genotypes, in a very short time, for several traits/genes at once and is also able to minimize the inhibited fear of linkage drag in wide crosses (Varshney et al. 2006). GAB, mainly marker-assisted breeding has achieved only limited success in peanut, and even that has been restricted to simply-inherited traits. The majority of agronomically important traits are complex in nature and governed by several genomic regions, which also show interactions with environments (G x E) and other genomic regions (epistasis). Hence, genomics tools along with modern decision making tools should be used along with proven conventional breeding approaches to understand the exact genetic nature of the target traits and for finding ways for their possible manipulation leading to genetic enhancement.

Currently, GAB could be used for crop improvement in three ways, i.e., marker-assisted backcrossing (MABC), Marker-Assisted Recurrent Selection (MARS) and Genomic Selection (GS). The first two approaches require QTL information, while the 3rd one does not. In practice, introgression of recessive genes and pyramiding of multiple genes is very difficult, costly, lengthy and error prone using conventional breeding methods. Marker-assisted selection (MAS) has proved its utility in several crops to overcome such problems and many genes can be pyramided either for the same trait or for different traits along with faster recurrent parent genome recovery through intense background selection (Varshney et al. 2006). In addition, MAS can be used to pyramid/introgress several recessive genes in less time and with more precision, which is almost impossible through conventional breeding. MAS has gained popularity due to its proven record in several crops and is easy to use even in smaller research stations that have low to moderate marker genotyping capabilities. In peanut, these tools have been integrated into the conventional breeding programs very late due to the lack of genomic resources such as molecular markers, genetic maps and most importantly tightly-linked markers for the most desirable traits in peanut. Nevertheless, some efforts have been made to use molecular markers in peanut breeding.

Root-knot nematode (*Meloidogyne arenaria*) resistance, the first trait for which linked molecular markers were identified, was introgressed from *A. cardenasii* through the amphidiploid pathway into cultivated peanut (Simpson 2001). This was relatively easy to identify due to sequence divergence between diploid and tetraploid genomes (Chu et al. 2007a; Nagy et al. 2010b). This effort led to the development of the first MAS product in peanut, named as NemaTAM (Simpson et al. 2003), the first peanut cultivar developed using MAS. MAS has shown several benefits in the development of "NemaTAM" such as selection of heterozygous and homozygous plants in early generations with very high precision at the seedling stage. Phenotyping for nematode resistance is prone to environmental fluctuations and more often leads to escapes (Simpson et al. 2003).

The RFLP marker system used to develop NemaTAM is very costly, requires DNA in large quantity, entails health risk due to the use of radioisotopes, also requires high technical expertise and has a long turnaround time for results. Since breeders require timely genotyping information to make backcrosses, efforts were made to develop more rapid and easy-to-assay markers for nematode resistance (Nagy et al. 2010b). Meanwhile, a tightly associated CAPS marker (1101/1048) became available for another important trait, i.e., high oleic acid (Chu et al. 2009). The associated markers for high oleic acid were deployed to backcross the high-oleate trait (*FAD2B*) into the nematode resistant cultivar, Tifguard (Holbrook et al. 2011). Homozygous recessive mutations in both *AhFAD2* homeologs are necessary to achieve high O/L. Since the frequency of a spontaneous loss-of-function allele of *AhFAD2A* is high in the ssp. *hypogaea* germplasm (Chu et al. 2007b) and fixed in most elite lines of US runner and Virginia market-type peanuts (Chu et al. 2009), therefore, MAS was required only to select the mutant allele of *AhFAD2B* for making Tifguard High O/L. Markers linked with nematode resistance were used to monitor flow of the nematode-resistant allele in backcross and selfed generations. These markers have been used during MABC to select desired DNA fragment carrying nematode resistance while simultaneously selecting for a recessive *AhFAD2B* allele necessary to recover lines with a high ratio of oleic:linoleic acid (O/L) leading to development of the 2nd MAS product in peanut namely, "Tifguard High O/L" (Chu et al. 2011).

Development of immortal populations and generation of multiseasonal phenotypic data resulted in the identification of stable QTLs and tightly-linked molecular markers for LLS and leaf rust (Khedikar et al. 2010; Sujay et al. 2012). The linked markers for leaf rust were deployed to introgress leaf rust resistance into the genetic background of three elite cultivars (ICGV 91114, JL 24 and TAG 24) through MABC at ICRISAT, India. An important result of this study was identification of SSR markers, which are easy to genotype even in smaller laboratories. Three codominant markers (GM2079, GM2301 and GN1536) and one dominant SSR marker (IPAHM103) were used to select heterozygous allele at backcrossed F_1 (BC_1F_1, BC_2F_1 and BC_3F_1) generations and homozygous allele at backcrossed F_2 (BC_2F_2 and BC_3F_2) generations. As a result, a total of 200 advanced generation introgression lines (117 BC_2F_5 and 83 BC_3F_5) were developed using the above markers for all the above three elite cultivars. Superior lines with desirable yield and higher resistance to leaf rust were selected based on replicated evaluation during the rainy seasons in 2011 and 2012 for further multiplication and multilocation trails (Varshney et al. 2013b). The initial screening has been very encouraging showing reduced disease symptoms and has led to the identification of several promising lines in all the three genetic backgrounds. However, in the case of drought tolerance, many QTLs were identified each contributing only small phenotypic variance (Varshney et al. 2009c; Ravi et al. 2011; Gautami et al. 2012a). In such cases, MABC approach may not be appropriate and hence, other modern breeding approaches such as MARS or GS might be better approaches (Bernardo and Yu 2007; Ribaut and Ragot 2007; Bernardo 2009; Heffner et al. 2009; Jannink et al. 2010).

Apart from three traits (nematode resistance, high oleate and leaf rust) discussed above, QTLs and linked markers for two more diseases namely LLS (Sujay et al. 2012) and TSWV (Qin et al. 2012) have been reported. These markers linked to LLS (GM1573/GM1009 and Seq8D09) and TSWV (IPAHM287 and Seq12F7) provide hope for marker-assisted improvement of resistance to these two diseases in the near future. The future of GAB in peanut may be more fruitful due to increased availability of linked markers to other important traits of peanut which will accelerate multiple trait improvement of existing high yielding cultivars and development of new cultivars through gene pyramiding.

5.6 Diversification and Enrichment of Primary Gene Pool

Tetraploidization has restricted gene flow from diploids to tetraploid (cultivated) which has created a serious genetic bottleneck. Efforts into making wide crosses through use of hexaploids, autotetraploids and allotetraploids have been plagued by serious problems with fertility barriers, linkage drag and difficulty in tracking introgressed alien genomic regions (Bertioli et al. 2011). Of these three barriers, two (linkage drag and tracking alien genomic regions) can be efficiently handled by integrating genomics into routine breeding programs to diversify the narrow peanut primary gene pool. GAB can help in tracking alien genomic regions and hence, linkage drag can be minimized. Several efforts have attempted to introgress wild genes into cultivated, most involving disease resistance (Simpson 1991; Singh 1996; Tansley and Nelson 1996; Stalker et al. 2002; Favero et al. 2006; Fonceka et al. 2009; Leal-Bertioli et al. 2011; Mallikarjuna et al. 2011).

Introgressing useful alleles from wild relatives can be done with higher precision using genomics and decision making tools. Molecular markers evenly distributed throughout genomes have been utilized for tracking genome recovery during backcrossing in several crops. While introgressing genes from wild relatives, stringent background selection is required using markers covering the full genome to avoid linkage drag from unwanted genomic segments from wild relatives. The lone effort towards alien genomic introgressions made using this approach in peanut was with the use of limited genomic resources by Fonceka et al. (2009). A synthetic amphidiploid (*A. duranensis* × *A. ipaënsis*) was used to cross with a cultivated variety (Fleur 11) followed by two backcrosses. Molecular markers were used to track alien genomic region introgressions in the genetic background of the cultivated genotype "Fleur 11" in backcross generations. This facilitated selection of several introgression lines with varied amounts of wild genomic segments for further study. With the availability of more genomic resources and high throughput genotyping platforms, it will become easier to broaden the genetic base of the primary gene pool by introgressing genomic segments from the wild species or synthetic amphidiploid genotypes with the help of molecular markers.

5.7 Towards Assembling the Genome Sequence

Recent advances in Next-Generation Sequencing (NGS) technology platforms have enabled much-needed faster sequence data generation along with advancements in informatics and assembly tools to manage and analyze NGS data (Varshney and May 2012). Before recent advances in technology whole-genome sequencing of crops with larger genome size and complex genomes was questionable. The main problem now lies in analyzing and transmission of information to apply for crop improvement through discovery of genes, and molecular markers associated with economically important traits (Edward and Baitley 2010). Using advanced technologies, whole genomes have been sequenced for several crop species but sequencing of the peanut genome has not been accomplished due to its large size, which is ~20-times larger than that of *Arabidopsis thaliana,* and 2-6-times larger than that of rice, sorghum or soybean. Nevertheless, sequencing for the peanut genome has been initiated by the Peanut Genome Consortium (PGC) http://www.peanutbioscience.com/peanutgenomeproject.html) for the tetraploid cultivar "Tifrunner". The Peanut Genome Project (PGP) is initiating sequencing of the peanut genome in collaboration with BGI-Shenzhen (China). It is, therefore, anticipated that a draft genome sequence along with extensive genome and trancriptome information will be available for the peanut research community within the near future. The genome sequence data will lead to the identification of several hundred molecular markers leading to the development of dense genetic maps, which will facilitate identification of linked/associated markers with economically important traits to use in genetic enhancement of cultivated peanut.

5.8 Summary and Future Prospects

GAB should accelerate genetic enhancement leading to improved productivity, oil quality and resistance/tolerance to stresses. Recent advances have resulted in the development of SSR markers and several genetic maps for different genomes (AA, BB, AABB genomes). The density of genetic maps in diploid (AA and BB) genomes was higher than the tetraploid genetic maps. Even though the tetraploid species has both the genomes, the genetic diversity observed in cultivated maps has been low. Therefore, only partial (<100 loci) to low-moderate (<300 loci) genetic maps could be constructed. One of the major challenges was to integrate as many markers as possible on a single genetic map, which was solved through successful development of a reference consensus genetic map with 897 marker loci based on 11 individual genetic maps. Now, the expectation lies with SNP markers to develop high density genetic maps but it will take few years before these markers are in routine use for breeding and genetic applications. Until that time, SSR markers are going to continue to be used in genetic and breeding applications in cultivated peanuts. Efforts with the available limited genomic resources led to the identification of linked markers for oil quality (high oleic acid) and disease resistance (nematode, rust, LLS and TSWV) traits in cultivated peanut through trait mapping. These developments also led to the deployment of linked markers to improve disease resistance and oil quality through MABC approaches. It is now feasible to pyramid resistance to all the four diseases along with the high oleic trait. Further attention is required towards other challenging areas such as drought stress along with aflatoxin/mycotoxin contamination, which has teratogenic and carcinogenic effects on humans and animals. The expected availability of genome sequence in the near future should provide huge genomic resources, which will hasten the efforts of the much-needed linking of phenotype with markers/genome sequences. However, it can only be achieved with precise and high-throughput phenotyping for complex traits. Recent advances in peanut genomics and molecular breeding efforts provide hope for efficient genetic enhancement of cultivated peanut to address different production as well as quality constraints.

Acknowledgements

The authors are grateful to the National Fund of the Basic and Strategic Research in Agriculture (NFBSRA) of the Indian Council of Agriculture Research (ICAR), New Delhi, India, Generation Challenge Program (http://www.generationcp.org) of CGIAR, Department of Biotechnology, Government of India for Center of Excellence (CoE) grant, The Peanut Foundation, USA, Georgia Peanut Commission, USA and National Peanut Board, USA for supporting work of several authors presented in this article.

Abbreviations

CAAS (China)	:	Chinese Academy of Agricultural Sciences
CENARGEN (Brazil)	:	National Research Center for Genetic Resources and Biotechnology
CRI (China)	:	Crops Research Institute
DGR (India)	:	Directorate of Groundnut Research
EMBRAPA (Brazil)	:	Brazilian Agricultural Research Corporation
GAAS (China)	:	Guangdong Academy of Agricultural Sciences

IBONE (Argentina)	:	Instituto de Botánica the Northeast
ICAR (India)	:	Indian Council of Agricultural Research
ICRISAT (India)	:	International Crops Research Institute for the Semi-Arid Tropics
INTA (Brazil)	:	Argentina Agronomic Institute of Campinas in Brazil; and Instituto Nacional de Agropecuaria
NBPGR (India)	:	National Bureau of Plant Genetic Resources
NCSU (USA)	:	North Carolina State University
OCRI (China)	:	Oil Crops Research Institute
TAMU (USA)	:	Texas A & M University
USDA (USA)	:	U.S. Department of Agriculture

References

Beavis WD, Grant D (1991) A linkage map based on information from four F_2 populations of maize (*Zea mays* L.). Theor Appl Genet 82: 636–644.

Bernardo R, Yu J (2007) Prospects for genome wide selection for quantitative traits in maize. Crop Sci 47: 1082–1090.

Bernardo R (2009) Genome wide selection for rapid introgression of exotic germplasm in maize. Crop Sci 49: 419–425.

Bertioli DJ, Seijo G, Freitas FO, Valls FM, Leal-Bertioli SCM, Moretzsohn (2011) An overview of peanut and its wild relatives. Plant Genetic Resour: Characteriz Utiliz 9: 134–149.

Bohra A, Saxena RK, Gnanesh BN, Saxena K, Byregowda M, Rathore A, Kavikishore PB, Cook DR, Varshney RK (2012) An intra-specific consensus genetic map of pigeonpea [*Cajanus cajan* (L.) Millspaugh] derived from six mapping populations. Theor Appl Genet 125: 1325–38.

Bravo JP, Hoshino AA, Angelici CMLCD, Lopes CR, Gimenes MA (2006) Transferability and use of microsatellite markers for the genetic analysis of the germplasm of some *Arachis* section species of the genus *Arachis*. Genet Mol Biol 29: 516–524.

Burow MD, Simpson CE, Paterson AH, Starr JL (1996) Identification of peanut (*Arachis hypogaea* L.) RAPD markers diagnostic for root-knot nematode (*Meloidogyne arenaria* (Neal) Chitwood) resistance. Mol Breed 2: 369–379.

Burow MD, Simpson CE, Starr JL, Paterson AH (2001) Transmission genetics of chromatin from a synthetic amphiploid in cultivated peanut (*A. hypogaea* L.): Broadening the gene pool of a monophyletic polyploid species. Genetics 159: 823–837.

Chen Z, Wang ML, Barkley NA, Pittman RN (2010) A simple allele-specific PCR assay for detecting *FAD2* alleles in both A and B genomes of the cultivated peanut for high-oleate trait selection. Plant Mol Biol Rep 28: 542–548.

Choi IY, Hyten DL, Matukumalli LK, Song Q, Chaky JM, Quigley CV, Chase K, Lark KG, Reiter RS, Yoon M-S, Hwang E-Y, Yi S-I, Young ND, Shoemaker RC, van Tassell CP, Specht JE, Cregan PB (2007) A soybean transcript map: gene distribution, haplotype and single-nucleotide polymorphism analysis. Genetics 176: 685–696.

Chu Y, Holbrook CC, Timper P, Ozias-Akins P (2007a) Development of a PCR-based molecular marker to select for nematode resistance in peanut. Crop Sci 47: 841–845.

Chu Y, Ramos L, Holbrook CC, Ozias-Akins P (2007b) Frequency of a loss-of-function mutation in oleoyl-PC desaturase (*ahFAD2A*) in the minicore of the U.S. peanut germplasm collection. Crop Sci 47: 2372–2378.

Chu Y, Holbrook CC, Ozias-Akins P (2009) Two alleles of *ahFAD2B* control the high oleic acid trait in cultivated peanut. Crop Sci 49: 2029–2036.

Chu Y, Wu CL, Holbrook CC, Tillman BL, Person G, Ozias-Akins P (2011) Marker-assisted selection to pyramid nematode resistance and the high oleic trait in peanut. Plant Genome 4: 110–117.

Cuc LM, Mace ES, Crouch JH, Quang VD, Long TD, Varshney RK (2008) Isolation and characterization of novel microsatellite markers and their application for diversity assessment in cultivated groundnut (*Arachis hypogaea*). BMC Plant Biol 8: 55.

Dwivedi SL, Gurtu S, Chandra S, Weng YJ, Nigam SN (2001) Assessment of genetic diversity among selected groundnut germplasm-I: RAPD analysis. Plant Breed 120: 345–349.

Edwards D, Batley J (2010) Plant genome sequencing: applications for crop improvement. Plant Biotechnol J 8: 2–9.

Falque M, Décousset L, Dervins D, Jacob AM, Joets J, Martinant J-P, Raffoux X, Ribière N, Ridel C, Samson D, Charcosset A, Murigneux A (2005) Linkage mapping of 1454 maize candidate gene loci. Genetics 70: 1957–1966.

Fávero AP, Simpson CE, Valls JFM, Valls NA (2006) Study of the evolution of cultivated peanut through crossability studies among *Arachis ipaënsis*, *A. duranensis*, and *A. hypogaea*. Crop Sci 46: 1546–1622.

Ferguson ME, Burow MD, Schultze SR, Bramel PJ, Paterson AH, Kresovich S, Mitchell S (2004) Microsatellite identification and characterization in peanut (*A. hypogaea* L.). Theor Appl Genet 108: 1064–1070.

Foncéka D, Hodo-Abalo T, Rivallan R, Faye I, Sall MN, Ndoye O, Fávero AP, Bertioli DJ, Glaszmann JC, Courtois B, Rami JF (2009) Genetic mapping of wild introgressions into cultivated peanut: a way toward enlarging the genetic basis of a recent allotetraploid. BMC Plant Biol 9: 103.

Froenicke L, Pandey M, Upadhyaya H, Moretzsohn MC, Guimarães P, Leal-Bertioli S, Varshney RK, Bertioli D, Michelmore RW (2011) Towards ultra-dense genetic maps of peanut generated by sequencing diploid and tetraploid RIL populations and a peanut diversity panel. In: Proc Advances in *Arachis* through Genomics and Biotechnology (AAGB), 13–15 June 2011, Brasília, Brazil.

Garcia GM, Stalker HT, Kochert G (1995) Introgression analysis of an interspecific hybrid population in peanuts (*Arachis hypogaea* L.) using RFLP and RAPD markers. Genome 38: 166–176.

Garcia GM, Stalker HT, Shroeder E, Kochert G (1996) Identification of RAPD, SCAR and RFLP markers tightly linked to nematode resistance genes introgressed from *Arachis cardenasii* to *A. hypogaea*. Genome 39: 836–845.

Garcia GM, Stalker HT, Schroeder E, Lyerly JH, Kochert G (2005) A RAPD-based linkage map of peanut based on a backcross population between the two diploid species *Arachis stenosperma* and *A. cardenasii*. Peanut Sci 32: 1–8.

Gautami B, Ravi K, Narasu ML, Hoisington DA, Varshney RK (2009) Novel set of groundnut SSR markers for germplasm analysis and inter-specific transferability. Int J Integr Biol 7: 100–106.

Gautami B, Pandey MK, Vadez V, Nigam SN, Ratnakumar P, Krishnamurthy L, Radhakrishnan T, Gowda MVC, Narasu ML, Hoisington DA, Knapp SJ, Varshney RK (2012a) QTL analysis and consensus genetic map for drought tolerance traits based on three RIL populations of cultivated groundnut (*Arachis hypogaea* L.). Mol Breed 32: 757–772.

Gautami B, Fonceka D, Pandey MK, Morezsohn MC, Sujay V, Qin H, Hong Y, Faye I, Chen X, Bhanuprakash A, Shah TM, Gowda MVC, Nigam SN, Liang X, Hoisington DA, Guo B, Bertioli DJ, Rami JF, Varshney RK (2012b) An international reference consensus genetic map with 897 marker loci based on 11 mapping populations for tetraploid groundnut (*Arachis hypogaea* L.). PLoS ONE 7: e41213.

Gentzbittel L, Vear F, Zhang YX, Breville A, Nicolas P (1995) Development of a consensus linkage RFLP map of cultivated sunflower (*Helianthus annus* L.). Theor Appl Genet 90: 1079–1086.

Gimenes MA, Hosino AA, Barbosa AVG, Palmieri DA, Lopes CR (2007) Characterization and transferability of microsatellite markers of cultivated peanut (*Arachis hypogaea*). BMC Plant Biol 7: 9.

Gobbi A, Teixeira C, Moretzsohn M, Guimarães P, Leal-Bertioli SC, Bertioli D, Lopes CR, Gimenes M (2006) Development of a linkage map to species of B genome related to the peanut (*Arachis hypogaea*-AABB). In: Proc Plant Anim Genome XIV Conf, 14–18 Jan 2006, San Diego, USA, p 679.

Guo BZ, Fedorova ND, Chen X, Wan CH, Wang W, Nierman WC, Bhatnagar D, Yu J (2011) Gene expression profiling and identification of resistance genes to *Aspergillus flavus* infection in peanut through EST and microarray startegies. Toxins 3: 737–753.

Guo Y, Khanal S, Tang S, Bowers JE, Heesacker AF, Khalilian N, Nagy ED, Zhang D, Taylor CA, Stalker HT, Ozias-Akins P, Knapp SJ (2012) Comparative mapping in intraspesific populations uncovers high degree of macrosynteny between A- and B-genome diploid species of peanut. BMC Genomics 13: 608.

Gupta PK, Kumar J, Mir RR, Kumar A (2010) Marker-assisted selection as a component of conventional plant breeding. Plant Breed Rev 33: 145–217.

Halward TM, Stalker HT, Kochert G (1993) Development of an RFLP linkage map in diploid peanut species. Theor Appl Genet 87: 379–384.

Hauge BM, Hanley SM, Cartinhour S, Cherry JM, Goodman HM (1993) An integrated genetic/RFLP map of the Arabidopsis thaliana genome. Plant J 3: 745–754.

He GH, Prakash CS (2001) Evaluation of genetic relationships among botanical varieties of cultivated peanut (*Arachis hypogaea* L.) using AFLP markers. Genet Resour Crop Evol 48: 347–352.

He GH, Meng R, Newman M, Gao G, Pittman RN, Prakash CS (2003) Microsatellites as DNA markers in cultivated peanut (*Arachis hypogaea* L.). BMC Plant Biol 3: 3.

Heffner EL, Sorrells ME, Jannink JL (2009) Genomic selection for crop improvement. Crop Sci 49: 1–12.

Herselman L (2003) Genetic variation among Southern African cultivated peanut (*Arachis hypogaea* L.) genotypes as revealed by AFLP analysis. Euphytica 133: 319–327.

Herselman LR, Thwaites FM, Kimmins B, Courtois PJA, Merwe VD, Seal SE (2004) Identification and mapping of AFLP markers linked to peanut (*Arachis hypogaea* L.) resistance to the aphid vector of groundnut rosette disease. Theor Appl Genet 109: 1426–1433.

Hilu KW, Stalker HT (1995) Genetic relationships between peanut and wild species of *Arachis* sect. *Arachis* (*Fabaceae*): evidence from RAPDs. Syst Evol 198: 167–178.

Holbrook CC, Dong W (2005) Development and evaluation of a mini core collection for the US peanut germplasm collection. Crop Sci 45: 1540–1544.

Holbrook CC, Ozias-Akins P, Chu Y, Guo B (2011) Impact of molecular genetic research on peanut cultivar development. Agronomy 1: 3–17.

Hong Y, Chen X, Liang X, Liu H, Zhou G, Li S, Wen S, Holbrook CC, Guo B (2010a) A SSR-based composite genetic linkage map for the cultivated peanut (*Arachis hypogaea* L.) genome. BMC Plant Biol 10: 17.

Hong YB, Chen XP, Liu HY, Zhou GY, Li SX, Wen SJ, Liang XQ (2010b) Development and utilization of orthologous SSR markers in Arachis through soybean (*Glycine max*) EST. Acta Agron Sin 36: 410–421 (in Chinese with English abstract).

Hopkins MS, Casa AM, Wang T, Mitchell SE, Dean RE, Kochert GD, Kresovich S (1999) Discovery and characterization of polymorphic simple sequence repeats (SSRs) in peanut. Crop Sci 39: 1243–1247.

Jannink JL, Lorenz AJ, Iwata H (2010) Genomic selection in plant breeding: from theory to practice. Brief Func Genom 9: 166–177.

Jesubatham AM, Burow MD (2006) PeanutMap: an online genome database for comparative molecular maps of peanut. BMC Bioinformatics 7: 375.

Jiang HF, Ren XP, Zhang XJ, Huang JQ, Lei Y, Yan LY, Liao BS, Upadhyaya HD, Holbrook CC (2010) Comparison of genetic diversity based on SSR markers between peanut mini core collections from China and ICRISAT. Acta Agron Sin 36: 1084–1091.

Khedikar YP, Gowda MVC, Sarvamangala C, Patgar KV, Upadhyaya HD, Varshney RK (2010) A QTL study on late leaf spot and rust revealed one major QTL for molecular breeding for rust resistance in groundnut (*Arachis hypogaea* L.). Theor Appl Genet 121: 971–984.

Khera P, Upadhyaya HD, Pandey MK, Roorkiwal M, Sriswathi M, Janila P, Guo Y, McKain MR, Nagy ED, Knapp SJ, Leebens-Mack J, Conner JA, Ozias-Akins P, Varshney RK (2013) SNP-based genetic diversity in the reference set of peanut (*Arachis* spp.) by developing and applying cost-effective KASPar genotyping assays. The Plant Genome 6, doi: 10.3835/plantgenome2013.

Kianian SF, Quiros CF (1992) Generation of a *Brassica oleracea* composite RFLP map: linkage arrangements among various populations and evolutionary implications. Theor Appl Genet 84: 544–554.

Kilian A (2008) DArT-based whole genome profiling and novel information technologies in support system of modern breeding of groundnut. In: Proc 3rd Int Conf for Peanut Genomics and Biotechnology on "Advances in *Arachis* through Genomics and Biotechnology (AAGB), 4–8 Nov 2008, Hyderabad, India.

Kochert G, Stalker HT, Gimenes M, Galgaro L, Lopes CR, Moore K (1996) RFLP and cytogenetic evidence on the origin and evolution of allotetraploid domesticated peanut, *Arachis hypogaea* (Leguminosae). Am J Bot 83: 1282–1291.

Koilkonda P, Sato S, Tabata S, Shirasawa K, Watanabe A, Wada T, Kishida Y, Tsuruoka H, Fujishiro T, Yamada M, Kohara M, Suzuki S, Hasegawa M, Kiyoshima H, Isobe S (2011) Large-scale development of expressed sequence tag-derived simple sequence repeat markers and diversity analysis in *Arachis* spp. Mol Breed DOI: 10.1007/s11032-011-9604-8.

Krapovickas A, Gregory WC (1994) Taxonomia del género *Arachis* (Leguminosae). Bonplandia 8: 1–186.

Leal-Bertioli SCM, José AC, Alves-Freitas DM, Moretzsohn MC, Guimarães PM, Nielen S, Vidigal BS, Pereira RW, Pike J, Fávero AP, Parniske M, Varshney RK, Bertioli DJ (2009) Identification of candidate genome regions controlling disease resistance in *Arachis*. BMC Plant Biol 9: 112.

Leal-Bertioli SCM, Moretzsohn MC, Guimarães PM, Santos SP, Nielen S, Araujo ACG, Brasileiro ACM, Morgante CM, Bertioli DJ (2011) Broadening the genetic base of peanut: Introgression of wild *Arachis* resistance genes using the Tetraploid Route with the aid of molecular and cytogenetic markers. In: Proc 5th Int Conf for Peanut Genomics and Biotechnology on "Advances in *Arachis* through Genomics and Biotechnology (AAGB), 13–15th June 2011, Brasília, Brazil.

Liang X, Zhou G, Hong Y, Chen X, Liu H, Li S (2009a) Overview of research progress on peanut (*Arachis hypogaea* L.) host resistance to aflatoxin contamination and genomics at the Guangdong Academy of Agricultural Sciences. Peanut Sci 36: 29–34.

Liang X, Chen X, Hong Y, Liu H, Zhou G, Li S, Guo B (2009b) Utility of EST-derived SSR in cultivated peanut (*Arachis hypogaea* L.) and *Arachis* wild species. BMC Plant Biol 9: 35.

Luo M, Dang P, Guo B, He G, Holbrook CC, Bausher MG, Lee RD (2005) Generation of expressed sequence tags (ESTs) for gene discovery and marker development in cultivated peanut. Crop Sci 45: 346–353.

Mace ES, Varshney RK, Mahalakshmi V, Seetha K, Gafoor A, Leeladevi Y, Crouch JH (2007) *In silico* development of simple sequence repeat markers within the aeschynomenoid/dalbergoid and genistoid clades of the leguminosae family and their transferability to *Arachis hypogaea*, groundnut. Plant Sci 174: 51–60.

Macedo SE, Moretzsohn MC, Leal-Bertioli SCM, Alves DMT, Gouvea EG, Azevedo VCR, Bertioli DJ (2012) Development and characterization of highly polymorphic long TC repeat microsatellite markers for genetic analysis of peanut. BMC Res Notes 5: 86.

Mallikarjuna N, Senthilvel S, Hoisington D (2011) Development of new sources of tetraploid *Arachis* to broaden the genetic base of cultivated groundnut (*Arachis hypogaea* L.). Genet Resour Crop Evol 58: 889–907.

Marcel TC, Varshney RK, Barbieri M, Jafary H, de Kock MJ, Graner A, Niks RE (2007) A high-density consensus map of barley to compare the distribution of QTLs for partial resistance to *Puccinia hordei* and of defence gene homologues. Theor Appl Genet 114: 487–500.

Milla SR (2003) Relationships and utilization of *Arachis* germplasm in peanut improvement. PhD Thesis, North Carolina State Univ, USA, p 150.

Moretzsohn MC, Hopkins MS, Mitchell SE, Kresovich S, Valls JFM, Ferreira ME (2004) Genetic diversity of peanut (*Arachis hypogaea* L.) and its wild relatives based on the analysis of hyper variable regions of the genome. BMC Plant Biol 4: 11.

Moretzsohn MC, Leoi L, Proite K, Guimaras PM, Leal-Bertioli SCM, Gimenes MA, Martins WS, Valls JFM, Grattapaglia D, Bertioli DJ (2005) A microsatellite-based, gene-rich linkage map for the AA genome of *Arachis* (*Fabaceae*). Theor Appl Genet 111: 1060–1071.

Moretzsohn MC, Barbosa AVG, Alves-Freitas DMT, Teixeira C, Leal-Bertioli SC (2009) A linkage map for the B-genome of *Arachis* (Fabaceae) and its synteny to the A-genome. BMC Plant Biol 9: 40.

Nagy E, Guo S, Khanal S, Taylor C, Ozias-Akins P, Stalker HT, Nielsen N (2010a) Developing a high density molecular map of the A-genome species *A. duranensis*. In: Proc American Peanut Research and Education Society (APRES), 12–15 July 2010, Florida, USA.

Nagy E, Chu Y, Guo Y, Khanal S, Tang S, Li Y, Dong W, Timper P, Taylor C, Ozias-Akins P, Holbrook C, Beilinson V, Nielsen N, Stalker T, Knapp S (2010b) Recombination is suppressed in an alien introgression in peanut harboring *Rma,* a dominant root-knot nematode resistance gene. Mol Breed 26: 357–370.

Naito Y, Suzuki S, Iwata Y, Kuboyama T (2008) Genetic diversity and relationship analysis of peanut germplasm using SSR markers. Breed Sci 58: 293–300.

Nelson MN, Phan HTT, Ellwood SR, Paula MM, Hane J, Williams A, Lone CE, Nyarko JF, Scobie M, Cakir M, Jones MGK, Bellgard M, Wolko B, Barker SJ, Oliver RP, Cowling WA (2006) The first gene-based map of *Lupinus angustifolius* L.—location of domestication genes and conserved synteny with *Medicago truncatula*. Theor Appl Genet 113: 225–238.

Palmieri DA, Hoshino AA, Bravo JP, Lopes CR, Gimenes MA (2002) Isolation and characterization of microsatellite loci from the forage species *Arachis pintoi* (Genus *Arachis*). Mol Ecol Notes 2: 551–553.

Palmieri DA, Bechara MD, Curi RA, Gimenes MA, Lopes CR (2005) Novel polymorphic microsatellite markers in section Caulorrhizae (*Arachis*, Fabaceae). Mol Ecol Notes 5: 77–79.

Pandey MK, Monyo E, Ozias-Akins P, Liang X, Guimarães P, Nigam SN, Upadhyaya HD, Janila P, Zhang X, Guo B, Cook DR, Bertioli DJ, Michelmore R, Varshney RK (2012a) Advances in *Arachis* genomics for peanut improvement. Biotechnol Adv 30: 639–651.

Pandey MK, Gautami B, Jayakumar T, Sriswathi M, Upadhyaya HD, Gowda MVC, Radhakrishnan T, Bertioli DJ, Knapp SJ, Cook DR, Varshney RK (2012b) Highly informative genic and genomic SSR markers to facilitate molecular breeding in cultivated groundnut (*Arachis hypogaea*). Plant Breed 131: 139–147.

Pandey MK, Feng S, Culbreath A, Varshney RK, Wang ML, Barkley NA, Holbrook CC, Guo B (2012c) Saturation of genetic maps for identification of QTLs controlling biotic resistance, morphological descriptors and oil quality in tetraploid peanut (*Arachis hypogaea*). In: Proc American Peanut Research and Education Society (APRES), 9–11 July 2012, Raleigh, USA, p 27.

Proite K, Leal-Bertioli SC, Bertioli DJ, Moretzsohn MC, da Silva FR, Martins NF, Guimarães PM (2007) ESTs from a wild *Arachis* species for gene discovery and marker development. BMC Plant Biol 7: 7.

Qin H, Feng S, Chen C, Guo Y, Knapp S, Culbreath A, He G, Wang ML, Zhang X, Holbrook CC, Ozias-Akins P, Guo B (2012) An integrated genetic linkage map of cultivated peanut (*Arachis hypogaea* L.) constructed from two RIL populations. Theor Appl Genet 124: 653–664.

Ravi K, Vadez V, Isobe S, Mir RR, Guo Y, Nigam SN, Gowda MVC, Radhakrishnan T, Bertioli DJ, Knapp SJ, Varshney RK (2011) Identification of several small-effect main QTLs and larg number of epistatic QTLs for drought tolerance in groundnut (*Arachis hypogaea* L.). Theor Appl Genet 122: 1119–1132.

Ribaut JM, Hoisington D (1998) Marker-assisted selection: new tools and strategies. Trends Plant Sci 3: 236–239.

Ribautl JM, Ragot M (2007) Marker-assisted selection to improve drought adaptation in maize: the backcross approach, perspectives, limitations, and alternatives. J Exp Bot 58: 351–360.

Sarvamangala C, Gowda MVC, Varshney RK (2011) Identification of quantitative trait loci for protein content, oil content and oil quality for groundnut (*Arachis hypogaea* L.). Field Crops Res 122: 49–59.

Selvaraj MG, Narayana M, Schubert AM, Ayers JL, Baring MR, Burow MD (2009) Identification of QTLs for pod and kernel traits in cultivated peanut by bulked segregant analysis. Elec J Biotechnol 12, DOI:10.2225/vol12.

Sharopova N, McMullen MD, Schultz L, Schroeder S, Sanchez-Villeda H, Gardiner J, Bergstrom D, Houchins K, Melia-Hancock S, Musket T, Duru N, Polacco M, Edwards K, Ruff T, Register JC, Brouwer C, Thompson R, Velasco R, Chin E, Lee M, Woodman-Clikeman W, Long MJ, Liscum E, Cone K, Davis G, Coe EH Jr (2002) Development and mapping of SSR markers for maize. Plant Mol Biol 48: 463–481.

Shirasawa K, Koilkonda P, Aoki K, Hirakawa H, Tabata S, Watanabe M, Hasegawa M, Kiyoshima H, Suzuki S, Kuwata C, Naito Y, Kuboyama T, Nakaya A, Sasamoto S, Watanabe A, Kato M, Kawashima K, Kishida Y, Kohara M, Kurabayashi A, Takahashi C, Tsuruoka H, Wada T, Isobe S (2012a) *In silico* polymorphism analysis for the development of simple sequence repeat and transposon markers and construction of linkage map in cultivated peanut. BMC Plant Biol 12: 80.

Shirasawa K, Hirakawa H, Tabata S, Hasegawa M, Kiyoshima H, Suzuki S, Sasamoto S, Watanabe A, Fujishiro T, Isobe S (2012b) Characterization of active miniature inverted-repeat transposable elements in the peanut genome. Theor Appl Genet 124: 1429–1438.

Shirasawa K, Bertioli DJ, Varshney RK, Moretzsohn MC, Leal-Bertioli SCM, Thudi M, Pandey MK, Rami J-F, Foncéka D, Gowda MVC, Qin H, Guo B, Hong Y, Liang X, Hirakawa H, Tabata S, Isobe S (2013) Integrated consensus map of cultivated peanut and wild relatives reveals structures of the A and B genomes of *Arachis* and divergence of the legume genomes. DNA Res, pp 1–12, doi:10.1093/dnares/dss042.

Simpson CE (1991) Pathways of introgression of pest resistance into *Arachis hypogaea* L. Peanut Sci 18: 22–26.

Simpson CE (2001) Use of wild *Arachis* species/introgression of genes into *A. hypogaea* L. Peanut Sci 28: 114–116.

Simpson CE, Starr JL, Church GT, Burow MD, Paterson AH (2003) Registration of 'NemaTAM' peanut. Crop Sci 43: 1561.

Singh KP, Raina SN, Singh AK (1996) Variation in chromosomal DNA associated with the evolution of *Arachis* species. Genome 39: 890–897.

Somers DJ, Isaac P, Edwards K (2004) A high-density microsatellite consensus map for bread wheat (*Triticum aestivum* L.). Theor Appl Genet 109: 1105–1114.

Song GQ, Li MJ, Xiao H, Wang XJ, Tang RH, Xia H, Zhao CZ, Bi YP (2010) EST sequencing and SSR marker development from cultivated peanut (*Arachis hypogaea* L.). Elec J Biotechnol 13, DOI:10.2225/vol13.

Song QJ, Marek LF, Shoemaker RC, Lark KG, Concibido VC, Delannay X, Specht JE, Cregan PB (2004) A new integrated genetic linkage map of the soybean. Theor Appl Genet 109: 122–128.

Stalker HT, Beute MK, Shew BB, Isleib TG (2002) Registration of five leaf spot-resistant peanut germplasm lines. Crop Sci 42: 314–316.

Subramanian V, Gurtu S, Nageswara Rao RC, Nigam SN (2000) Identification of DNA polymorphism in cultivated groundnut using random amplified polymorphic DNA (RAPD) assay. Genome 43: 656–660.

Sujay V, Gowda MVC, Pandey MK, Bhat RS, Khedikar YP, Nadaf HL, Gautami B, Sarvamangala C, Lingaraju S, Radhakrishan T, Knapp SJ, Varshney RK (2012) QTL analysis and construction of consensus genetic map for foliar diseases resistance based on two RIL populations in cultivated groundnut (*Arachis hypogaea* L.). Mol Breed 32: 773–788.

Tanksley SD, Nelson JC (1996) Advanced backcross QTL analysis: a method for the simultaneous discovery and transfer of valuable QTLs from unadapted germplasm into elite breeding lines. Theor Appl Genet 92: 191–203.

Upadhyaya HD, Ortiz R, Bramel PJ, Singh S (2003) Development of a groundnut core collection using taxonomical, geographical and morphological descriptors. Genet Resour Crop Evol 50: 139–148.

Varshney RK, Graner A, Sorrells ME (2005) Genomics-assisted breeding for crop improvement. Trends Plant Sci 10: 621–630.

Varshney RK, Hoisington DA, Tyagi AK (2006) Advances in cereal genomics and applications in crop breeding. Trends Biotechnol 24: 490–499.

Varshney RK, Marcel TC, Ramsay L, Russell J, Röder MS, Stein N, Waugh R, Langridge P, Niks RE, Graner A (2007) A high density barley microsatellite consensus map with 775 SSR loci. Theor Appl Genet 114: 1091–1103.

Varshney RK, Close TJ, Singh NK, Hoisington DA, Cook DR (2009a) Orphan legume crops enter the genomics era. Curr Opin Plant Biol 12: 202–210.

Varshney RK, Hoisington DA, Nayak SN, Graner A (2009b) Molecular plant breeding: Methodology and achievements. In: Somers D, Langridge P, Gustafson PJ (eds) Methods in Molecular Biology: Plant Genomics. Humana Press, Totowa, NJ, USA, pp 283–304.

Varshney RK, Bertioli DJ, Moretzsohn MC, Vadez V, Krishnamurthy L, Aruna R, Nigam SN, Moss BJ, Seetha K, Ravi K, He G, Knapp SJ, Hoisington DA (2009c) The first SSR-based genetic linkage map for cultivated groundnut (*Arachis hypogaea* L.). Theor Appl Genet 118: 729–739.

Varshney RK, Thudi M, May GD, Jackson SA (2010a) Legume genomics and breeding. Plant Breed Rev 33: 257–304.

Varshney RK, Glaszmann JC, Leung H, Ribaut JM (2010b) More genomic resources for less-studied crops. Trends Biotechnol 28: 452–460.

Varshney RK, May GD (2012) Next-generation sequencing technologies: opportunities and obligations in plant genomics. Brief Funct Genom 11: 1–2.

Varshney RK, Mohan SM, Gaur PM, Gangarao NVPR, Pandey MK, Bohra A, Sawargaonkar S, Kimurto PK, Janila P, Saxena KB, Fikre A, Sharma M, Pratap A, Tripathi S, Datta S, Chaturvedi SK, Anuradha G, Babbar A, Chaudhary RG, Mhase MB, Bharadwaj CH, Mannur DM, Harer PN, Guo B, Liang X, Nadarajan N, Gowda CLL (2013a) Achievements and prospects of genomics-assisted breeding in three legume crops of the semi-arid tropics. Biotechnol Adv http://dx.doi.org/10.1016/j.biotechadv.2013.01.001.

Varshney RK, Pandey MK, Janila P, Nigam SN, Sudhini H, Gowda MVC, Sriswathi M, Radhakrishnan T, Manohar SS, Nagesh P (2013b) Marker-assisted introgression of a QTL region improved rust resistance in three varieties of peanut (*Arachis hypogaea* L.). Theor Appl Genet, under review.

Wang CT, Yang XD, Chen DX, Yu SL, Liu GZ, Tang YY, Xu JZ (2007) Isolation of simple sequence repeats from groundnut. Elec J Biotechnol 10: 3.

Wang H, Penmetsa RV, Yuan M, Gong L, Zhao Y, Guo B, Farmer AD, Rosen BD, Gao J, Isobe S, Bertioli DJ, Varshney RK, Cook DR, He G (2012) Development and characterization of BAC-end sequence derived SSRs, and their incorporation into a new higher density genetic map for cultivated peanut (*Arachis hypogaea* L.). BMC Plant Biol 12: 10.

Wang H, Pandey MK, Qiao L, Qin H, Culbreath AK, He G, Varshney RK, Guo B (2013) Genetic mapping and QTL analysis for disease resistance using F_2 and F_5 generation-based genetic maps derived from Tifrunner × GT-C20 in peanut (*Arachis hypogaea* L.). The Plant Genome 6, doi: 10.3835/plantgenome2013.05.0018.

Xu Y, Crouch JH (2008) Marker-assisted selection in plant breeding: from publications to practice. Crop Sci 48: 391–407.

Young ND, Weeden NF, Kochert G (1996) Genome mapping in legumes (Family *Fabaceae*). In: Paterson AH (ed) Genome Mapping in Plants. Landes Biomedical Press, Texas, USA, pp 211–277.

Yuan M, Gong L, Meng R, Li S, Dang P, Guo B, He G (2010) Development of trinucleotide $(GGC)_n$ SSR markers in peanut (*Arachis hypogaea* L.). Elec J Biotechnol DOI:10.2225/vol13-issue6-fulltext-6.

Zhao Y, Prakash CS, He G (2012) Characterization and compilation of polymorphic simple sequence repeat (SSR) markers of peanut from public database. BMC Res Notes 5: 362.

6

An Overview of Peanut Genome Structure

DJ Bertioli,[1] *ACG Araujo,*[2,*] *S Nielen,*[3] *P Heslop-Harrison,*[4] *PM Guimarães,*[5] *T Schwarzacher,*[6] *S Isobe,*[7] *K Shirasawa*[8] and *SCM Leal-Bertioli*[9]

ABSTRACT

Cultivated peanut (*Arachis hypogaea* L.) is a Papilionoid grain legume crop, important throughout the tropics. It is an allotetraploid of recent origin with an AB type genome ($2n = 4x = 40$) and has very low DNA polymorphism, a characteristic that has hampered genetic studies. The A and B genomes are of similar size and are composed mostly of metacentric chromosomes. The A genome is characterized by a pair of small chromosomes and the presence of strong centromeric heterochromatic bands, in contrast, B chromosomes are all of similar size and have much weaker centromeric bands. The genome of peanut is estimated at about 2.8 Gb and with a high repetitive DNA content. Its most probable diploid ancestors are *A. duranensis* and *A. ipaënsis*, donors of the A and B genomes, respectively. These two subgenomes diverged from a common ancestor about three and a half million years ago, more recently than the subgenomes of cotton or soybean. Consequently, homeologous A and B genic sequences have very high sequence identity. Genetically, cultivated peanut acts as a diploid, the two subgenomes have very high genetic synteny, and do not appear to have undergone major structural rearrangements after polyploidization. Indeed, the peanut subgenomes even have detectable genetic synteny with legumes that diverged during evolution about 55 Mya. The patterns of synteny indicate that the A and B genomes are highly diploidized, and that gene-space is likely to be ordered into about 10

Authors' affiliations given at the end of the chapter.

conserved blocks. In contrast to their conserved genetic synteny, the repetitive DNA components of the subgenomes are very significantly diverged. This may be substantially explained by the activity of a few retrotransposons since the time of genome divergence. In addition, the peanut genome harbors many miniature inverted-repeat transposable elements that have been active since polyploidization. This activity has probably contributed to the phenotypic variability of peanut.

Keywords: Allotetraploid, Polyploidization, Genome structure, Karyotypes, Genomic shock, Domestication, Genomic *in situ* hybridization (GISH), Long terminal repeat (LTR)

6.1 Introduction

It may seem a difficult task to write about peanut genome structure when the sequencing of the entire DNA sequence genome is imminent (The International Peanut Genome Initiative—IPGI) (http://www.peanutbioscience.com). At present, we know only a restricted amount of DNA sequence and what is known is yet unordered and unanchored. Fortunately however, many broad features of the peanut genome are already apparent, and the knowledge of these features will be of help to put into context and even to generate, assemble and order an entire genome sequence.

Many characteristics of a genome structure can be observed independently of the DNA sequencing. Cytogenetics, genetics and linkage mapping are prime examples that have given useful information on the large-scale organization of the *Arachis* genome. In addition, a relatively small amount of DNA information is very informative of genome structure because, in eukaryotes, typically a few very abundant transposons make up a substantial proportion of the genome. Already, some of the most important properties of the transposon components of the peanut genome are apparent, and some of the most abundant transposons are now well defined.

6.1.1 The Position of the Genus Arachis within the Legumes and Base Chromosome Number

Peanut, like most other economically important legumes is within the subfamily Papilionoid. Most of these important legumes fall within two subclades of this subfamily that diverged from each other about 50 Million years ago (Mya): the Phaseoloids and Galegoids (Lewis et al. 1995; Lavin et al. 2005). The Phaseoloids, which include soya and common bean, have a base chromosome number of $2x = 20$ or 22. The Galegoids, which include pea, lentil, *Lotus* and clover have a base chromosome number of

$2x = 10–16$. However, *Arachis* falls in a different clade, the Dalbergioids. This clade diverged from the Galegoids and Phaseoloids about 55 Mya (Cronk et al. 2006). Dalbergioids are predominantly from the New World tropical areas and have an ancestral chromosome number of $2x = 20$. Accordingly, most species of *Arachis* have 20 chromosomes and those that have different chromosome numbers can be assumed to be derived states. Of particular interest here is the cultivated peanut that differs from most of its wild relatives in having an allotetraploid genome ($2n = 4x = 40$).

6.2 The Chromosomes of Wild Arachis and Cultivated Peanut

Peanut has mostly metacentric chromosomes of similar size. The diploid wild *Arachis* species that are most closely related to it, and that have symmetrical karyotypes, have been assigned to two genome types, A and B, and belong to the botanical section *Arachis*. Species with the A genome have a small pair of chromosomes, "the A chromosomes" (Husted 1936; Smartt et al. 1978). Those species with symmetric karyotypes but without A chromosomes were traditionally considered B genome species more properly called non-A species. Recently, based on heterochromatin distribution and rDNA loci localization, these non-A species have been divided into three groups: B *sensu stricto*, F and K (Seijo et al. 2004; Robledo and Seijo 2010). The B genome *sensu stricto* has undetectable or much weaker centromeric heterochromatin, whilst F and K genomes have heterochromatic bands on most chromosomes, but differ in the amount and distribution. Phylogenies based on DNA sequence data strongly support the validity of this division (Moretzsohn et al. 2004; Milla et al. 2005; Tallury et al. 2005; Bravo et al. 2006; Bechara et al. 2010).

Cytological analysis showing heterochromatic distribution, rDNA loci and genomic *in situ* hybridization (GISH) indicated that the karyotype of cultivated peanut is equivalent to the sum of the karyotypes of *A. duranensis* (A genome) and *A. ipaënsis* (B genome) or very closely related species (Seijo et al. 2004, 2007). This origin is supported by DNA sequence data, the analysis of which has also revealed a very limited genetic variability in peanut and the wild tetraploid *A. monticola* (Halward et al. 1991; Kochert et al. 1996; Raina et al. 2001; Milla et al. 2005). Indeed, these two species are most probably the same species. This very limited DNA variability also indicates that *A. hypogaea* and *A. monticola* most probably had their origin in a single or very few hybridizations followed by chromosome duplication. It is not known if this origin occurred in the wild, or spontaneously when the two diploids were cultivated in close proximity by ancient inhabitants of South America. In either case, archaeological studies indicate the presence of *A. hypogaea* in the Huarmey Valley in Peru as long as 5,000 BP (Bonavia 1982).

Polyploids are sufficiently common among cultivated plants suggesting that they may have an advantage during artificial selection by man (Hilu 1993). This may be in part because of greater vigor due to 1) heterosis, and 2) increased size of harvested organs (Gepts 2003). In addition, events caused by polyploidy such as 3) changes in gene expression through the increased possibilities offered by: higher gene dosage, differential silencing, creation of new diversity through "genomic shock"; activation of transposons, and 4) relaxed selection on duplicated genes granting the acquisition of new gene functions are also likely to be important factors in allowing greater adaptability to cultivation (Soltis and Soltis 1995; Wessler and Carrington 2005). Whatever are the exact molecular mechanisms involved, it is remarkable that allotetraploid peanut, which has a very narrow genetic base, was transformed by domestication into one of the world's most important crops: completely distinct in plant architecture, seed size and pod form from its wild ancestors. In contrast, the much more genetically diverse diploid species, which have been cultivated for at least the same amount of time, only gave rise to a couple of proto-domesticate species cultivated on a very limited scale by indigenous people to this day (JFM Valls, pers. comm.; Freitas 2004).

6.3 Some Aspects of the Genetic Behavior of Wild and Cultivated Peanuts

The center of diversity of the genus *Arachis* is in the Cerrado biome, a savannah-like vegetation that experiences highly seasonal rainfall with very distinct wet and dry seasons (Krapovickas and Gregory 1994; Valls and Simpson 2005). Perhaps as an evolved adaptation to these climatic conditions, all peanut species have an unusual reproductive biology. They bear their fruits under the ground, or in other words, they are geocarpic. The flowers are borne above ground, and, after fertilization the young fruit is pushed into the ground through an elongated portion of the fruit structure ("peg") (Pelegrin et al. 2013) that has a specialized resistant tip. It penetrates the soil, a few centimeters for cultivated peanut, and much further for the wild species and the pod develops underground (Smith 1950). Deposited below the soil surface, peanut seeds are afforded protection from many pests and predators, favorable conditions for germination, and privileged access to soil moisture. However, a buried seed cannot be efficiently dispersed, and in natural conditions dispersal is mostly limited to the area covered by the maternal plant. More rarely, seeds may be deposited further afield by water-driven soil erosion, or animals (including man). In these cases, a single or very few seeds then found a new population, which results in natural populations that are "patches" with typically only 10s to 100s of individuals.

The combination of multiple recurrent severe genetic bottlenecks, small population sizes and a typically high rate of self-fertilization have provided the perfect conditions for genetic drift and the evolution of genetic isolation. In the classic Bateson–Dobzhansky–Muller model for the evolution of sexual incompatibility, diverging lineages evolve by mutations at different loci that are innocuous in their native genomic context, but interact negatively in hybrids (Bomblies and Weigel 2007). In the context of *Arachis*, its reproductive biology provides the perfect scenario for the fixation of weakly deleterious mutations that affect reproduction or indeed any other aspect of the plant's biology. Once a mutation is fixed within a population, any compensatory mutation in the same or an interacting gene will be positively selected, thus driving forward species divergence and genetic isolation. These mechanisms may have caused the remarkable degree of sexual incompatibility observed between different collections classified as the same species of wild *Arachis* (Krapovikas and Gregory 1994). At a genomic level, we may expect the genomes of wild species to harbor the signatures of genetic bottlenecks, isolation, inbreeding and genetic drift.

A very severe genetic bottle was imposed at the origin of peanut, and with cultivation came different population dynamics. With man actively transporting seeds, populations became mobile, and would be likely to experience more genetic mixing. However, considering the low densities of itinerant farmers in prehistory, genetic bottlenecks could still easily have occurred. Also, a new genetic phenomenon may have accompanied domestication: the selective sweep. Here, through strong artificial selection, agronomically favorable alleles and their surrounding genomic regions rapidly spread through populations and are genetically fixed. Such sweeps leave characteristic signals in genome sequences, although they may be difficult to detect in a background of very low DNA polymorphism.

6.4 The A and B Genomes of Peanut

Although peanut is an allotetraploid, chromosome pairing during meiosis is almost entirely bivalent (Smartt 1990). This presumably indicates that the A and B chromosomes have significantly diverged at the molecular level. Recently, the date of evolutionary divergence of the A and B genomes has been estimated for the first time. Sets of orthologous sequences were obtained from four intron regions, from *A. duranensis* and *A. ipaënsis, Lotus* and both palaeopolyploid components of soybean. Calibrating a molecular clock of DNA sequence divergence using the known divergence dates of the two genomic components of soybean (13 Mya), and the Galegoid, Phaseoloid and Dalbergioid clades, the divergence of the *Arachis* A and B genomes was estimated at 3.5 Mya (Nielen et al. 2011). Although such estimations must always be used with caution, we can confidently say that A and B peanut genomes are much more similar than the soya subgenomes, and almost certainly more similar than the allotetraploid cotton subgenomes, which have an estimated divergence of 5–7 million years.

In terms of evolution, 3.5 million years is a relatively short time. However, it is ample time for very significant transposon activity. Indeed, most easily dated transposons in plant genomes are less than three million years old. Older elements tend to be degraded by mutation or eliminated by unequal crossing-over and illegitimate recombination (Vicient et al. 1999; Devos et al. 2002; Pereira 2004). A substantial divergence in the repetitive component of the two genome components of peanut is consistent with *in situ* hybridization experiments where chromosome spreads were probed with *A. duranensis* BAC clones (Fig. 6-1) (Guimarães et al. 2008; Araujo et al. 2012), or with GISH using whole genomic DNA of its most probable ancestral diploids *A. duranensis* and *A. ipaënsis* (Seijo et al. 2007). The genomic probes do not hybridize exclusively, but predominantly to the chromosomes of their respective genome components. This shows that the repetitive components of the ancestral species diverged substantially during their separate evolutionary journeys traced since the time of their most recent common progenitor. Also, that any movement of repetitive DNA between the A and B genomes that occurred since the formation of the allotetraploid species was not sufficient to homogenize their repetitive DNA contents. Furthermore, the absence of any significant mosaic or chimeric hybridization patterns indicates that no large translocations between the A and B chromosomes have occurred since polyploidization. Indeed, the genome of peanut observed by GISH (Fig. 6-1a) is not distinguishable from the genome of a synthetic allotetraploid made from *A. ipaënsis* and *A. duranensis* (Fávero et al. 2006). Overall, it seems that there have been no obvious major structural changes in the diploid ancestral genomes following polyploidization.

The details of the GISH hybridization patterns could also be informative as to the distribution of repetitive DNA within the chromosomes. The strongest hybridization signals are at the interstitial chromosome regions. Hybridization is not detectable at the centromeres or the chromosome ends (Seijo et al. 2007; Nielen et al. 2010). Additionally, the smallest A chromosome pair, which has the most pronounced heterochromatic band, exhibits only very weak hybridization signals. From this, it is tempting to conclude that the repetitive DNA content of centromere and terminal regions of chromosomes are distinct from the interstitial regions and that the A chromosome pairs have distinct repetitive DNA profiles. However, it is also possible that the different hybridization patterns may, at least in part, be due to different states of chromatin condensation that influence the access of probes.

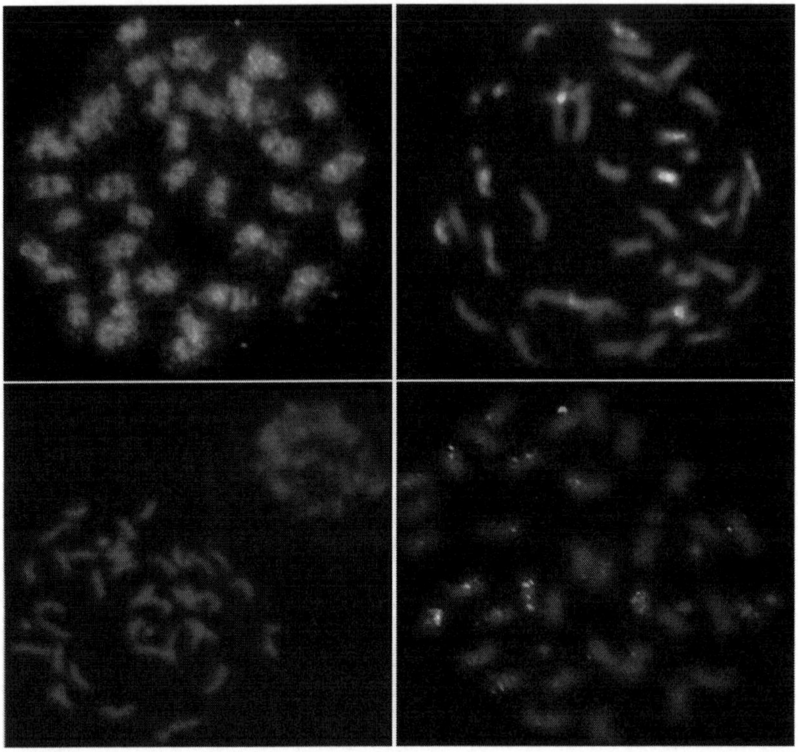

Figure 6-1 *In situ* hybridization on metaphase spreads of *Arachis* spp. with DAPI counterstaining. a) GISH on the synthetic amphidiploid (*Arachis duranensis* × *A. ipaënsis*)^{4x} with both parentals genomic DNA probes—green signals with *A. duranensis* on half of the chromosomes (A genome) and red signals with *A. ipaënsis* in the other half chromosomes (B genome) with signals overlapping part of some chromosomes; b) BAC-FISH with ADH79O23 (F12_Sl2_6OVER) probe with red signals over half of the chromosomes (A genome) and some dots on B genome chromosomes, more concentrated labeling but at different intensity depending on the chromosome. Hybridization signals were absent at centromere and telomere regions; c) BAC-FISH with ADH51I17 probe with diffused red signals in the pericentromere regions only on A genome chromosomes (F12_Sl4_5OVER); d) BAC-FISH with ADH179B13 probe with spotted green signals on A and B genome chromosomes but stronger on A chromosomes. Red signals correspond to the rDNA 5S sites (F17_Sl2_4OVER1). Scale bar: 5 µm.

Color image of this figure appears in the color plate section at the end of the book.

Fluorescent *in situ* hybridization (FISH) patterns are also informative as to the distribution of genes and repetitive DNA within genome. Hybridization signals in *A. hypogaea* metaphase chromosomes using some gene-poor clones from *A. duranensis* Bacterial Artificial Chromosome (BAC) library (A genome) as probes, were scatteredly distributed at the interstitial chromosome regions mainly in A chromosomes and eventually also in B (Fig. 6-1b), suggesting that the repetitive content of the A and B genomes is diverged. Patterns of hybridization such as intensity, diffused or spotted, varied according to the clone used as probe (Fig. 6-1b–d). Hybridization was generally at the pericentromere region but not at telomeres (Fig. 6-1b–d). Preliminary sequence analysis indicated that most of the repetitive sequences present in these clones could be accounted for by multiple copies of just few Long Terminal Repeat (LTR) retrotransposons and part of them (Bertioli et al. 2013). This will help the understanding of the peanut genome evolution and to delineate strategies to help the assemblage of its full genome sequence.

6.5 The Size and Repetitive Content of the Peanut Genome

Estimates of genome size vary depending on the methodology and standards used for its determination. The first estimate of genome size of peanut (Singh et al. 1996) was later re-evaluated substantially downwards to an equivalent of about 2.8 Gbp (Temsch and Greilhuber 2000, 2001). The size estimates for the most probable ancestral A and B genome donors, *A. duranensis* and *A. ipaënsis,* have been reported as being roughly similar to each other, which is consistent with observed chromosome sizes. The overall repetitive structure of the peanut genome has been studied by the kinetics of renaturation (Dhillon et al. 1980; Table 6-1). The high estimated percentage of repetitive DNA is within the range expected for genomes of this size.

Eukaryotic genomes harbor different types of repetitive DNA. Here we shall discuss ribosomal DNAs (rDNAs) and transposons.

Table 6-1 Repetitive structure of the peanut genome according to renaturation kinetics (data from Dhillon et al. 1980).

% of Genome	Class	Average copy number
11.90%	High repeat	38,000
14.80%	Intermediate repeat	6,700
37.40%	Rarely repeated	200
36.00%	Single copy	1

6.5.1 rDNAs in Peanut and Its Diploid Ancestors

The most thorough study of the location and number of rDNAs was conducted by Seijo and collaborators (2004) using FISH. The study showed, as previously mentioned, that the number, size, and distribution of rDNA clusters in *A. hypogaea* are virtually equivalent to the sum of those present in *A. duranensis* and *A. ipaënsis*. A single pair of 5S sites is present on each of the A and B chromosome complements, and two pairs of 18S–25S sites on the A chromosomes and three pairs on the B. The only exception to this equivalence is that in both of the diploid species, 18S–25S sites bear a thread-like constriction indicating intense transcriptional activity (forming the SAT chromosome; Fernandez and Krapovickas 1994). However, in the allotetraploid the constrictions are observed only on the A genome. This indicates that the transcriptional activity of the B genome rDNAs has been silenced, a common event in polyploids called nucleolar dominance (Cermeno et al. 1984; Preuss and Pikaard 2007).

6.5.2 Transposons and Their Evolution in the A and B Genomes

Transposons can be divided into two classes depending on whether their transposition intermediate is RNA (class 1, or retrotransposons) or DNA (class 2, or DNA transposons). Archetypal members of each group encode the protein products required for their transposition and are autonomous in function. However in a *reductio ad absurdum* of the rhyme:

"Big fleas have little fleas,
upon their backs to bite 'em,
and little fleas have lesser fleas,
and so on, *ad infinitum...*,"

even transposons are not free of parasites! In both classes of transposons there exist "parasitic" members with incomplete, degraded or completely absent coding regions. These transposons are non-autonomous, and depend on the proteins encoded by other elements for transposition. Plant genomes harbor a great diversity of transposons. However, two types, Miniature Inverted-repeat Transposable Elements (MITEs) and Long Terminal Repeat (LTR) retrotransposons have made particularly notable contributions to plant genome organization and evolution (Feschotte et al. 2002).

MITEs are non-autonomous DNA transposons of less than 600 bp in length. In peanut, Patel and collaborators (Patel et al. 2004) reported that, following treatment with a chemical mutagen, a MITE insertion caused functional disruption of the fatty-acid desaturase-encoding gene *ahFAD2B*, one of the homeologous genes controlling the very important quality trait of high oleic/linoleic fatty acid ratio in peanut seeds. This MITE did not belong to the most common Tourist or Stowaway families but showed similarities to the Bigfoot family in *Medicago* (Charrier et al. 1999). Later, AhMITE1, a transposon with sequence similarities to the previously reported MITE, was observed to excise from a single locus in spontaneous and artificially induced mutants (Gowda et al. 2010, 2011).

Evidence of activity and a tendency to transpose into genes or their flanking regions (Feschotte et al. 2002) stimulated further interest in MITEs, and recently a large-scale analysis in peanut has been completed (Shirasawa et al. 2012). Using enriched genomic libraries, 504 unique AhMITE1 sequences and their flanking genomic regions were obtained and shown to group into six families. Intriguingly, southern blots showed multiple AhMITE1 copies in the genomes of *A. magna* (a wild diploid B genome species very closely related to *A. ipaënsis*) and *A. hypogaea*, but not in the genome of *A. duranensis*, the most probable A genome donor to peanut. This suggests that AhMITE1 elements amplified in the B genome, but not in the A genome after their divergence about 3.5 Mya. Surveying of AhMITE1 insertion sites in cultivated varieties by PCR showed 13% polymorphism within a small sample of Virginia Runner type and 30% polymorphism between three Virginia cultivars and a Spanish type. This clearly indicates large-scale activity of AhMITE1 elements since the formation of the cultivated peanut and indicated the possibility that transposition events from the B to the A genome may have occurred in this tetraploid. The distribution of AhMITE1 markers in all the linkage groups of the most dense linkage map for peanut produced to date support that this migration has happened (Shirasawa et al. unpubl. data). This conclusion is compatible with the apparent equivalence in GISH patterns of peanut and synthetic allotetraploid mentioned above, because MITEs are small and their movement would not be expected to significantly change genomewide chromosome hybridization patterns.

The insertion rate of AhMITE1 into BLASTX detectable genes (10.5%) is much more frequent than would be expected by chance, and indicates that this family of transposons is likely to have affected the expression of numerous genes since the formation of the tetraploid, and may have had an important role in the generation of present-day morphological diversity of cultivated peanuts (Shirasawa et al. 2012).

The first comprehensively characterized peanut retrotransposon was an autonomous Ty3-*gypsy* type element of about 11,200 bp named FIDEL (Nielen et al. 2010). FISH analysis of peanut, dot blots and BAC-end sequences from *A. duranensis* and *A. ipaënsis* indicate that this element is more frequent in the A than in the B genome, with copy numbers of about 3,000 and 820 per haploid genome respectively (0.7% of the tetraploid genome). Phylogenetic analysis of reverse transcriptase sequences showed distinct evolution of FIDEL in the A and B genomes and indicated that FIDEL most probably underwent two major events of transposition and multiplication in the A genome after its evolutionary divergence from the B genome.

In contrast to AhMITE1, FIDEL is less frequent near single copy genes, a tendency that was observed using paired sequences from BAC clones (Nielen et al. 2010). Interestingly, this tendency could not be demonstrated with resistance gene homologs (see more on the association of transposons and resistance gene homologs later in this chapter). On a chromosome scale, the distribution of FIDEL; strongest hybridization in the interstitial regions of chromosome arms, and absence of detectable signal from centromeres, telomeric regions and nucleolar organizer region, closely resemble that of whole genomic DNA probes. This indicates that FIDEL may be an important component of the divergence of the repetitive DNA of the A and B genomes (Nielen et al. 2010).

Whilst FIDEL is common in the peanut genome, most copies are likely to be in a mutated and/or epigenetic silenced state. This is suggested by stop codons in coding regions and by the Ts/Tv ratio of >1.5:1 in the LTR sequences (Nielen et al. 2010). However, searching Expressed Sequenced Tag (EST) data does reveal some activity, and more intriguingly, transcription from FIDEL seems to be up-regulated under drought and disease stress (Brasileiro et al. 2012). Whether this evidence for activation indicates the generation of functional proteins and transposition remains to be answered. Whichever, FIDEL's localization in euchromatic regions suggests that it may have modified the expression of other genes, through insertional inactivation, or through the promotor activity of its LTRs.

More recently, a 6,179 bp autonomous Ty1-*copia* retrotransposon from the *Bianca* lineage named *Matita* has been characterized in peanut (Nielen et al. 2011). *Matita* is much less abundant than FIDEL with an estimated 520 copies in the haploid cultivated peanut genome. Also, in contrast to FIDEL, *Matita* is mainly located on the distal regions of chromosome arms and is of approximately equal frequency on both A and B chromosomes. Furthermore, phylogenetic analysis and molecular dating of transposition events suggest that although *Matita* has been active since the divergence of the A and B genomes, it underwent its last major burst of transposition activity at around the same time as the evolutionary divergence of peanut's diploid ancestors. By probing BAC libraries it was shown that *Matita* is also not randomly distributed in the genome but exhibits a significant tendency of being more abundant near resistance gene homologs than near single copy genes.

These studies have given a glimpse of the importance of transposons in the evolution of the peanut genome, in terms of their influence on gene expression and on genome structure. It is a common theme in plant genome structure that a relatively few transposon species are present at high number, and many more are present in low numbers. Peanut seems to follow this pattern, as the sequence analysis of 12 A-genome BACs, spanning about 1.25 Mb has recently shown (Araujo et al. 2012; Bertioli et al. 2013). Within these BAC sequences, most of the repetitive sequences could be accounted for by multiple copies of just seven LTR retrotransposons, their solo LTRs and remnants (Bertioli et al. 2013). Interestingly, only three of these elements were autonomous and four were non-autonomous, with one of the non-autonomous elements having FIDEL-like LTRs. Most of the datable transpositions were less than three Mya, indicating that much of the divergence of the A and B gene-space may be accounted by the activity of a few species of LTR retrotransposons. These elements are frequent in gene-space, but may be even more so in regions of the genome with few or no genes (see the example in Fig. 6-2).

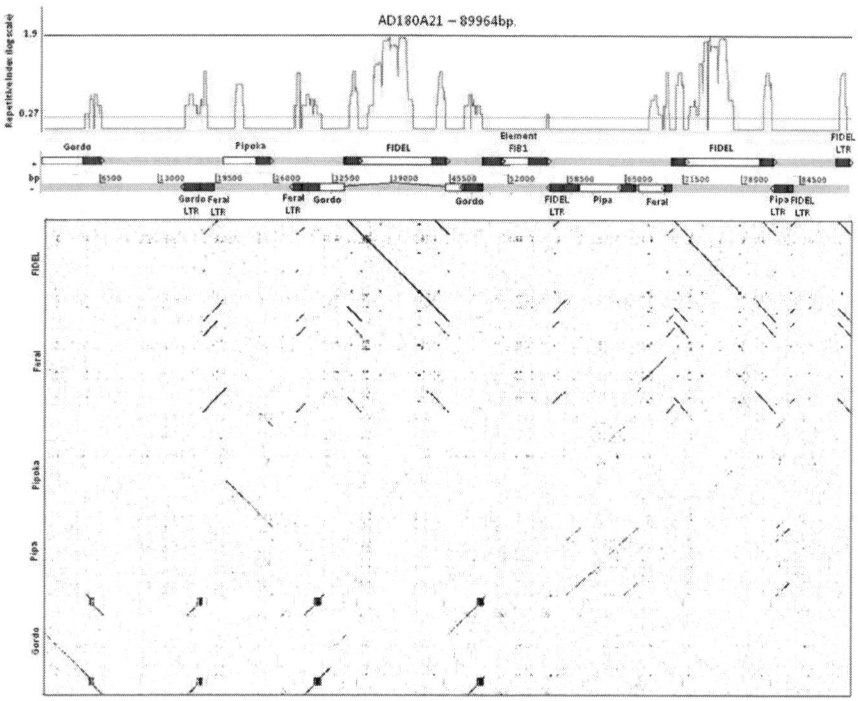

Figure 6-2 Representation of BAC clone AD180A21 consisting of two contigs, one of 84,046 bp positioned at left hand side, and one of 5,920 bp positioned at right. Top: Repetitive Index graph; Middle: Annotation scheme, and Bottom: dot plot. Repetitive Index is a score for repeat content based on BLASTN against 41,856 *A. duranensis* BAC-end sequences. The score is calculated using the formula Repetitive Index = log10(N+1), where N is the number of BLASTN homologies with an evalue of 1e-20 or less. The highest peak represented here is 1.9, which is equivalent to 88 BLASTN homologies; the lowest peak is 0.3, which is equivalent to a single BLASTN homology. The annotation scheme represents long terminal repeats (LTRs) in blue and internal regions of transposons in white. Transposons in positive orientation are represented on upper strand, and those in negative on the lower. The dot plot is of the BAC sequence (horizontal) against whole representative sequences of the transposons FIDEL, Feral, Pipoka, Pipa and Gordo (vertical).

This BAC clone consists almost entirely of LTR retrotransposons their solo elements and remnants, and does not contain any non-transposon gene. The sequence contains two complete FIDELs, one complete Pipa, and a complete Gordo interrupted by one of the FIDEL elements (names in bold type), a lower copy LTR transposon (Element-FIB1), plus retrotransposon fragments. All highly repetitive sequences in the BAC are derived from five retrotransposons: FIDEL, Feral, Pipoka, Pipa and Gordo.

Color image of this figure appears in the color plate section at the end of the book.

6.6 Comparison of the A and B Genomes in Low Copy Regions of the Genome

Whilst the consideration of repetitive DNA above emphasizes the differences between the A and B genomes, comparisons of genetic maps (see later in this chapter) and lower copy DNA emphasize the similarities. At the moment there is little data, but sequence identity is on average 94% between two 45 kb A and B homeologous genome sequences characterized by the authors (Bertioli et al. unpubl. data). This degree of sequence similarity in noncoding regions is compatible with the estimated date of evolutionary divergence (3.5 Mya) and a substitution rate of 1.3×10^{-8} per site per year (Ma et al. 2004). This is likely to be representative of many regions of the genome. In coding regions that are under conservative selection, we may expect DNA sequence identity to be even higher. However, these regions of high identity will be "broken" by fast evolving repetitive DNA. We may expect that many of these breaks can be accounted by the activity of a rather few species of transposons (Araujo et al. 2012; Bertioli et al. 2013).

We have already seen that the distributions of AhMITE1, FIDEL and *Matita* are not spatially random. Over evolutionary time, nonrandom patterns of transposon activity and elimination from the genome create a genomic landscape with identifiable broad features. Next we will see what light genetic mapping has shed on this.

6.7 Genetic Maps and Genome Structure

Sturtevant and Morgan's insight that the percentage of meiotic recombinants between two loci can be used as a measure of the distance between them allowed the construction of genetic maps. The first genetic maps were made using phenotypic characteristics, but traits controlled by single locus are scarce, and modern genetic maps are based on DNA markers. In peanut, which has very low DNA polymorphism, the generation of informative genetic markers has been very difficult, and this has been a fundamental limitation to peanut genetics.

The development of molecular markers for peanut has followed the technical trends of the times. The first studies were based on isozymes and proteins (Krishna and Mitra 1988; Grieshammer and Wynne 1990; Lu and Pickersgill 1993), followed by Restriction Fragment length polymorphism —RFLPs (Kochert et al. 1991, 1996; Paik-Ro et al. 1992), random Amplified Polymorphic DNA—RAPDs (Halward et al. 1991, 1992; Hilu and Stalker 1995; Subramanian et al. 2000), Amplified Fragment Length Polymorphism —AFLPs (He and Prakash 1997, 2001; Gimenes et al. 2002; Herselman 2003; Ferguson et al. 2004; Milla et al. 2005; Tallury et al. 2005), more recently microsatellite markers (Hopkins et al. 1999; Palmieri et al. 2002; He et al.

2003, 2005; Moretzsohn et al. 2004, 2005, 2009; Palmieri et al. 2005; Bravo et al. 2006; Budiman et al. 2006; Mace et al. 2006; Gimenes et al. 2007; Proite et al. 2007; Wang et al. 2007; Cuc et al. 2008; Guo et al. 2008; Naito et al. 2008; Liang et al. 2009; Yuan et al. 2010; Koilkonda et al. 2011) and molecular markers based on MITE markers (Shirasawa et al. 2012 and unpubl. data). Generally, these markers have shown a trend towards becoming more informative, and now microsatellites, being codominant and easy to score in the tetraploid genome, are considered the molecular marker of choice, with MITE markers also showing much potential.

Markers based on Single Nucleotide Polymorphism (SNP) have proved very difficult to apply for peanut. This is because they are very rare and difficult to detect against the background of false A-B polymorphism. If we consider that even very diverse peanut cultivars have diverged only a few thousand years ago, whilst the A and B genomes diverged a few Mya, then we can expect true A-A and B-B SNP rates to be in the region of 1,000 times less frequent than false A-B SNP rate. In addition to this difficulty of discovering SNPs, the considerable problem of scoring them in a tetraploid genome complicates it even more.

The obstacles to mapping in peanut meant that the first maps were generated using crosses involving wild species. Some of the maps also used the simpler diploid genetics of wild diploid species. Halward and collaborators (1991) produced such a diploid map, based on RFLPs and an F_2 population derived from two diploid A genome species *A. stenosperma* and *A. cardenasii*. Another RFLP-based map was published for a BC_1 tetraploid population derived from a synthetic allotetraploid [*A. batizocoi* x (*A. cardenasii* x *A. diogoi*)]^{4x} crossed with peanut (Burow et al. 2001). In this latter map, 370 RFLP loci were mapped onto 23 linkage groups, spanning a total of 2,207 cM. This tetraploid map was particularly informative to genome structure because it allowed the assignment of marker alleles to A or B genomes by reference to the known genomes of the diploid parents of the synthetic allotetraploid. In this way homologous linkage groups could be aligned. This showed that marker order was highly conserved between the A and B genomes.

The first map based on microsatellites was derived from a cross between *A. duranensis* and *A. stenosperma* (Moretzsohn et al. 2005). This map consisted of 11 linkage groups covering 1,230 cM. Subsequently, a microsatellite map of the B genome based on a cross of *A. ipaënsis* and the closely related *A. magna* map had 10 linkage groups, with 149 loci spanning a very similar total map distance of 1,294 cM. The comparison of 51 shared markers between these two maps revealed high levels of synteny, with all but one of the B linkage groups showing a single main correspondence to an A linkage group. This seems largely consistent with the observations for the previously mentioned tetraploid map. The main differences being: in the tetraploid study, one large B linkage group shows no marker correspondences to the A genome, whilst comparisons of the diploid maps showed no "orphan" linkage groups. Furthermore, in the diploid comparison, two B linkage groups correspond to one A, a situation not observed in the tetraploid map (Moretzsohn et al. 2009).

Further markers were subsequently included in the diploid A genome map. The most recently published version has 369 markers in 10 linkage groups (Leal-Bertioli et al. 2009). The total genetic distance covered by this map was more than 2,990 cM. Considering comparisons with subsequently published maps (Foncéka et al. 2009), we can conclude that this distance is overestimated several fold, probably due to the mixing of different dominant and codominant marker types. Nevertheless, this overestimate does not significantly reduce the information content of the map, because marker order seems correct and virtually all markers were sequence-characterized. Furthermore, many of the markers were particularly informative: 102 were genome comparative markers developed from intron regions of low copy genes (Leg anchor markers; Fredslund et al. 2006), and 35 were resistance gene homologs.

The sequence-characterized markers, and high proportion of low or single copy gene markers allowed the map to be aligned to the fully sequenced genomes of *Lotus japonicus* and *Medicago truncatula* (Sato et al. 2008; www.medicago.org). These were represented as "genome plots" (Fig. 6-3; Bertioli et al. 2009). Inspection of these plots shows surprising degrees of synteny considering the time of species divergence (estimated 55 Mya). Although there are some regions of double affinities between *Arachis* and these model legumes, most synteny blocks have a single main affinity and not multiple affinities interleaved. This is an important observation. Genome evolution, for instance, chromosomal translocations and inversions, progressively breaks down syntenic relationships between species over evolutionary time. However, in addition, whole genome duplications occur periodically during plant evolution, followed by progressive diploidization. In this chapter until now we have referred to diploid and tetraploid as if they were absolute states. In fact, genome duplication in plant evolution is sufficiently frequent that almost no plant is fully diploid, but in varying states of diploidization, following the most recent polyploidy event (Adams and Wendel 2005; Cui et al. 2006). From the single pairwise affinities in the *Arachis* genome plots two main conclusions can be made. The common ancestral genome of *Arachis*, that existed some 55 Mya was already substantially diploidized during the last universal legume whole genome duplication, which predated the divergence of *Arachis* from the Galegoids and Phaseoloids.

Secondly, that the so-called diploid *Arachis* genomes are therefore truly substantially diploid; their internal duplication is likely to be in the same range as for *Lotus* and *Medicago* 6.8 and 9.7%, respectively (Cannon et al. 2006).

Most economically important legumes and the two most important model legumes, *Medicago* and *Lotus,* belong to the Galegoid or Phaseoloid clades. *Arachis* is an outgroup, and so comparisons are particularly informative for making evolutionary inferences. For this reason, the *Arachis* vs. *Lotus*/*Medicago* plots were drawn with equivalent chromosomal orders as a previously published comparison between *Lotus* and *Medicago* genomes (Bertioli et al. 2009; Cannon et al. 2006; Fig. 6-3). Then, all possible *Arachis-Lotus-Medicago* species-by-species analysis could be observed in a comparable format. In this way, 10 distinct conserved synteny blocks and also nonconserved regions could be observed in all genomes. This clearly implies that certain legume genomic regions are consistently more stable during evolution than others. It is notable that these regions are large scale, and apparently in some cases consist of entire chromosomal arms.

An explanation for these observations was found by analyzing transposon distributions in *Lotus* and *Medicago.* Retrotransposons are very unevenly distributed in both the model legumes and it was observed that the retrotransposon-rich regions tend to correspond to variable regions, intercalating with the synteny blocks, which are relatively retrotransposon poor. This tendency is particularly evident for *Medicago,* but somewhat less so for *Lotus.* Furthermore, while the variable regions generally have lower densities of single copy genes than the more conserved regions, some harbor high densities of the fast evolving disease resistance genes (Bertioli et al. 2009; Fig. 6-3). For *Arachis* it was notable that LGs 2 and 4, which harbor the most prominent clusters of Resistance Gene Homologues (RGHs) and Quantitative Trait Loci (QTLs), showed shattered synteny with both *Lotus* and *Medicago.* In a different study, resistance to root-knot nematode was mapped to LGA 9 (Nagy et al. 2010). The upper region of LGA 9 is a synteny block, but its lower region appears to be a variable region. The region that confers nematode resistance is derived from the wild diploid *A. cardenasii,* and is particularly genetically interesting because it displays strongly suppressed recombination with the A genome of *A. hypogaea* and appears to cover about one-third to a half of a chromosome.

Through large scale screening of Simple Sequence Repeat (SSR) markers, a sufficient number of polymorphic markers were identified for the generation of the first genetic linkage maps based on cultivated x cultivated crosses (Varshney et al. 2009; Hong et al. 2008, 2010). These maps are very useful for breeding because they incorporate QTLs for agronomically important traits, such as disease resistance and drought-related traits. For the creation of the highest density map of peanut to date, markers screening

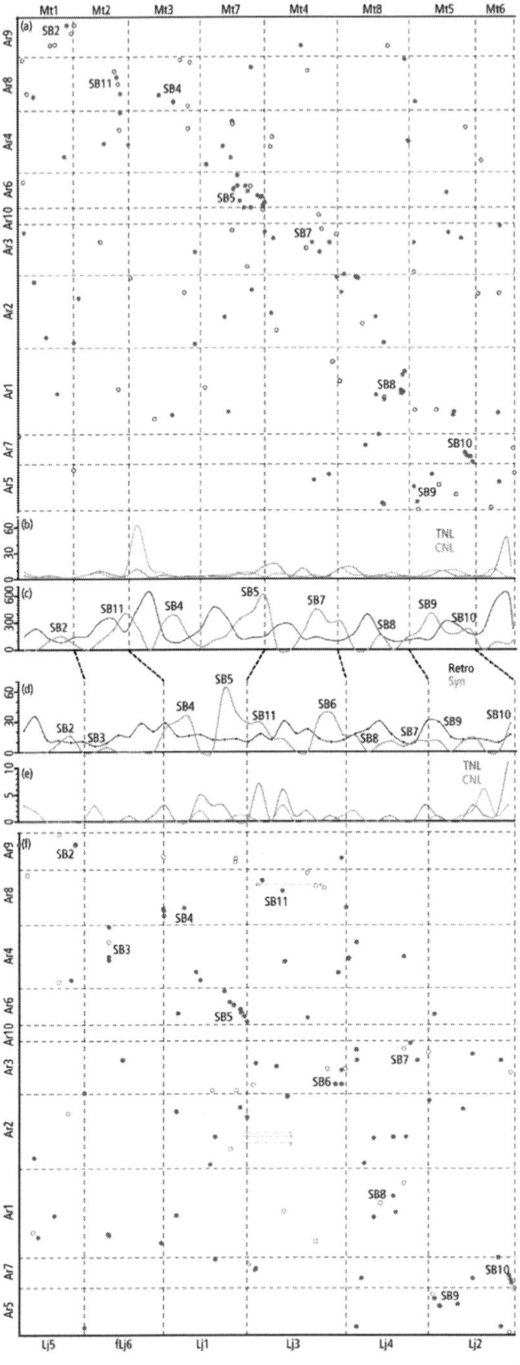

Figure 6-3 contd....

was done by *in silico* analysis of the parents. The map has 1,114 markers and is 2,166 cM in length. Interestingly it has 21 linkage groups, two of which are much lower density than the others (Shirasawa et al. unpubl. data).

6.8 Sequencing the Peanut Genome

In a new development, as an initial phase in the IPGI (http://www.peanutbioscience.com), the very large capacity of Illumina sequencing is being used for the generation of high density genetic maps. The data for the generation of these maps is obtained essentially by using low coverage sequencing as a method of high density genotyping. The approach is being used in diploid and tetraploid peanut mapping populations and a peanut diversity panel (Froenicke et al. 2011, 2012). The genetic maps generated are expected to be especially useful in the ordering of contigs and scaffolds in the Peanut Genome Project.

The estimated size of the cultivated peanut's genome is about 2.8 Gbp, almost as large as the human genome. Sequencing and properly ordering of such a large genome is a challenging task. The main obstacle is not the size itself but the repeat structures present within large genomes. Very significant portions of large genomes consist of almost identical copies of DNA repeated multiple times. During sequence assembly, the placement

Figure 6-3 contd.

Figure 6-3 Genome plots of *Arachis* vs. *Medicago* and *Arachis* vs. *Lotus*, integrated with each other and graphs of synteny with *Arachis*, and retrotransposon, and resistance gene homolog distributions for *Medicago* and *Lotus* (original figure is from Bertioli et al. 2009). Chromosome orders and numbering of synteny blocks are the same as a *Medicago* vs. *Lotus* plot in Cannon et al. 2006, allowing direct comparisons. Equivalent conserved regions (synteny blocks) and variable regions are present in all possible combinations of species comparisons *Arachis-Lotus-Medicago*. This shows that some genomic regions (synteny blocks) are consistently more stable during evolution than others.

(a) Genome Plot of *Arachis* vs. *Medicago*.
(b) Density of blast detected resistance gene homologs of the TNL (red line) and CNL (green line) subclasses plotted along the *Medicago* genome. High densities of resistance gene homologs and retrotransposons coincide.
(c) Black line: density of blast detected retrotransposons plotted along the *Medicago* genome. Cyan-blue line: scaled synteny score of *Medicago* with *Arachis*. Synteny blocks occur in regions of low retrotransposon density.
(d) Black line: percentage genome coverage of retrotransposons plotted along the *Lotus* genome. Cyan-blue line: scaled synteny score of *Lotus* with *Arachis*. Synteny blocks tend to occur in regions of low retrotransposon coverage.
(e) Density of resistance gene homolog encoding sequences, TNL (red) and CNL (green), plotted along the Lotus genome. Clusters of resistance gene homologs and retrotransposons coincide.
(f) Genome Plot of *Arachis* vs. *Lotus*. Markers mapped to intervals are plotted as horizontal lines.

Color image of this figure appears in the color plate section at the end of the book.

of sequence reads derived from these repeats into the wrong position can prevent assembly being completed, or worse, induce the assembly to be completed in the wrong way. The peanut genome is no exception and harbors numerous repeat structures. It is know that each retrotransposition event can create a new repeat structure and as discussed above, there seems to have been significant evolutionary recent activity of transposons in the *Arachis* genome, which is a potential problem for assembly.

The most problematic transposition events are very recent ones where mutation has not had sufficient time to reduce sequence identity. The size of transposons, and even many solo LTRs, usually substantially exceeds the size of individual sequence reads. Therefore, paired sequence reads at different scales will be particularly important for spanning transposons and other repeat structures of varying scales and enabling assembly.

Although the overall repetitive profile of peanut seems compatible with whole genome shotgun sequencing, the allotetraploid genome with relatively recently diverged A and B components will be especially problematic. Assembly of such a genome may encounter two frequent problems, the first one being breaks in contigs because of misassembles at ends of contigs (A reads at ends of B contigs or vice versa) and the second, the generation of mixed A and B (chimeric) contigs. These problems are likely to be worse with shorter sequence reads because, for instance, identical 100 bp A and B homeologous regions will be much more common than identical 500 bp regions. Strategies will be necessary to overcome these difficulties, especially if the project is to take advantage of Illumina sequencing, which produces massive amounts of data, but short sequence reads. Two possible options are the sequencing of the diploid progenitors to provide templates for a tetraploid assembly, and a multiplexed BAC-by-BAC strategy.

6.9 Conclusions

Although, at the time of writing this chapter, there is relatively little ordered or anchored genomic DNA sequence available for peanut, many general features of the genome are apparent. The knowledge of these features is useful in the design of sequencing strategies, and should be useful to guide assembly methods, and to generate and test hypotheses when an assembled genome is available.

References

Adams KL, Wendel JF (2005) Polyploidy and genome evolution in plants. Curr Opin Plant Biol 8: 135–141.

Araujo A, Nielen S, Vidigal B, Moretzsohn M, Leal-Bertioli S, Ratnaparkhe M, Kim C, Bailey J, Paterson A, Guimarães P, Schwarzacher T, Heslop-Harrison P, Bertioli DJ (2012) An analysis of the repetitive component of the peanut genome in the evolutionary context of the Arachis A-B genome divergence. In: Plant Anim Genome XX Conf, San Diego CA, USA, P 0460. https://pag.confex.com/pag/xx/webprogram/Paper3749.html.

Bechara M, Moretzsohn M, Palmieri D, Monteiro J, Bacci M, Martins J, José FM, Valls J, Lopes C, Gimenes M (2010) Phylogenetic relationships in genus *Arachis* based on ITS and 5.8S rDNA sequences. BMC Plant Biol 10: 255.

Bertioli D, Moretzsohn M, Madsen L, Sandal N, Leal-Bertioli S, Guimaraes P, Hougaard B, Fredslund J, Schauser L, Nielsen A, Sato S, Tabata S, Cannon S, Stougaard J (2009) An analysis of synteny of *Arachis* with *Lotus* and *Medicago* sheds new light on the structure, stability and evolution of legume genomes. BMC Genomics 10: 45.

Bertioli D, Vidigal B, Nielen S, Ratnaparkhe M, Lee T-H, Leal-Bertioli S, Kim C, Guimarães P, Seijo G, Schwarzacher T, Paterson A, Heslop-Harrison P, Araujo A (2013) The repetitive component of the A genome of peanut (*Arachis hypogaea*) and its role in remodelling intergenic sequence space since its evolutionary divergence from the B genome. Ann Bot 112: 545–559.

Bomblies K, Weigel D (2007) Hybrid necrosis: autoimmunity as a potential gene-flow barrier in plant species. Nat Rev Genet 8: 382–393.

Bonavia D (1982) Precerámico peruano, Los Gavilanes, oásis en la história del hombre. Corporación Financiera de Desarrollo S.A. COFIDE e Instituto Arqueológico Alemán. Lima, Peru.

Brasileiro AC, Morgante CV, Leal-Bertioli SC, Santos CM, Araujo ACG, Pappas GJ, Bonfim O, da Silva FR, Silva AK, Martins ACQ, Bertioli DJ, Patricia G (2012) Transcriptome survey on wild peanut relatives for discovery of drought-responsive genes. In: Plant Anim Genome XX Conf, San Diego, CA, USA, P 0462. https://pag.confex.com/pag/xx/webprogram/Paper4256.html.

Bravo JP, Hoshino AA, Angelici C, Lopes CR, Gimenes MA (2006) Transferability and use of microsatellite markers for the genetic analysis of the germplasm of some *Arachis* section species of the genus *Arachis*. Genet Mol Biol 29: 516–524.

Budiman M, Jones J, Citek R, Warek U, Bedell J, Knapp S (2006) Methylation-filtered and shotgun genomic sequences for diploid and tetraploid peanut taxa: http://www.ncbi.nlm.nih.gov/ (Accessed Aug 2013).

Burow MD, Simpson CE, Starr JL, Paterson AH (2001) Transmission genetics of chromatin from a synthetic amphidiploid to cultivated peanut (*Arachis hypogaea* L.) broadening the gene pool of a monophyletic polyploid species. Genet 159: 823–837.

Cannon SB, Sterck L, Rombauts S, Sato S, Cheung F, Gouzy J, Wang X, Mudge J, Vasdewani J, Schiex T, Spannagl M, Monaghan E, Nicholson C, Humphray SJ, Schoof H, Mayer KF, Rogers J, Quétier F, Oldroyd GE, Debellé F, Cook DR, Retzel EF, Roe BA, Town CD, Tabata S, Van de Peer Y, Young ND (2006) Legume genome evolution viewed through the *Medicago truncatula* and *Lotus japonicus* genomes. Proc Natl Acad Sci USA 103: 14959–14964.

Cermeno MC, Orellana J, Santos JL, Lacadena JR (1984) Nucleolar activity and competition (Amphiplasty) in the genus *Aegilops*. Heredity 53: 603–611.

Charrier B, Foucher F, Kondorosi E, D'Aubenton-Carafa Y, Thermes C, Kondorosi A, Ratet P (1999) Bigfoot: a new family of MITE elements characterized from the *Medicago* genus. Plant J 18: 431–441.

Cronk Q, Ojeda I, Pennington R (2006) Legume comparative genomics: progress in phylogenetics and phylogenomics. Curr Opin Plant Biol 9: 99–103.

Cuc LM, Mace ES, Crouch JH, Quang VD, Long TD, Varshney RK (2008) Isolation and characterization of novel microsatellite markers and their application for diversity assessment in cultivated groundnut (*Arachis hypogaea* L.). BMC Plant Biol 8: 55.

Cui L, Wall PK, Leebens-Mack JH, Lindsay BG, Soltis DE, Doyle JJ, Soltis PS, Carlson JE, Arumuganathan K, Barakat A, Albert VA, Ma H, dePamphilis CW (2006) Widespread genome duplications throughout the history of flowering plants. Genome Res 16: 738–749.

Devos KM, Brown JKM, Bennetzen JL (2002) Genome size reduction through illegitimate recombination counteracts genome expansion in *Arabidopsis*. Genome Res 12: 1075–1079.

Dhillon SS, Rake AV, Miksche JP (1980) Reassociation kinetics and cytophotometric characterization of peanut (*Arachis hypogaea* L.) DNA. Plant Physiol 65: 1121–1127.

Fávero AP, Simpson CE, Valls FMJ, Velo NA (2006) Study of evolution of cultivated peanut trough crossability studies among *Arachis ipaënsis*, *A. duranensis* and *A. hypogaea*. Crop Sci 46: 1546–1552.

Ferguson ME, Burow MD, Schulze SR, Bramel PJ, Paterson AH, Kresovich S, Mitchell S (2004) Microsatellite identification and characterization in peanut (*A. hypogaea* L.). Theor Appl Genet 108: 1064–1070.

Fernandez A, Krapovickas A (1994) Cromosomas y evolucion en *Arachis* (Leguminosae). Bonplandia 8: 187–200.

Feschotte C, Jiang N, Wessler SR (2002) Plant transposable elements: where genetics meets genomics. Nat Rev Genet 3: 329–341.

Foncéka D, Hodo-Abalo T, Rivallan R, Faye I, Sall M, Ndoye O, Fávero A, Bertioli D, Glaszmann J, Courtois B, Rami J (2009) Genetic mapping of wild introgressions into cultivated peanut: a way toward enlarging the genetic basis of a recent allotetraploid. BMC Plant Biol 9: 103.

Fredslund J, Madsen L, Hougaard B, Nielsen A, Bertioli D, Sandal N, Stougaard J, Leif Schauser L (2006) A general pipeline for the development of anchor markers for comparative genomics in plants. BMC Genomics 7: 207.

Freitas FO (2004) Ampliação da área de ocorrência da espécie *Arachis villosulicarpa* Embrapa Comunicado Técnico- Embrapa, Brasilia-Brazil.

Froenicke L, Pandey MK, Upadhyaya H, Moretzsohn MC, Guimaraes PM, Leal-Bertioli S, Varshney RK, Bertioli D, Michelmore RW (2011) Towards ultra-dense genetic maps of peanut generated by sequencing diploid and tetraploid RIL populations and a peanut diversity panel. In: 5th Int Conf of the Peanut Research Community on Advances in *Arachis* through Genomics and Biotechnology (AAGB-2011), Brasilia, Brazil, 13–15 June 2011. Embrapa, Brazil, p 57.

Froenicke L, Beitel C, Pandey MK, Upadhyaya HD, Moretzsohn MC, Guimaraes PM, Leal-Bertioli SCM, Varshney RK, Bertioli DJ, Michelmore R (2012) Applications of genome sequencing for peanut improvement. In: Plant Anim Genome XX Conf, San Diego, CA, USA, p W355. https://pag.confex.com/pag/xx/webprogram/Paper4430.html.

Gepts P (2003) Ten thousand years of crop evolution. In: Chrispeels M, Sadava D (eds) Plants, Genes, and Crop Biotechnology. Jones and Bartlett Publishers, Sudbury, MA, USA, pp 328–359.

Gimenes MA, Lopes CR, Galgaro ML, Valls JF, Kochert G (2002) RFLP analysis of genetic variation in species of section Arachis, genus *Arachis* (Leguminosae). Euphytica 123: 421–429.

Gimenes MA, Hoshino AA, Barbosa AVG, Palmieri DA, Lopes CR (2007) Characterization and transferability of microsatellite markers of the cultivated peanut (*Arachis hypogaea*). BMC Plant Biol 7: 9.

Gowda M, Bhat R, Sujay V, Kusuma P, Varshakumari V, Bhat S, Varshney RK (2011) Characterization of *AhMITE1* transposition and its association with the mutational and evolutionary origin of botanical types in peanut (*Arachis* spp.). Plant Syst Evol 291: 153–158.

Gowda MVC, Bhat RS, Motagi BN, Sujay V, Kumari V, Bhat S (2010) Association of high-frequency origin of late leaf spot resistant mutants with *AhMITE*1 transposition in peanut. Plant Breed 129: 567–569.

Grieshammer U, Wynne JC (1990) Isozyme variability in mature seeds of US peanut cultivars and collections. Peanut Sci 17: 72–75.

Guimarães P, Garsmeur O, Proite K, Leal-Bertioli S, Seijo G, Chaine C, Bertioli D, D'Hont A (2008) BAC libraries construction from the ancestral diploid genomes of the allotetraploid cultivated peanut. BMC Plant Biol 8: 14.

Guo BZ, Chen XP, Dang P, Scully BT, Liang XQ, Holbrook C, Yu J, Culbreath A (2008) Peanut gene expression profiling in developing seeds at different reproduction stages during *Aspergillus parasiticus* infection. BMC Dev Biol 8: 12, doi:10.1186/1471-213x-8-12.

Halward T, Stalker T, LaRue E, Kochert G (1992) Use of single-primer DNA amplifications in genetic studies of peanut (*Arachis hypogaea* L.). Plant Mol Biol 18: 315–325.

Halward TM, Stalker HT, Larue EA, Kochert G (1991) Genetic variation detectable with molecular markers among unadapted germ-plasm resources of cultivated peanut and related wild species. Genome 34: 1013–1020.

He G, Prakash CS (1997) Identification of polymorphic DNA markers in cultivated peanut (*Arachis hypogaea* L.). Euphytica 97: 143–149.

He G, Prakash C (2001) Evaluation of genetic relationships among botanical varieties of cultivated peanut (*Arachis hypogaea* L.) using AFLP markers. Genet Resour Crop Evol 48: 347–353.

He G, Meng R, Newman M, Gao G, Pittman RN, Prakash C (2003) Microsatellites as DNA markers in cultivated peanut (*Arachis hypogaea* L.). BMC Plant Biol 3: 3.

He G, Meng R, Gao H, Guo B, Newman M, Pittman R, Prakash C (2005) Simple sequence repeat markers for botanical varieties of cultivated peanut (*Arachis hypogaea* L.). Euphytica 142: 131–136.

Herselman L (2003) Genetic variation among Southern African cultivated peanut (*Arachis hypogaea* L.) genotypes as revealed by AFLP analysis. Euphytica 133: 319–327.

Hilu KW (1993) Polyploidy and the evolution of domesticated plants. Am J Bot 80: 2521–2528.

Hilu KW, Stalker HT (1995) Genetic relationships between peanut and wild species of *Arachis* sect. Arachis (Fabaceae): Evidence from RAPDs. Plant Syst Evol 198: 167–178.

Hong YB, Liang XQ, Chen XP, Liu HY, Zhou GY, Li SX, Wen SJ (2008) Construction of genetic linkage map based on SSR markers in peanut (*Arachis hypogaea* L.). Agri Sci China 7: 915–921.

Hong YB, Chen XP, Liang XQ, Liu HY, Zhou GY, Li S, Wen S, Holbrook CC, Guo B (2010) A SSR-based composite genetic linkage map for the cultivated peanut (*Arachis hypogaea* L.) genome. BMC Plant Biol 10: 1–13.

Hopkins M, Casa A, Wang T, Mitchell S, Dean R, Kochert G, Kresovich S (1999) Discovery and characterization of polymorphic simple sequence repeats (SSRS) in peanut. Crop Sci 39: 1243–1247.

Husted L (1936) Cytological Studies on The Peanut. *Arachis*. II—Chromosome number, morphology and behavior, and their application to the problem of the origin of the cultivated forms. Cytol 7: 396–423.

Kochert G, Halward T, Branch WD, Simpson CE (1991) RFLP variability in peanut (*Arachis hypogaea* L.) cultivars and wild species. Theor Appl Genet 81: 565–570.

Kochert G, Stalker HT, Gimenes M, Galgaro L, Lopes CR, Moore K (1996) RFLP and cytogenetic evidence on the origin and evolution of allotetraploid domesticated peanut, *Arachis hypogaea* (Leguminosae). Am J Bot 83: 1282–1291.

Koilkonda P, Sato S, Tabata S, Shirasawa K, Hirakawa H, Sakai H, Sasamoto S, Watanabe A, Wada T, Kishida Y, Tsuruoka H, Fujishiro T, Yamada M, Kohara M, Suzuki S, Hasegawa M, Kiyoshima H, Isobe S (2011) Large-scale development of expressed sequence tag-derived simple sequence repeat markers and diversity analysis in *Arachis* spp. Mol Breed 30: 125–138.

Krapovickas A, Gregory WC (1994) Taxonomia del genero *Arachis* (Leguminosae). Bonplandia 8: 1–186.

Krishna TG, Mitra R (1988) The probable genome donors to *Arachis hypogaea* L. based on arachin seed storage protein. Euphytica 37: 47–52.

Lavin M, Herendeen PS, Wojciechowski MF (2005) Evolutionary rates analysis of Leguminosae implicates a rapid diversification of lineages during the Tertiary. Syst Biol 54: 575–594.

Leal-Bertioli SCM, Jose ACVF, Alves-Freitas DMT, Moretzsohn MC, Guimaraes PM, Nielen S, Vidigal BS, Pereira RW, Pike J, Fávero AP, Parniske M, Varshney RK, Bertioli D (2009) Identification of candidate genome regions controlling disease resistance in *Arachis*. BMC Plant Biol 9: 112.

Lewis GP, Schrire B, Mackinder B, Lock M (1995) Legumes of the world. Royal Botanic Gardens, Kew, Surrey, UK, p 8.

Liang X, Chen X, Hong Y, Liu H, Zhou G, Li S, Guo B (2009) Utility of EST-derived SSR in cultivated peanut (*Arachis hypogaea* L.) and *Arachis* wild species. BMC Plant Biol 9: 35.

Lu J, Pickersgill B (1993) Isozyme variation and species relationships in peanut and its wild relatives (*Arachis* L. Leguminosae). Theor Appl Genet 85: 550–560.

Ma J, Devos KM, Bennetzen JL (2004) Analyses of LTR-retrotransposon structures reveal recent and rapid genomic DNA loss in rice. Genome Res 14: 860–869.

Mace E, Phong D, Upadhyaya H, Chandra S, Crouch J (2006) SSR analysis of cultivated groundnut (*Arachis hypogaea* L.) germplasm resistant to rust and late leaf spot diseases. Euphytica 152: 317–330.

Milla SR, Isleib TG, Stalker HT (2005) Taxonomic relationships among *Arachis* sect. *Arachis* species as revealed by AFLP markers. Genome 48: 1–11.

Moretzsohn M, Leoi L, Proite K, Guimarães P, Leal-Bertioli S, Gimenes M, Martins W, Valls JF, Grattapaglia D, Bertioli DJ (2005) A microsatellite-based, gene-rich linkage map for the AA genome of *Arachis* (Fabaceae). Theor Appl Genet 111: 1060–1071.

Moretzsohn MC, Hopkins MS, Mitchell SE, Kresovich S, Valls JFM, Ferreira ME (2004) Genetic diversity of peanut (*Arachis hypogaea* L.) and its wild relatives based on the analysis of hypervariable regions of the genome. BMC Plant Biol 4: 11.

Moretzsohn MC, Barbosa AVG, Alves-Freitas DMT, Teixeira C, Leal-Bertioli SCM, Guimarães PM, Pereira RW, Lopes CR, Cavallari MM, Valls JF, Bertioli DJ, Gimenes MA (2009) A linkage map for the B-genome of *Arachis* (Fabaceae) and its synteny to the A-genome. BMC Plant Biol 9: 40.

Nagy ED, Chu Y, Guo Y, Khanal S, Tang S, Li Y, Dong WB, Timper P, Taylor C, Ozias-Akins P, Holbrook CC, Beilinson V, Nielsen NC, Stalker HT, Knapp SJ (2010) Recombination is suppressed in an alien introgression on chromosome 5A of peanut harboring *Rma*, a dominant root-knot nematode resistance gene. Mol Breed 26: 357–370.

Naito Y, Suzuki S, Iwata Y, Kuboyama T (2008) Genetic diversity and relationship analysis of peanut germplasm using SSR markers. Breed Sci 3: 293–300.

Nielen S, Campos-Fonseca F, Leal-Bertioli S, Guimarães P, Seijo G, Town C, Arrial R, Bertioli D, (2010) FIDEL—a retrovirus-like retrotransposon and its distinct evolutionary histories in the A- and B-genome components of cultivated peanut. Chrom Res 18: 227–246.

Nielen S, Vidigal BS, Leal-Bertioli SC, Ratnaparkhe M, Paterson AH, Garsmeur O, D'Hont A, Guimarães PM, Bertioli D (2011) Matita, a new retroelement from peanut: characterization and evolutionary context in the light of the Arachis A-B genome divergence. Mol Genet Genom 287: 21–38.

Paik-Ro OG, Smith RL, Knauft DA (1992) Restriction fragment length polymorphism evaluation of six peanut species within the *Arachis* section. Theor Appl Genet 84: 201–208.

Palmieri DA, Hoshino AA, Bravo JP, Lopes CR, Gimenes MA (2002) Isolation and characterization of microsatellite loci from the forage species *Arachis pintoi* (Genus *Arachis*). Mol Ecol Notes 2: 551–553.

Palmieri DA, Bechara MD, Curi RA, Gimenes MA, Lopes CR (2005) Novel polymorphic microsatellite markers in section *Caulorrhizae* (*Arachis*, Fabaceae). Mol Ecol Notes 5: 77–79.

Patel M, Jung S, Moore K, Powell G, Ainsworth C, Abbott A (2004) High-oleate peanut mutants result from a MITE insertion into the *FAD2* gene. Theor Appl Genet 108: 1492–1502.

Pelegrin CMG, Valls JFM, Mariath JEA (2013) Pericarp ontogenesis in wild species of *Arachis* L. (Leguminosae; Papilionoideae). 208: 118–127.

Pereira V (2004) Insertion bias and purifying selection of retrotransposons in the *Arabidopsis thaliana* genome. Genome Biol 5: R79.

Preuss S, Pikaard CS (2007) rRNA gene silencing and nucleolar dominance: insights into a chromosome-scale epigenetic on/off switch. Biochim Biophys Acta 1769: 383–392.

Proite K, Leal-Bertioli S, Bertioli D, Moretzsohn M, da Silva F, Martins N, Guimarães P (2007) ESTs from a wild *Arachis* species for gene discovery and marker development. BMC Plant Biol 7: 7.

Raina SN, Rani V, Kojima T, Ogihara Y, Singh KP, Devarumath M (2001) RAPD and ISSR fingerprints as useful genetic markers for analysis of genetic diversity, varietal identification, and phylogenetic relationships in peanut (*Arachis hypogaea*) cultivars and wild species. Genome 44: 763–772.

Robledo G, Seijo G (2010) Species relationships among the wild B genome of *Arachis* species (section *Arachis*) based on FISH mapping of rDNA loci and heterochromatin detection: a new proposal for genome arrangement. Theor Appl Genet 121: 1033–1046.

Sato S, Nakamura Y, Kaneko T, Asamizu E, Kato T, Nakao M, Sasamoto S, Watanabe A, Ono A, Kawashima K, Fujishiro T, Katoh M, Kohara M, Kishida Y, Minami C, Nakayama S, Nakazaki N, Shimizu Y, Shinpo S, Takahashi C, Wada T, Yamada M, Ohmido N, Hayashi M, Fukui K, Baba T, Nakamichi T, Mori H, Tabata S (2008) Genome structure of the legume, *Lotus japonicus*. DNA Res 15: 227–239.

Seijo G, Lavia GI, Fernandez A, Krapovickas A, Ducasse DA, Bertioli D, Moscone E (2007) Genomic relationships between the cultivated peanut (*Arachis hypogaea*, Leguminosae) and its close relatives revealed by double GISH. Am J Bot 94: 1963–1971.

Seijo JG, Lavia GI, Fernandez A, Krapovickas A, Ducasse D, Moscone E (2004) Physical mapping of the 5S and 18S-25S rRNA genes by FISH as evidence that *Arachis duranensis* and *A. ipaënsis* are the wild diploid progenitors of *A. hypogaea* (leguminosae). Am J Bot 91: 1294–1303.

Shirasawa K, Hirakawa H, Tabata S, Hasegawa M, Kiyoshima H, Suzuki S, Sasamoto S, Watanabe A, Fujishiro T, Isobe S (2012) Characterization of active miniature inverted-repeat transposable elements in the peanut genome. Theor Appl Genet 124: 1429–38.

Singh K, Raina S, Singh A (1996) Variation in chromosomal DNA associated with the evolution of *Arachis* species. Genome 39: 890–897.

Smartt J, Gregory WC, Gregory MP (1978) The genomes of *Arachis hypogaea*. 1. Cytogenetic studies of putative genome donors. Euphytica 27: 665–675.

Smartt J (1990) The groundnut, *Arachis hypogaea* L. In: Smartt J (ed) Grain Legumes: Evolution and Genetic Resources. Cambridge Univ Press, Cambridge, UK, pp 30–84.

Smith B (1950) *Arachis hypogaea*, aerial flower and subterranean fruit. Am J Bot 37: 802–850.

Soltis DE, Soltis PS (1995) The dynamic nature of polyploid genomes. Proc Natl Acad Sci USA 92: 8089–8091.

Subramanian V, Gurtu S, Nageswara Rao RC, Nigam SN (2000) Identification of DNA polymorphism in cultivated groundnut using random amplified polymorphic DNA (RAPD) assay. Genome 43: 656–660.

Tallury SP, Hilu KW, Milla SR, Friend SA, Alsaghir M, Stalker HT, Quandt D (2005) Genomic affinities in *Arachis* section *Arachis* (Fabaceae): molecular and cytogenetic evidence. Theor Appl Genet 111: 1229–1237.

Temsch EM, Greilhuber J (2000) Genome size variation in *Arachis hypogaea* and *A. monticola* re-evaluated. Genome 43: 449–451.

Temsch EM, Greilhuber J (2001) Genome size in *Arachis duranensis*: a critical study. Genome 44: 826–830.

Valls JFM, Simpson CE (2005) New species of *Arachis* L. (Leguminosae) from Brazil, Paraguay and Bolivia. Bonplandia 14: 35–63.

Varshney RK, Bertioli DJ, Moretzsohn MC, Vadez V, Krishnamurty L, Aruna R, Nigam SN, Moss BJ, Seetha K, Ravi K, He G, Knapp SJ, Hoisington DA (2009) The first SSR based genetic linkage map for cultivated groundnut (*Arachis hypogaea* L.). Theor Appl Genet 118: 729–739.

Vicient CM, Suoniemi A, Anamthawat-Jansson K, Tanskanen J, Beharav A, Nevo E, Schulman AH (1999) Retrotransposon *BARE*-1 and its role in genome evolution in the genus *Hordeum*. Plant Cell Online 11: 1769–1784, doi:10.1105/tpc.11.9.1769.

Wang CT, Yang XD, Chen D, Lin Yu SX, Liu GZ, Tang YY, Xu JZ (2007) Isolation of simple sequence repeats from groundnut. Elec J Biotechnol 10: 473–480.

Wessler SR, Carrington JC (2005) Genome studies and molecular genetics: The consequences of gene and genome duplication in plants. Curr Opin Plant Biol 8: 119–121.

Yuan M, Gong L, Meng R, Li S, Dang P, Guo B, He G (2010) Development of trinucleotide (GGC)n SSR markers in peanut (*Arachis hypogaea* L.). Elec J Biotechnol 13: 5–6.

[1]Institute of Biological Sciences, Genetics Department, University of Brasilia, Brazil, 70910-900 Brasília-DF, Brazil.
Email: *davidbertioli@unb.br*

[2]Embrapa Genetic Resources and Biotechnology. PqEB W5 Norte Final, 70770-917 Brasília-DF, Brazil. CEP 71.770-900.

[3]Plant Breeding and Genetics Section, International Atomic Energy Agency, Vienna, Austria.
Email: *s.nielen@iaea.org*

[4]Department of Biology, University of Leicester, Leicester, LE1 7RH, UK.
Email: *phh4@le.ac.uk*

[5]Embrapa Genetic Resources and Biotechnology. PqEB W5 Norte Final, 70770-917 Brasília-DF, Brazil. CEP 71.770-900.
Email: *patricia.guimaraes@embrapa.br*

[6]Department of Biology, University of Leicester, Leicester, LE1 7RH, UK.
Email: *ts32@leicester.ac.uk*

[7]Department of Plant Genome Research, Kazusa DNA Research Institute, Chiba, Japan.
Email: *sisobe@kazusa.or.jp*

[8]Department of Plant Genome Research, Kazusa DNA Research Institute, Chiba, Japan.
Email: *shirasaw@kazusa.or.jp*

[9]Embrapa Genetic Resources and Biotechnology. PqEB W5 Norte Final, 70770-917 Brasília-DF, Brazil. CEP 71.770-900.
Email: *soraya.bertioli@embrapa.br*

*Corresponding author: *ana-claudia.guerra@embrapa.br*

7

Peanut Transcriptomics

Xiaoping Chen and *Xuanqiang Liang**

ABSTRACT

Transcriptome designates a specific subset or complete set of messenger RNA (mRNA) expressed in a particular cell, tissue, organ or organism, and their quantity for a given developmental stage or physiological condition. Transcriptomics or genomewide transcriptional profiling allows simultaneous examination of transcriptome. It has been increasingly used to describe transcripts of a range of peanut tissues at different developmental stages under various environmental stresses. Here, we review commonly used technologies for transcriptome studies, their representation for three important peanut tissues (pod/seed, root and leaf), for stress response in peanut, as well as their use for marker development. It is now clear that various transcriptomics strategies are complementary and synergistic and the increasing availability of newly developed methods opens new opportunities for peanut transcriptome analysis, contributing to the understanding of genetic mechanism underlying important agronomic traits for peanut improvement.

Keywords: Messenger RNA (mRNA), Transcriptome, Gene expression, Microarray analysis, Protein-encoding genes, RNA-seq, "omics" technologies

Crops Research Institute, Guangdong Academy of Agricultural Sciences, Guangzhou 510640, China.
Email: *xpchen1011@gmail.com*
*Corresponding author: *Liang-804@163.com*

7.1 Introduction

Peanut or groundnut (*Arachis hypogaea* L.) is widely used as a source of edible oil and protein. It is cultivated mainly in tropical, subtropical and warm regions around the world. It is also the second most important seed legume of the world following soybean, with a total global production of 36.46 million tons, ranking among the top five oilseeds cultivated in the world (FAO 2009). China, India and the US have been the leading producers of peanut for decades and consume approximately 65% of the world´s production (Gunstone 2011). Despite the importance of peanut, progress in peanut breeding, genetics and genomics still lag far behind that of major crops. This is in large part because of the relatively few researchers involved in this crop, the low financial resources allocated to fundamental study as well as the complex genome of peanut (allotetraploidy, large genome size of 2,800 Mb; Temsch et al. 2000). Peanut is an economically important food crop worldwide, and thus its quality has become increasingly important and received more attention from the food industry and consumers, while growers are more interested on high-yielding varieties for higher income. Peanut well known for its allergens, is affected by a wide range of diseases such as those caused by *Aspergillus*, leaf fungi, nematodes and also various abiotic environmental stresses including drought. These constraints have led to yield loss as well as quality deterioration, e.g., allergenicity and aflatoxin contamination (Li et al. 2000; Holbrook et al. 2003; Ratnaparkhe et al. 2011). Peanut is morphologically diverse, however, many agronomical traits are difficult to select by conventional breeding, as they are quantitatively inherited. For this reason, it is quite hard to diminish the rate of adverse reactions and to enhance resistance to abiotic and biotic stresses through conventional breeding systems. The application of genomic tools in the breeding programs would greatly facilitate the genetic enhancement of cultivated peanut in a relatively rapid way (Livingstone et al. 2005; Varshney et al. 2009; Knoll et al. 2011). Unfortunately, understanding the genetic mechanism underlying complex traits in peanut is hindered by the fact that peanut possesses a narrow genetic diversity and its genome has not yet been sequenced and transcriptome resources are still limited (Pandey et al. 2011).

For a complex agronomic trait, the traditional approach to elucidate the underlying mechanism is to divide it into smaller, simpler and thus more tractable units for studying (Romero et al. 2006). If a divided unit or trait was dominated by a single gene/Quantitative Trait Locus (QTL), the gene/locus can be isolated and investigated using various molecular techniques. This strategy can assign a phenotype/trait to specific genes or QTLs, on the basis of which, however, a complete understanding of genetic mechanism underlying complex traits is not tenable (Strange 2005). Such complexity can become tractable with the use of "omics" technologies, referring to simultaneously studying DNA, mRNA, proteins, and metabolites in a cell, tissue, organ or organism (Romero et al. 2006). The first and still most well-known of "omics" technologies is transcriptomics or genomewide transcriptional profiling, allowing simultaneous examination of mRNA or transcript abundance and variations for much or all of genome with the goal of understanding genes and pathways involved in the biological processes (Gomase et al. 2008). "Omics" sciences have been taken to be a general term of reference to studies of entities in aggregate (Weinstein 1998; Evans 2000), and hence the term "transcriptomics" is designated to study the transcriptome that is a specific subset or complete set of mRNA expressed in a cell, tissue, organ or organism, and their quantity, for a given developmental stage or physiological condition (Hegde et al. 2003; Gomase et al. 2008; Wang et al. 2009b). It has been widely applied in plant biology, both in model species such as *Arabidopsis*, and also in crop plants including rice and maize.

This chapter will provide an overview of peanut transcriptomics including commonly used strategies for genomewide analysis, its current status in peanut, its representation for three major tissues (pod/seed, hereafter refer to as seed, root and leaf) and in stress response, its use for marker development, as well its future perspectives.

7.2 Transcriptomics Strategies

In plants, transcriptome technologies are often used to identify the genes whose expression is differentially regulated at different developmental stages or in response to various biotic and abiotic stresses. There are many strategies that have been developed in this particular area for examination of large-scale gene expression profiling. Most of these strategies can be classified into two broad categories (Shackel et al. 2006): (1) "closed architecture systems" requiring existing knowledge of gene sequences and no novel sequence information generated for this type of strategies, such as microarrays (Schena et al. 1995); (2) "open architecture systems" not requiring a priori knowledge of gene sequences and potentially producing new sequences, including initial cDNA sequencing (expression sequence tags, ESTs; Adams et al. 1991) and RNA sequencing (RNA-seq) recently developed and becoming increasingly important in transcriptomics (Wang et al. 2009b).

Array technology is a good representative of the "closed architecture system". It has revolutionized our ability to monitor global changes in gene expression at a genomewide level and has become the preferred technology of rapidly monitoring differential gene expression in hundreds to thousands of mRNA transcripts in a single experiment (Schena et al.

1998; Lipshutz 2000; Suarez et al. 2009). From its inception in the 1990s, a number of derivative technologies for expression profiling proliferated from this methodology (Schena et al. 1998; Lipshutz 2000; Govind et al. 2009). There are currently two types of arrays used to monitor gene expression in plants: cDNA-based and oligonucleotide arrays. cDNA arrays are based upon the deposition of pre-assembled cDNA probes including the nylon-based cDNA macroarray (Lennon et al. 1991; Pietu et al. 1996) and cDNA microarray (Schena et al. 1995). Oligonucleotide arrays are made through *in situ* synthesis of oligonucleotide probes. According to the length of probe, oligonucleotide arrays can be further divided into short oligonucleotide of around 20 nucleotides (Affymetrix arrays or gene chips) and long oligonucleotide with 50–70 mers (Agilent Technologies and Roche NimbleGen; Kane et al. 2000; Kasuga et al. 2005). Regardless of the density, probe length and preparation and support of all arrays, their underlying principle remains the same on the basis of interactions between complementary strands of DNA (Southern 1975). The advent of array technologies has made a quantum leap in expression profiling analysis across diverse biological conditions. However, array technologies suffer from two main limitations: (1) reliance upon a priori knowledge of gene sequences, and thus transcriptomics is limited to genes that are represented on the array; and (2) the lack of standards presenting and exchanging data, hindering the integration of data from different experiments.

In contrast to array technologies, sequence-based methods do not require existing knowledge about transcript sequences, and directly determine the cDNA sequence. There are a large number of sequencing-based technologies used for transcriptome analysis including EST sequencing (Adams et al. 1991), Serial Analysis of Gene Expression (SAGE) (Velculescu et al. 1995), Cap-Analysis Gene Expression (CAGE) (Shiraki et al. 2003), Massively Parallel Signature Sequencing (MPSS) (Brenner et al. 2000), Polony Multiplex Analysis of Gene Expression (PMAGE) (Kim et al. 2007), Differential Display Reverse Transcription Polymerase Chain Reaction (DDRT-PCR) (Liang et al. 1992; Cho et al. 2001), Suppressive Subtractive Hybridization (SSH) (Diatchenko et al. 1996), cDNA Amplified Fragment Length Polymorphism (cDNA-AFLP) (Kivioja et al. 2005), as well as High Coverage Expression Profiling (HiCEP) (Fukumura et al. 2003). Compared to the "closed architecture system", they generate tags independent of knowledge of gene sequences (Forrest et al. 2009). Most importantly, they can produce novel sequences with the potential for gene discovery. However, most are based on expensive and relatively low-throughput Sanger sequencing technology, making it infeasible to annotate transcriptome at a high resolution. Recently, on the basis of Next Generation Sequencing (NGS) technologies (Ansorge 2009), RNA-seq is becoming a revolutionary tool for transcriptomics, like the appearance of microarray technologies for expression profiling in the mid-1990s. It has revolutionized our view of the extent and complexity of transcriptomes as well as provided more precise measurement of levels of transcripts and their isoforms than the previous methods (Wang et al. 2009b).

Despite the advent of EST sequencing (Adams et al. 1991) and the microarray technology (Schena et al. 1995) nearly 20 years ago, the application of transcriptomics in peanut has not happened until a decade ago, far lagging behind the major crops such as rice and soybean. Microarray and EST sequencing are commonly used strategies for peanut transcriptomics, followed by other technologies such as SSH, DD-PCR and cDNA-AFLP. Transcriptome analyses in peanut using these strategies provided a certain amount of transcriptional information for a variety of peanut tissues at different developmental stages under various environmental stresses.

7.3 Peanut Transcriptome

Because a complete reference genome is not available for peanut, consensus transcriptome assembled from EST sequences have been developed as an alternative reference for designing micorarray probes and aligning the transcript reads generated by NGS technologies. Most of the peanut transcriptome assemblies are based on dbEST and nucleotide divisions in NCBI (www.ncbi.nlm.nih.gov). The earliest available assembly for peanut was provided in 2007 by the Institute of Genome Research (TIGR, now the J. Craig Venter institute, http://plantta.jcvi.org/cgi-bin/plantta_release.pl). The assembly had only 6,965 unigenes with 1,491 contigs and 5,474 singletons since the TIGR plant transcript assemblies were updated in July 2007 when there were 13,174 peanut ESTs in GenBank. Another assembly for peanut is PlantGDB-assembled Unique Transcripts (PUT, version 171a, http://www.plantgdb.org), which was released in June 2009. A total of 85,614 sequences filtered from 125,531 ESTs and cDNAs in GenBank were used for assembly, resulting in 30,319 unigenes. Most recently, GeneBank Unigene Build #2 consisting of 33,015 unigene clusters was developed based on 117,331 sequences from dbESTs and nucleotide divisions through 9 June 2011. In addition, a peanut transcriptome assembly containing 32,619 contigs was reported by the so-called peanutDB that currently focused on the transcriptomics analysis of *Arachis hypogaea* (Schmidt et al. 2011). On the basis of sequenced transcriptome ESTs from 17 tetraploid genotypes and publicly-available sequences, a comprehensive transcriptome assembly has been developed, which is comprised of 211,244 unigenes with 3,907 bp of the largest contig and an average length of 563 bp (Pandey et al. 2011).

With the NGS like 454 pyrosequencing and Illumina SBS, more and more transcriptome sequences are being generated for peanut. Currently, there are 2.7 Gb sequence data generated using 454 Titanium and Illumina GA platforms. In addition, a total of 225,264 peanut ESTs are available in dbEST (release 120111) comprised of 150,922 ESTs from cultivated peanut (*Arachis hypogaea*) as well as 35,291, 32,787, and 6,264 from three wild relatives *A. duranensis*, *A. ipaensis*, and *A. stenosperma*, respectively. Most of these were produced from seed, leaf and root tissues at different developmental stages under various environmental conditions. Detailed information is listed in Table 7-1. The length of these ESTs ranged from 37 bp to 2,038 bp with an average length of 535 bp. They have been assembled into 60,328 unigenes with 20,346 contigs and 39,982 singletons. Of these, 38,614, 16,776, 16,831 and 4,774 unigenes were expressed in *A. hypogaea*, *A. duranensis*, *A. ipaensis*, and *A. stenosperma*, respectively (Fig. 7-1A).

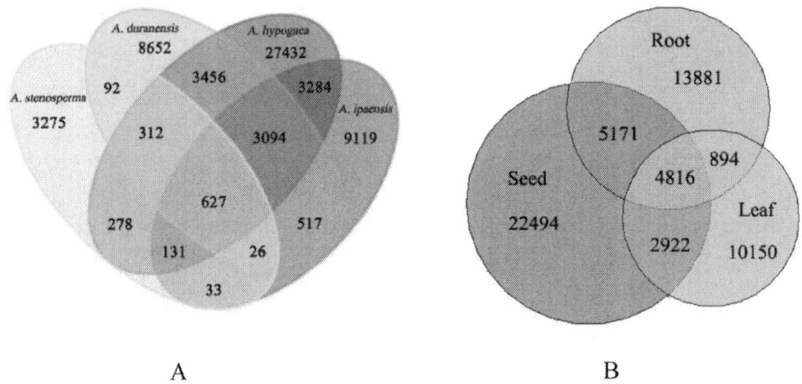

Figure 7-1 Peanut transcriptome assembled from ESTs in dbEST. (A) Distribution of transcripts among cultivated species (*A. hypogaea*) and three wild relatives (*A. duranensis*, *A. ipaensis* and *A. stenosperma*). (B) Distribution of transcripts among seed, leaf and root tissues.

Color image of this figure appears in the color plate section at the end of the book.

7.3.1 The Seed Transcriptome

Seed, the unique harvesting organ mainly accumulating oil and protein, is undisputedly the organ of the greatest importance in peanut from an agronomic perspective. However, little is known pertaining to the molecular mechanisms underlying seed development. Defining its transcriptome and understanding global expression profiling will provide critical information for improvement of nutritional composition and enhancement of resistance to pests of peanut. Gene expression profiling is essential to identify seed-specific genes with preferential expression at specific developmental stages. Because peanut has not a complete genome sequence and a limited number of transcripts are publicly available, little is known with respect to the number of protein-encoding genes and transcripts derived from

Table 7-1 List of expressed sequence data (EST) in GenBank dbEST generated from different tissues at different developmental stages and under various stresses in peanut.

Tissue	Species	Developmental stage	Conditions/ Stress	# of ESTs	GenBank ID	Last update	Submitter/Reference
Pod/ seed	*A. hypogaea*	30–50 days after flowering	N/A	20596	JK146921-JK167516	09-Jul-2011	Huang et al. Unpublished
		Mid-maturation, Developing embryo	N/A	20628	GO322902-GO343529	24-Mar-2009	Nagy et al. Unpublished
		9 to 50 days after pegging	N/A	770	ES490643-ES491295 EY396003-EY396119	19-Nov-2007	Zhuang et al. Unpublished
		20–60 days after pegging	N/A	7454	EE123340-EE127745 EG372473-EG374035 EG374037-EG374270 EG529454-EG529477 EG529479-EG530705	23-Oct-2006	Bi et al. (Bi et al. 2010)
		20–60 days after pegging	N/A	1115	GW714098-GW714528 GW774918-GW774921 GW775566-GW775567 HO115456-HO116133	14-Jul-2011	Wang et al. (Bi et al. 2010)
		Mid-maturation, Cotyledon	N/A	2569	CO897496-CO897530 CX127898-CX128247 EG028562-EG030729 EG357719-EG357733 EG357735	13-Oct-2006	Huang et al. (Yan et al. 2005)
		R6	N/A	853	CD037991-CD038843	23-Dec-2010	Guo et al. (Luo et al. 2005b)
		R5, R6 and R7	*A. parasiticus* drought	21778	ES702769-ES724546	01-Oct-2008	Guo et al. (Guo et al. 2008)
		Developing seeds	N/A	8	GW391722-GW391729	25-Nov-2010	Yan et al. Unpublished

Table 7-1 contd....

Table 7-1 contd.

Tissue	Species	Developmental stage	Conditions/ Stress	# of ESTs	GenBank ID	Last update	Submitter/Reference
		Developing seeds	N/A	1	AF366559	03-Feb-2001	Simth et al. (Paik-Ro et al. 2002)
	A. duranensis	3–8 weeks developing seeds	N/A	20667	GW939348-GW960014	24-May-2010	Bowers et al. Unpublished
	A. ipaensis	3-8 weeks developing seeds	N/A	16984	GW975818-GW992801	24-May-2010	Bower et al. Unpublished
Leaf	*A. hypogaea*	21 day-old seedlings	Salt	5	FL641010-FL641014	25-Aug-2008	Sharma et al. Unpublished
		30 day-old seedling	Drought	25	FL640985-FL641009	25-Aug-2008	Sharma et al. Unpublished
		first and second opened leaves	*C. personatum*	745	EL966274-EL967017 ES216626	06-Mar-2011	Nobile et al. 2008 (Nobile et al. 2008)
		21 day-old seedlings	Drought	336	DT044264-DT044476 EC391278-EC391382 EC589996-EC590005 CO868713-CO868719	23-Jun-2006	Sudhakar et al. Unpublished
		20 day-old seedlings	Drought	845	EC268400-EC268655 EC365167-EC365455 EC366281-EC366550	15-Jun-2006	Udayakumar et al. Unpublished
		late vegetative	Moisture	2	DQ256366 DQ256368	12-Nov-2005	Muthappa et al. Unpublished
		seedling	Drought	1	DQ119294	02-Aug-2005	Devaiah et al. Unpublished
		Unbloomed plant	N/A	20	CK326108-CK326116 CK394220-CK394229	10-Jan-2011	Matand et al. Unpublished

		newly expanded leaves 100 days after planting	No	492	CD037499-CD037990	23-Dec-2010	Guo et al. (Luo et al. 2005b)
		Cotyledon and young leaves	N/A	12327	GO256999-GO269325	13-Mar-2009	Nagy et al. Unpublished
		100 days after planting	N/A	16931	ES751523-ES768453	02-Mar-2011	Guo et al. (Guo et al. 2009)
		N/A	*R. solanacearum*	21751	JK167517-JK178109 JK188935-JK200092	09-July-2011	Huang et al. Unpublished
		N/A	Oxidative stress	44	GW672393-GW672436	09-Mar-2010	Gallo et al. Unpublished
		Three-month old plants mature leaf	Healthy leaf	240	EH041934-EH048107	25-Feb-2011	Proite et al. 2007
Root	*A. hypogaea*	plantlet	N/A	10825	JK178110-JK188934	09-Jul-2011	Huang et al. Unpublished
		plantlet	*R. solanacearum*	10062	JK200093-JK210154	09-Jul-2011	Huang et al. Unpublished
		Nodulation Stage	*B. radyrhizobium*	439	GT735098-GT735536	03-Mar-2010	Khanderao et al. Unpublished
		N/A	N/A	197	CX018023-CX018219	20-Dec-2010	Paterson et al. Unpublished
		Seedling	*M. earenaria*	103	GW275984-GW276086	11-Mar-2011	Gallo et al. (Tirumalaraju et al. 2011)
	A. duranensis	2 weeks roots ()	N/A	14624	GW924724-GW939347	24-May-2010	Bowers et al. Unpublished
	A. ipaensis	2 weeks roots ()	N/A	15803	GW960015-GW975817	24-May-2010	Bowers et al. Unpublished
	A. stenosperma	Three-month old plants	Healthy root	6024	EH041936-EH048196	24-Feb-2011	Proite et al. 2007

alternative splicing in peanut. This aggravates the already difficult task that is to estimate how many transcripts are expressed in peanut seed tissue. Until now, there are no comprehensive studies of seed transcriptome and rare resources for identifying and comparing organ transcriptomes. Knowledge of seed transcriptome is inferred from transcript assemblies, which are mainly compiled from ESTs. Based on the current available ESTs in dbEST, a total of 35,403 unigenes were found to express in seed tissue of peanut (Fig. 7-1B).

An initial *de novo* generation of ESTs resulted in 1,056 sequences from immature pods (Luo et al. 2005b). Yan et al. (2005) reported more than 400 ESTs obtained from mid-maturation stage cotyledons of peanut. The largest seed EST collection of 21,777 high-quality EST sequences were generated from six cDNA libraries derived from developing peanut seeds at three reproduction stages (R5, R6 and R7). These ESTs represented 8,689 unigenes in seed transcriptome. More than 20 functional categories were found among these transcripts, providing some insights into the complexity of the seed transcriptome (Guo et al. 2008). This study also showed that gene expression patterns in seed were significantly different not only at various development stages but also between genotypes (Guo et al. 2008). Comparative transcriptome analysis indicated that peanut transcriptome was more closely related to legume species than to cereal crops, and more similar to dicot than to monocot plant species (Guo et al. 2008).

Microarray analysis is another widely-used method to examine seed transcriptome and to provide insights into its complexity. Payton et al. (2009) designed an oligonucleotide microarray containing 15,744 unique probes and employed it to profile gene expression in various tissues of peanut, with the purpose of developing a tool for expression profiling studies in diverse tissues of peanut and of identifying tissue-specific genes. A total of 108 putatively pod-specific/abundant genes were identified. From these, almost half represented unknown genes that were possibly peanut-specific genes (Payton et al. 2009b). In the same year, Kottapalli et al. (2009) used 8 x 15K micoarrays to monitor changes in the transcriptome of peanut seeds at six developmental stages from R2 to R8, aiming to investigate regulatory processes and mechanisms underlying the development of peanut seeds. Several clusters of gene profiles were identified with different time-scales and models would be proposed to demonstrate how novel pathways may impinge on the molecular mechanism of seed development in peanut (Kottapalli et al. 2009). More recently, a cDNA microarray was developed based on 17,000 ESTs from immature seeds at different developmental stages (Bi et al. 2010). It was used to analyze gene expression profiles in a range of tissues as well as at different seed developmental stages. This study showed that genes for seed protein and late embryogenesis proteins accounted for more than one-fourth of the total transcripts at immature developmental stages. A large proportion of seed-preferentially expressed genes encoded storage proteins and fatty acid metabolism proteins. Bi et al. (2010) identified 277 transcripts that were more highly expressed in seed in comparison to other tissues including root, stem, leaf, flower and gynophore.

NGS has also been used for the understanding of peanut transcriptome. Most recently, a study (Burow et al. 2011) using Illumina (Solexa) technology was conducted on the high oleic Texas AgriLife cultivar Olin, generating 28.8 million short reads, of which 21.3 million can be mapped to the UGA Tifrunner reference set (Pandey et al. 2011). A total of 36,201 putative genes were assembled from these reads, of which 19,000 genes contained 72,586 polymorphisms through a comparison to Tifrunner. Currently, to provide deep insights into the pod transcriptome and comprehensively characterize its expression dynamics throughout development, we conducted RNA-seq using RNA isolated from 20 separated seed and shell samples representing 11 distinct stages of pod development, producing approximately 100 Gb of sequence data, more than 30 times the size of peanut genome (Chen and Liang unpubl. data).

7.3.2 The Leaf Transcriptome

Leaf, as a source organ, is also of great importance in peanut. Understanding the leaf transcriptome is very important for the elucidation of transcription profiles in leaf responding to diverse conditions. A large number of ESTs were generated from healthy and stressed leaves to date and submitted in GenBank (Table 7-1). Using 40,000 publicly available peanut ESTs, Payton et al. (2009a) developed a high-density oligonucleotide microarray (Agilent Technologies) and investigated gene expression in two lines. A total of 623 transcripts in leaf showed genotype-specific expression patterns (Payton et al. 2009a). In another study, they identified 1,204 Differentially Expressed Genes (DEGs) showing tissue-specific expression patterns in leaf tissue (Payton et al. 2009b). In addition to microarrays, non normalized EST sequencing data have also been used to study the leaf transcriptome and global gene expression profiling in leaf. An initial study reported 769 ESTs generated from leaf tissue, of which energy-related genes accounted for 27.3% (Luo et al. 2005b). A larger collection of ESTs from leaves was submitted by Guo et al. (2009). They reported the generation of 17,376 ESTs from leaf tissues of two peanut cultivars, Tifrunner and GT-C20. These ESTs represented 6,888 tentative consensus transcripts expressed in leaf tissue (Guo et al. 2009). Out of these unigenes, only 948 were shared by the two libraries, indicating significantly differential expression profiles present between the two genotypes. In a slightly later study, the use of SSH generated nearly 700 genes that were identified to be enriched in subtractive cDNA library from gradual process of drought stress adaptation (Govind et al. 2009). Most recently, normalized cDNA was produced from leaf tissue, generating 22,356 long-read ESTs using the Sanger technology as well as 509,180 short-read ESTs using NGS technology (Nagy et al. 2010). In addition, a study of leaf tissue of a wild species, *Arachis stenosperma*, was also reported, providing sequence resource and expression information from another aspect (Proite et al. 2007).

7.3.3 The Root Transcriptome

On the basis of transcriptome assembly, approximately 41% of transcripts (representing 24,762 clusters) were identified in root tissue, in comparison to seed (58%) and leaf (31%) (Fig. 7-1B). More than half of the ESTs used for root transcriptome assembly came from wild relatives of cultivated peanut. The first study on the analysis of root transcriptome of a wild species was conducted by Proite et al. (2007). Four cDNA libraries were constructed using RNA extracted from root tissues, producing a total of 8,785 ESTs, of which 6,024 (71.3%) had high quality with 3,500 clusters composed of 963 contigs and 2,537 singletons. SSH was also used to explore the root transcriptome challenged by root-knot nematode (RKN) in both resistant and susceptible genotypes (Tirumalaraju et al. 2011). A total of 960 differentially expressed ESTs were sequenced from two SSH libraries, of which, 70 ESTs were classified into several functional categories, representing a range of expressed genes in response to root knot nematode infection. Payton et al. (2009) found that only 78 out of 4,046 DEGs showed root-specific expression patterns using a microarray. This is because most of the sequences used for probe designing were generated from pod and leaf tissues (Payton et al. 2009b). Additionally, a large collection of root ESTs were submitted in July, 2011 (Table 7-1), which was not yet used by any study. Nagy (2010) used both Sanger and NGS technologies to sequence root transcriptome, aiming to develop a high-density molecular map of A-genome species, generating 21,487 long-read ESTs using the Sanger technology as well as 501,820 short-read ESTs using NGS technology (Nagy et al. 2010).

7.4 Transcriptome Response to Stress in Peanut

Various stresses are major yield- and quality-limiting factors for peanut improvement. Peanut suffers from a number of diseases caused by fungi, virus and bacteria pathogens (Jackson 1969; Holbrook et al. 1994; Branch et al. 1999; Tirumalaraju et al. 2011), and is also influenced by abiotic stresses such as drought and heat (Rucker et al. 1995; Vara Prasada et al. 1998).

The transcriptome profile of an organ can be an exquisite sensitive indicator of stress. Global analysis of mRNA abundance via transcriptomics strategies such as microarray and *de novo* generation of sequence tags is one of the most important approaches to discover important genes to organisms undergoing environmental stress (Feder et al. 2005). Moreover, identification of plant genes responding to environmental stresses is a critical step leading to the elucidation of molecular mechanisms underlying plant/stress interaction (Yuksel et al. 2005). This is of principal importance in devising strategies to cope with challenges such as biotic and/or abiotic stresses.

7.4.1 Biotic Stress

Relatively little genetic diversity in peanut increases its vulnerability to nematode, fungi, bacteria and virus pathogens. A large number of studies were conducted to investigate gene expression patterns resulting from pathogens invasion, aiming to understand the resistance mechanisms at the transcriptome level. Results obtained from these studies will contribute to elucidate the defense mechanisms in peanut and provide the framework for the generation of pathogen-resistant peanut cultivars.

Many studies were conducted to identify candidate genes for resistance to *Aspergillus* spp., causative of aflatoxin contamination, which is a great concern in peanut production worldwide. Phenotyping studies suggested that peanut cultivars with drought tolerance may have less aflatoxin contamination (Holbrook et al. 2000). An initial cDNA microarray with 384 cDNA clones that were selected from 1,825 ESTs was used to characterize *A. parasiticus* infection-induced changes in gene expression under drought stress (Luo et al. 2005b). Forty-two upregulated genes were identified in response to both *A. parasiticus* challenge and drought stress. This study identified a range of genes encoding pathogenesis-related proteins or with high homology to disease resistance genes. Guo et al. (2008) constructed six non normalized cDNA libraries using RNA from resistant and susceptible peanuts and identified 10 differentially-expressed resistant genes such as PR10 protein and the putative defensin 2.1 precursor. These studies identified a number of common resistance genes, but failed to recognize specific genes involved in the resistance to *Aspergillus* infection.

Nematodes are very damaging to peanut in some regions, and their infection causes devastating yield penalties with considerable economic losses annually around the world. The root-knot nematodes (RKN), *M. arenaria* and *M. javanica,* are important pathogens of peanut (Starr et al. 2006). SSH was used to investigate differential gene expression between the RKN resistant cultivar NemaTAM and the near-isogenic susceptible cultivar Florunner (Tirumalaraju et al. 2011). Differential screening of 960 clones from SSH libraries identified 140 clones from the forward library showing a higher level of expression in NemaTAM, and another 123 clones from the reverse library being highly expressed in Florunner. These ESTs were assembled into 70 unigenes, which were annotated and categorized into seven GO functional categories. The largest subset of differentially expressed sequences in both cultivars represented signal transduction. Discreet gene expression of Pathogenesis-Related (PR) genes, patatin-like proteins and universal stress related proteins, as well as those implicated in alleviation of oxidative stress were primarily represented in RKN-infected NemaTAM roots, which reflects a basal level of resistance operative against invading nematodes (Tirumalaraju et al. 2011). Wild relatives are also an important source of resistant genes, which could be integrated into cultivated peanut via wide crossing or genetic transformation. A total of 8,785 ESTs were produced from libraries inoculated with nematodes (*M. arenaria*) and non inoculated cDNA libraries from leaves and roots of *A. stenosperma* (Proite et al. 2007). These ESTs were assembled into 3,500 clusters and classified into 23 different functional categories. Numerous sequences related to disease resistance were identified. *In silico* analysis identified three DEGs Auxin Repressed protein (AsARP), Cytokinin Oxidase (AsCKX), Metallothionein Type 2 (AsMET2) that were also found to have similar differential expression profiles in macroarray analysis in the resistant and susceptible species both after, and sometimes even before, challenge with nematodes (Guimarães et al. 2010). These genes are possibly related to resistance through their roles in plant hormone balance or in the hypersensitive response.

Early leaf spot and late leaf spot are the major destructive diseases of peanut worldwide (Backman et al. 1984). An initial transcriptome analysis for leaf spot was performed using a cDNA microarray with 384 cDNA probes picked up from an EST sequencing study (Luo et al. 2005a; Luo et al. 2005b). Based on the microarray analysis, there were 56 upregulated genes and 65 downregulated genes in resistant genotype (C34-24), whereas susceptible genotype (GT-YY20) had 27 upregulated genes and 16 downregulated genes (Luo et al. 2005a). These genes were expressed at higher levels in the resistant genotype, as a response to *C. personatum* challenge, than in the susceptible one. Recently, a study was performed to investigate the molecular components of the initial stages of the resistance to late leaf spot using SSH and differential screening of cDNA macroarray techniques (Nobile et al. 2008). More than 700 unigenes were involved in defense signaling pathways and cell cycle at the early stages of *C. personatum* pathogenesis. More recently, the use of *de novo* generation of ESTs from leaf tissues of peanut cultivars resistant and susceptible to tomato spotted wilt virus and leaf spots suggested that gene expression profiles between resistant and susceptible genotypes were significantly different, implying the relative importance of specific transcripts to the degree of disease resistance (Guo et al. 2009). Like the nematode study, a wild species, *Arachis diogoi*, was used as a highly resistant genotype to perform differential gene expression analysis in leaf tissues challenged by late leaf spot pathogen (Kumar et al. 2011). Sixty partial cDNAs were cloned and sequenced from *Arachis diogoi* in response to fungal inoculation, which were upregulated within 48 hour post-inoculation. Temporal expression patterns were determined for cloned genes involved in phenylpropanoid pathway and lignification process. A pathogen-induced cyclophilin (AdCyp) was identified to reduce susceptibility towards late leaf spot pathogen and enhance resistance to *Ralstonia solanacearum* in transgenic tobacco and the resistance was associated with higher transcript levels of various defense-related genes (Kumar et al. 2011).

Bacterial Wilt (BW), caused by *R. solanacearum,* is also a primary constraint to peanut production in several Asian and African countries. The cDNA-AFLP technique was used to analyze differential expression of the genes related to BW resistance, aiming to illustrate the molecular mechanism of peanut resistant to BW (Peng et al. 2011). A total of 12,596 transcript-derived fragments were amplified with 256 primer combinations, of which 709 fragments generated from 119 primer combinations showed differential expression. Functional analysis indicated various pathways, e.g., defense, signal transduction, transcription and abiotic stress involved in the resistance to BW in peanut.

7.4.2 Abiotic Stress

More than 70% of peanut production worldwide falls under tropical, subtropical and warm regions, which are drought-prone areas (Holbrook et al. 2003). Thus, drought stress is one of the most limiting factors for peanut production. Several studies have been conducted to investigate DEGs that were involved in desiccation tolerance using a range of transcriptomics strategies including DDRT-PCR (Jain et al. 2001), cDNA microarray (Luo et al. 2005c), oligo microarray (Payton et al. 2009a), and SSH (Govind et al. 2009).

One of the first studies of peanut related to drought stress was carried out using DDRT-PCR (Jain et al. 2001). They identified 1,235 and 950 differential display products in irrigated and stressed peanut plants respectively, with 21 primer combinations. Further analysis indicated that stress could suppress RNA synthesis both qualitatively and quantitatively, and thus reducing overall protein synthesis (Jain et al. 2001). Because drought stress increases the predisposition of peanut to *Aspergillus* invasion and subsequent aflatoxin contamination (Wilson et al. 1983), studies on resistance to *Aspergillus* were often conducted in plants also submitted to drought stress. Luo et al. (2005) used a cDNA microarray to characterize gene expression profiling in peanut infected by *Aspergillus* under drought stress. This study identified 42 upregulated genes in response to both *A. parasiticus* and drought stress and 52 upregulated genes only responding to drought stress. From these, 25 cross-talking genes were found, which may be common genes related to both treatments (Luo et al. 2005). Similarly, a high-density oligonucleotide microarray was developed to investigate gene expression in tolerant and susceptible genotypes under water-deficit stress conditions. A total of 623 transcripts showed genotype-specific expression patterns in peanut plants under water deficit stress (Payton et al. 2009a). A study combining transcriptomics and proteomics revealed that several genes encoding lipoxygenase, fatty acid biosynthesis enzyme, acetyl-CoA carboxylase carboxyl transferase, lectin and aldolases could contribute to a physiological advantage under drought stress, these genes are potential candidates for marker-assisted breeding (Kottapalli et al. 2008). Recently, the use of SSH generated nearly 700 genes, of which, approximately 50% were not characterized, implying the underlying mechanism in peanut tolerant to drought stress was complex (Govind et al. 2009). Most recently, expression profiling studies of leaf, root and seed under deficit irrigation and in response to heat stress were conducted through screening the US mincore peanut germplasm collection. A number of stress-responsive genes showed genotype-, tissue- and time-specific response patterns and two types of DEGs were observed: constitutive DEGs between the tolerant and susceptible genotypes, and DEGs resulting from condition and physiological changes towards optima. Comparing the two subset genes, they supposed that pre-acclimation stress followed by a return to optimal conditions can confer the susceptible genotype again "constitutive" DEGs (Payton et al. 2011). Though it was not possible to identify specific genes or regulated networks involved in peanut tolerance to drought stress, insights gained from these studies will provide the foundation for further studies to understand the question of how peanut plants are able to adapt to naturally occurring harsh drought conditions.

7.5 Transcriptome Mining for Markers in Peanut

Marker-Assisted Selection (MAS) could accelerate the conventional breeding process and provide a possible approach for precision breeding. Relatively narrow genetic diversity requires more efficient markers to generate enormous polymorphisms in cultivated peanut. It is generally accepted that the development of genome-derived Simple Sequence Repeat (SSR) markers is expensive, labor intensive and time-consuming (He et al. 2003). Transcriptome sequencing, especially the advent of RNA-seq, facilitates the development of SSRs and Single Nucleotide Polymorphism (SNP) markers in peanut in a short time and at low cost. These markers would not only be useful for gene discovery and genetic mapping, but could also be used as anchor points in comparative mapping among diverse populations. A large number of EST-derived SSR (EST-SSR) have been developed from cultivated and wild *Arachis* species by various groups to date. Diversity analysis revealed that fairly limited polymorphisms existed among cultivated peanut, while high levels of polymorphisms were found within wild species (Proite et al. 2007; Liang et al. 2009; Song et al. 2010).

The first report for peanut EST sequencing and subsequent SSR development was a preliminary study that produced only 44 EST-SSRs from 400 unigenes (Luo et al. 2005b). A larger EST collection was used for EST-SSR development, identifying 856 EST-SSRs among 6,888 unique ESTs (Guo et al. 2009). Similarly, on the basis of 3,500 unigenes from a wild genotype a total of 206 microsatellites with di- and tri-nucleotide motifs being 119 and 72, respectively, were identified, of which 188 SSR markers have been developed (Proite et al. 2007). Polymorphism analysis indicated that 21 of these exhibited polymorphism for the AA population, and four for cultivated peanut. Recently, a study for utility of EST-SSR generated 881 microsatellites with tri-nucleotide (63.9%) being the most abundant, of which 290 SSR markers have been developed and were used for validation of the amplification and assessment of the polymorphism among 22 genotypes of cultivated peanut and 16 accessions of wild species, revealing that 26 of these were polymorphic among cultivated peanut, and 221 for wild species (Liang et al. 2009). A slightly later study identified 3,104 microsatellites from 28,023 non redundant unique sequences (Wang et al. 2009a). More recently, Song et al. (2010) found 610 ESTs that contained one or more microsatellites from 12,000 peanut ESTs. The most abundant SSRs were also tri-nucleotide motifs (66.3%). A total of 94 SSRs have been developed, of which 33 were used for polymorphism assessment among 73 cultivated peanut and 25 wild species, showing only five were polymorphic for cultivated peanut, while a high level of polymorphism was observed in wild species (Song et al. 2010). To circumvent low polymorphisms owing to narrow genetic diversity within the cultivated peanut, a total of 2,134 SSR markers developed from an *A. hypogaea* EST database were evaluated for polymorphism in the two progenitor diploid species. Of them 2,319 markers were mapped into 10 linkage groups, including 971 SSRs, 221 single stranded DNA conformation polymorphism (SSCP) markers, and 1,127 SNPs (Nagy et al. 2010). The linkages identified in this study would be a resource for sorting the A and B genomes and linkage relationships in the cultivated species. Most recently, a total of 3,187 markers were generated from 10,102 potential non redundant EST sequences (Koilkonda et al. 2011). Of these 1,571 EST-SSR markers showed clear polymorphisms among 24 *Arachis* accessions. Further polymorphic analysis with Fluoro-fragment Analyzer implied that 1,281 had polymorphisms among 16 *Arachis* accessions including cultivated peanut and wild species harboring A or B genome, as well as 366 were polymorphic among the 12 cultivated peanuts. Taken together, these EST-SSR markers provide a valuable resource for genetic mapping, QTL analysis, comparative genetic studies as well as MAS.

7.6 Perspectives

Transcriptomics in peanut is still in its infancy in comparison to that in model plants and major crops. Nevertheless, it has already made contributions to our understanding of peanut transcriptome and global expression variations in response to environmental conditions. Regarding strategies, microarray and EST sequencing have dominated peanut transcriptomics during the past several years. Although it has been speculated that microarray will be phased out owing to the rapid development of NGS, in the short term, microarray will continue to be used until sequence-based technologies have become cost-effective and easily used. Moreover, no transcriptome technologies to date are optimal to any experiments, and thus there is no single "correct" approach or methodology to examine the transcriptome. Therefore, many transcriptomics technologies should exist at the same time for sufficing different purposes of research and become complementary and synergistic. No matter which strategy is used, it is clear now that transcriptomics is becoming an indispensable tool for research in peanut functional genomics. Furthermore, with the help of computational biology, transcriptomics should not be an end in itself and new theoretical frameworks built for understanding the genetic mechanisms underlying various agronomic traits may invoke further research.

References

Adams MD, Kelley JM, Gocayne JD, Dubnick M, Polymeropoulos MH, Xiao H, Merril CR, Wu A, Olde B, Moreno RF (1991) Complementary DNA sequencing: expressed sequence tags and human genome project. Science 252: 1651–1656.

Ansorge WJ (2009) Next-generation DNA sequencing techniques. N Biotechnol 25: 195–203.

Backman PA, Crawford MA (1984) Relationship between yield loss and severity of early and late leaf spot diseases of peanut. Phytopathology 74: 1101–1103.

Bi YP, Liu W, Xia H, Su L, Zhao CZ, Wan SB, Wang XJ (2010) EST sequencing and gene expression profiling of cultivated peanut (*Arachis hypogaea* L.). Genome 53: 832–839.

Branch WD, Brenneman TB (1999) Stem rot disease evaluation of mass-selected peanut populations. Crop Prot 18: 127–130.

Brenner S, Johnson M, Bridgham J, Golda G, Lloyd DH, Johnson D, Luo S, McCurdy S, Foy M, Ewan M, Roth R, George D, Eletr S, Albrecht G, Vermaas E, Williams SR, Moon K, Burcham T, Pallas M, DuBridge RB, Kirchner J, Fearon K, Mao J, Corcoran K (2000) Gene expression analysis by massively parallel signature sequencing (MPSS) on microbead arrays. Nat Biotechnol 18: 630–634.

Burow G, Xin Z, Chopra R, Swaroop S, Burow M (2011) Next generation transcriptome sequencing of the high oleic peanut Cultivar OLin and Identification of SNPs between cultivars. American Peanut Research and Education Society Proceedings (APRES). Volume 43, p53, July 11–14, 2011, San Antonio, TX, USA.

Cho YJ, Meade JD, Walden JC, Chen X, Guo Z, Liang P (2001) Multicolor fluorescent differential display. Biotechniques 30: 562–568, 570, 572.

Diatchenko L, Lau YF, Campbell AP, Chenchik A, Moqadam F, Huang B, Lukyanov S, Lukyanov K, Gurskaya N, Sverdlov ED, Siebert PD (1996) Suppression subtractive hybridization: a method for generating differentially regulated or tissue-specific cDNA probes and libraries. Proc Natl Acad Sci USA 93: 6025–6030.

Evans GA (2000) Designer science and the "omic" revolution. Nat Biotechnol 18: 127.

FAO (2009) Food and Agricultural Organization of the United Nations, FAO Statistical Database. http://faostat.fao.org.

Feder ME, Walser JC (2005) The biological limitations of transcriptomics in elucidating stress and stress responses. J Evol Biol 18: 901–910.

Forrest AR, Carninci P (2009) Whole genome transcriptome analysis. RNA Biol 6: 107–112.

Fukumura R, Takahashi H, Saito T, Tsutsumi Y, Fujimori A, Sato S, Tatsumi K, Araki R, Abe M (2003) A sensitive transcriptome analysis method that can detect unknown transcripts. Nucleic Acids Res 31: e94.

Gomase VS, Tagore S (2008) Transcriptomics. Curr Drug Metab 9: 245–249.

Govind G, Harshavardhan VT, Patricia JK, Dhanalakshmi R, Senthil Kumar M, Sreenivasulu N, Udayakumar M (2009) Identification and functional validation of a unique set of drought induced genes preferentially expressed in response to gradual water stress in peanut. Mol Genet Genomics 281: 591–605.

Guimarães PM, Brasileiro AC, Proite K, Araújo AC, Leal-Bertioli SCM, Pic-Taylor A, Silva FR, Morgante CV, Ribeiro SG, Bertioli DJ (2010) A study of gene expression in the nematode resistant wild peanut relative, *Arachis stenosperma*, in response to challenge with *Meloidogyne arenaria*. Tropical Plant Biol 3: 183–192.

Gunstone FD (2011) Vegetable Oils in Food Technology: Composition, Properties and Uses. Front Matter Wiley-Blackwell, Oxford, UK.

Guo B, Chen X, Dang P, Scully BT, Liang X, Holbrook CC, Yu J, Culbreath AK (2008) Peanut gene expression profiling in developing seeds at different reproduction stages during *Aspergillus parasiticus* infection. BMC Dev Biol 8: 12.

Guo B, Chen X, Hong Y, Liang X, Dang P, Brenneman T, Holbrook C, Culbreath A (2009) Analysis of gene expression profiles in leaf tissues of cultivated peanuts and development of EST-SSR markers and gene discovery. Int J Plant Genomics 2009: 715605.

He G, Meng R, Newman M, Gao G, Pittman RN, Prakash CS (2003) Microsatellites as DNA markers in cultivated peanut (*Arachis hypogaea* L.). BMC Plant Biol 3: 3.

Hegde PS, White IR, Debouck C (2003) Interplay of transcriptomics and proteomics. Curr Opin Biotechnol 14: 647–651.

Holbrook CC, Matheron ME, Wilson DM, Anderson WF, Elizabeth Will M, Norden AJ (1994) Development of a large-scale field system for screening peanut for resistance to preharvest aflatoxin contamination. Peanut Sci. 21: 20–22.

Holbrook CC, Kvien CK, Rucker KS, Wilson DM, Hook JE (2000) Preharvest aflatoxin contamination in drought tolerant and intolerant peanut genotypes. Peanut Sci 27: 45–48.

Holbrook CC, Stalker HT (2003) Peanut breeding and genetic resources. John Wiley & Sons, Inc., Oxford, UK.

Jackson CR, Bell DK (1969) Diseases of peanut (groundnut) caused by fungi. Univ Ga Coll Agric Exp Stn Res Bull 56: 1–5.

Jain AK, Basha SM, Holbrook CC (2001) Identification of drought-responsive transcripts in peanut (*Arachis hypogaea* L.). Electron J Botechnol 4.

Kane MD, Jatkoe TA, Stumpf CR, Lu J, Thomas JD, Madore SJ (2000) Assessment of the sensitivity and specificity of oligonucleotide (50mer) microarrays. Nucleic Acids Res 28: 4552–4557.

Kasuga T, Townsend JP, Tian C, Gilbert LB, Mannhaupt G, Taylor JW, Glass NL (2005) Long-oligomer microarray profiling in Neurospora crassa reveals the transcriptional program underlying biochemical and physiological events of conidial germination. Nucleic Acids Res 33: 6469–6485.

Kim JB, Porreca GJ, Song L, Greenway SC, Gorham JM, Church GM, Seidman CE, Seidman JG (2007) Polony multiplex analysis of gene expression (PMAGE) in mouse hypertrophic cardiomyopathy. Science 316: 1481–1484.

Kivioja T, Arvas M, Saloheimo M, Penttila M, Ukkonen E (2005) Optimization of cDNA-AFLP experiments using genomic sequence data. Bioinformatics 21: 2573–2579.

Knoll JE, Ramos ML, Zeng Y, Holbrook CC, Chow M, Chen S, Maleki S, Bhattacharya A, Ozias-Akins P (2011) TILLING for allergen reduction and improvement of quality traits in peanut (*Arachis hypogaea* L.). BMC Plant Biol 11: 81.

Koilkonda P, Sato S, Tabata S, Shirasawa K, Hirakawa H, Sakai H, Sasamoto S, Watanabe A, Wada T, Kishida Y, Tsuruoka H, Fujishiro T, Yamada M, Kohara M, Suzuki S, Hasegawa M, Kiyoshima H, Isobe S (2011) Large-scale development of expressed sequence tag-derived simple sequence repeat markers and diversity analysis in *Arachis* spp. Mol Breed, doi:10.1007/s11032-011-9604-8.

Kottapalli K, Payton P, Puppala N, Burow M (2009) Transcript profiling of developing peanut seeds. In: Proceedings of The American Peanut Research and Education Society, Vol 41, July 14–17, 2009. Raleigh, NC, USA.

Kottapalli KR, Burrow MD, Puppala N, Payton P (2008) Transcriptomics and proteomics of drought tolerance in peanuts. The FASEB Journal 22: 841–845.

Kumar KR, Kirti PB (2011) Differential gene expression in Arachis diogoi upon interaction with peanut late leaf spot pathogen, Phaeoisariopsis personata and characterization of a pathogen induced cyclophilin. Plant Mol Biol 75: 497–513.

Li XM, Serebrisky D, Lee SY, Huang CK, Bardina L, Schofield BH, Stanley JS, Burks AW, Bannon GA, Sampson HA (2000) A murine model of peanut anaphylaxis: T- and B-cell responses to a major peanut allergen mimic human responses. J Allergy Clin Immunol 106: 150–158.

Liang P, Pardee AB (1992) Differential display of eukaryotic messenger RNA by means of the polymerase chain reaction. Science 257: 967–971.

Liang X, Chen X, Hong Y, Liu H, Zhou G, Li S, Guo B (2009) Utility of EST-derived SSR in cultivated peanut (*Arachis hypogaea* L.) and Arachis wild species. BMC Plant Biol 9: 35.

Lipshutz RJ (2000) Applications of high-density oligonucleotide arrays. Novartis Found Symp 229: 84–90; discussion 90–93.

Livingstone DM, Hampton JL, Phipps PM, Grabau EA (2005) Enhancing resistance to Sclerotinia minor in peanut by expressing a barley oxalate oxidase gene. Plant Physiol 137: 1354–1362.

Luo M, Dang P, Bausher MG, Holbrook CC, Lee RD, Lynch RE, Guo BZ (2005a) Identification of transcripts involved in resistance responses to leaf spot disease caused by *Cercosporidium personatum* in peanut (*Arachis hypogaea*). Phytopathology 95: 381–387.

Luo M, Dang P, Guo BZ, He G, Holbrook CC, Bausher MG, Lee RD (2005b) Generation of expressed sequence tags (ESTs) for gene discovery and marker development in cultivated peanut. Crop Sci 45: 346–353.

Luo M, Liang XQ, Dangd P, Holbrooke CC, Bausherd MG, Leea RD, Guo BZ (2005c) Microarray-based screening of differentially expressed genes in peanut in response to Aspergillus parasiticus infection and drought stress. Plant Sci. 169: 695–703.

Nagy E, Guo Y, Khanal S, Taylor C, Knapp S, Oziasakins P, Stalker HT, Nielsen N (2010) Developing a High-Density Molecular Map of the A-Genome Species A. duranensis American Peanut Research and Education Society Proceedings (APRES). Volume 42, p25, July 14–15, 2010, Clearwater, FL, USA.

Nobile PM, Lopes CR, Barsalobres-Cavallari C, Quecini V, Coutinho LL, Hoshino AA, Gimenes MA (2008) Peanut genes identified during initial phase of *Cercosporidium personatum* infection. Plant Sci 74: 78–87.

Paik-Ro OG, Seib JC, Smith RL (2002) Seed-specific, developmentally regulated genes of peanut. Theor Appl Genet 104: 236–240.

Pandey MK, Monyo E, Ozias-Akins P, Liang X, Guimaraes P, Nigam SN, Upadhyaya HD, Janila P, Zhang X, Guo B, Cook DR, Bertioli DJ, Michelmore R, Varshney RK (2011) Advances in *Arachis* genomics for peanut improvement. Biotechnol Adv.

Payton P, Kottapalli K, Rakwal R, Shibato J, Puppala N, Burow M (2009a) Transcriptome and proteome response to water-deficit stress in peanut (*Arachis* sp.). In: Plant and Animal Genome Conference, January 10–14, 2009, San Diego, USA.

Payton P, Kottapalli KR, Puppala N, Burow M (2011) A systems approach to study abiotic stress response in groundnut (*Arachis hypogaea*). In: Plant and Animal Genome Conference, W002, January 15–19, 2011, San Diego, CA, USA.

Payton P, Kottapalli KR, Rowland D, Faircloth W, Guo B, Burow M, Puppala N, Gallo M (2009b) Gene expression profiling in peanut using high density oligonucleotide microarrays. BMC Genomics 10: 265.

Peng WF, Lv JW, Ren XP, Huang L, Zhao XY, Wen QG, Jiang HF (2011) Differential expression of genes related to bacterial wilt resistance in peanut (*Arachis hypogaea* L.). Yi Chuan 33: 389–396 (in Chinese with English abstract).

Pietu G, Alibert O, Guichard V, Lamy B, Bois F, Leroy E, Mariage-Sampson R, Houlgatte R, Soularue P, Auffray C (1996) Novel gene transcripts preferentially expressed in human muscles revealed by quantitative hybridization of a high density cDNA array. Genome Res 6: 492–503.

Proite K, Leal-Bertioli SC, Bertioli DJ, Moretzsohn MC, da Silva FR, Martins NF, Guimaraes PM (2007) ESTs from a wild *Arachis* species for gene discovery and marker development. BMC Plant Biol 7: 7.

Ratnaparkhe MB, Rainville LK, Lemke C, Kim C, Wang X, Compton RO, Paterson AH (2011) Comparative and evolutionary studies of major peanut allergen genes. Plant and Animal Genome Conference. P421, January 15–19, 2011, San Diego, CA, USA.

Romero R, Espinoza J, Gotsch F, Kusanovic JP, Friel LA, Erez O, Mazaki-Tovi S, Than NG, Hassan S, Tromp G (2006) The use of high-dimensional biology (genomics, transcriptomics, proteomics, and metabolomics) to understand the preterm parturition syndrome. Bjog 113 Suppl 3: 118–135.

Rucker KS, Kvien CK, Holbrook CC, Hook JE (1995) Identification of peanut genotypes with improved drought avoidance traits. Peanut Sci. 22: 14–18.

Schena M, Shalon D, Davis RW, Brown PO (1995) Quantitative monitoring of gene expression patterns with a complementary DNA microarray. Science 270: 467–470.

Schena M, Heller RA, Theriault TP, Konrad K, Lachenmeier E, Davis RW (1998) Microarrays: biotechnology's discovery platform for functional genomics. Trends Biotechnol 16: 301–306.

Schmidt E, Duan X, Zhang J, Liang C (2011) PeanutDB: integrated bioinformatics web portal for *Arachis hypogaea* transcriptomics. Rustbelt RNA Meeting.

Shackel NA, Seth D, Haber PS, Gorrell MD, McCaughan GW (2006) The hepatic transcriptome in human liver disease. Comp Hepatol 5: 6.

Shiraki T, Kondo S, Katayama S, Waki K, Kasukawa T, Kawaji H, Kodzius R, Watahiki A, Nakamura M, Arakawa T, Fukuda S, Sasaki D, Podhajska A, Harbers M, Kawai J, Carninci P, Hayashizaki Y (2003) Cap analysis gene expression for high-throughput analysis of transcriptional starting point and identification of promoter usage. Proc Natl Acad Sci USA 100: 15776–15781.

Song GQ, Li MJ, Xiao H, Wang XJ, Tang RH, Xia H, Zhao CZ, Bi YP (2010) EST sequencing and SSR marker development from cultivated peanut (*Arachis hypogaea* L.). Electron J Biotechnol 13: DOI: 10.2225/vol13-issue3-fulltext-10.

Southern EM (1975) Detection of specific sequences among DNA fragments separated by gel electrophoresis. J Mol Biol 98: 503–517.

Starr JL, Simpson CE (2006) Improving the utility of nematode resistance in groundnut. Commun Agric Appl Biol Sci 71: 647–651.

Strange K (2005) The end of "naive reductionism": rise of systems biology or renaissance of physiology? Am J Physiol Cell Physiol 288: C968–974.

Suarez E, Burguete A, McLachlan GJ (2009) Microarray data analysis for differential expression: a tutorial. P R Health Sci J 28: 89–104.

Temsch EM, Greilhuber J (2000) Genome size variation in Arachis hypogaea and A. monticola re-evaluated. Genome 43: 449–451.

Tirumalaraju SV, Jain M, Gallo M (2011) Differential gene expression in roots of nematode-resistant and -susceptible peanut (*Arachis hypogaea*) cultivars in response to early stages of peanut root-knot nematode (*Meloidogyne arenaria*) parasitization. J Plant Physiol 168: 481–492.

Vara Prasada PV, Craufurd PQ, Summerfield RJ (1998) Sensitivity of peanut to timing of heat stress during reproductive development. Crop Sci 39: 1352–1357.

Varshney RK, Nayak SN, May GD, Jackson SA (2009) Next-generation sequencing technologies and their implications for crop genetics and breeding. Trends Biotechnol 27: 522–530.

Velculescu VE, Zhang L, Vogelstein B, Kinzler KW (1995) Serial analysis of gene expression. Science 270: 484–487.

Wang J, Pan L, Yang Q, Yu S (2009a) Development and characterization of EST–SSR markers from NCBI and cDNA library in cultivated peanut (*Arachis hypogaea* L.). Mol Plant Breed 7: 806–810.

Wang Z, Gerstein M, Snyder M (2009b) RNA-Seq: a revolutionary tool for transcriptomics. Nat Rev Genet 10: 57–63.

Weinstein JN (1998) Fishing expeditions. Science 282: 628–629.

Wilson DM, Stansell JR (1983) Effect of irrigation on aflatoxin contamination of peanut pods. Peanut Sci 10: 54–56.

Yan YS, Lin XD, Zhang YS, Wang L, Wu KQ, Huang SZ (2005) Isolation of peanut genes encoding arachins and conglutins by expressed sequence tags. Plant Sci 169: 439–445.

Yuksel B, Estill JC, Schulze SR, Paterson AH (2005) Organization and evolution of resistance gene analogs in peanut. Mol Genet Genomics 274: 248–263.

8

Advances in Proteomics Research for Peanut Genetics and Breeding

Ramesh Katam,[1,*] *Virginia Gottschalk,*[1] *Prashanth Survajahala,*[2] *George Brewster,*[1] *Payton Paxton,*[3] *Katsumi Sakata,*[4] *Ching Lee,*[5] *Crayger Williams,*[1] *Baozhu Guo*[6] and *Lekan M Latinwo*[1]

ABSTRACT

Crop trait improvement aimed at increased yield and quality relies on an understanding of the biology of the plant, particularly interactions occurring across hierarchical scales of organization. In this regard, the application of "-omics" techniques combined with field-level agronomy is poised to deliver novel insight into previously unknown or apparently unrelated interactions associated with developmental and environmental cues that combine to give the final plant phenotype. The challenge for the peanut research community, and all crop species, will be to ensure that "-omics" capabilities and data are generated, interpreted and integrated towards crop improvement. Peanuts are a globally important renewable source of oil, protein and carbohydrate for edible and industrial applications. Production areas range from subtropical, water-abundant regions to semi-arid regions around the world. The primary limitations to production vary across regions and the major areas of production research address: 1) allergenicity and human

[1]Department of Biological Sciences, Florida A&M University, Tallahassee, FL USA.
[2]Bioclues.org.
[3]USDA-ARS Cropping Systems Research Laboratory, Lubbock, TX USA.
[4]Maebashi Institute of Technology, Maebashi, Japan.
[5]University of Wisconsin, Madison, WI USA.
[6]USDA ARS, Crops Protection and Management Research Unit, Tifton, GA USA.
*Corresponding author: *ramesh.katam@gmail.com*

nutrition, 2) abiotic stress, primarily water-deficit stress, and 3) biotic stress. In this chapter, we describe the current knowledge regarding the peanut proteome, new technologies, limitations to proteomics and future omics approaches using example studies addressing allergenicity, thermal and water deficit stress and *Aspergillus* infection. Additionally, perspectives are presented on potential applications of these technologies to molecular breeding, metabolic engineering and elucidation of stress-response pathways.

Keywords: Structural genomics, Gene expression, Transcriptional modification, cDNA libraries, Peanut allergens, Protein-protein interactions (PPI)

8.1 Introduction

Legumes are the second crop group contributing to human sustenance, following only to cereal grains as a source of calories (FAO 2006). With a total global production exceeding 30 million tons, peanuts (*Arachis* species), in particular, are an important source of dietary protein in many developing nations, and provide an important source of oil for cooking and fuel. Biological fixation of nitrogen makes legumes an important crop for improving soil fertility and an important part of many crop rotations.

Cultivated peanut originated in South America, with putative progenitor species identified in Bolivia and Brazil (Simpson et al. 2001). Hybridization of A and B genome species was followed by chromosome doubling to the tetraploid level and subsequent domestication of the tetraploid species. European exploration resulted in the movement of peanuts to Africa and Asia which now, along with South America, are all centers of diversity. Peanut production is now widespread and occurs from tropical regions to semi-arid subtropical production areas. Hence, a wide array of adaptive responses to biotic and abiotic stress have been selected, making peanut an important experimental organism and model crop. Current peanut research using genomics techniques have focused on *Aspergillus*-resistance and other biotic stresses, abiotic stress tolerance and allergenicity, using a wide array of cutting-edge experimental techniques including structural genomics (Rogers et al. 2008), functional genomics (Luo et al. 2005a,b; Payton et al. 2009), and proteomics (Kottapalli et al. 2008). To illustrate the importance of peanuts as an experimental model, we will outline important research on the development and elucidation of the possible mechanisms involved in associated with this extremely important food source.

A number of phytochemical compounds are present in peanuts with potential antioxidant capacity including polyphenolics tocopherols, and proteins (Talcott et al. 2005). Numerous researchers have focused their studies on the role of phenolic antioxidants in preventing a great number of pathological disturbances associated to the generation of free radicals. Raw peanut seeds contain endogenous non enzymatic antioxidants such as a-tocopherol that is an excellent chain breaking antioxidant (Cobb and Johnson 1973). The antioxidative component has been identified as luteolin (Yen and Duh 1996) in peanut hull. Aflatoxin contamination in peanut is the major safety issue worldwide (Sampson 2004). Aflatoxin contamination can occur at any time from preharvest to storage. Peanuts are used in a variety of food products that are widely consumed, amplifying the health risks of aflatoxin contamination (Sicherer et al. 2007). Although there are genotypic differences in susceptibility to fungal invasion and aflatoxin production, the resistance in the germplasm and breeding lines is low. Drought-stressed plants lose moisture from pods, which lead to the reduction in the seeds' physiological activity, thereby increasing the susceptibility to fungal invasion (Dorner et al. 1989). Drought stress is also known to alter nutritional quality of peanut seeds—oil and protein and oil quality (Dwivedi et al. 1996) and phytoalexin/resveratrol levels (Tang et al. 2010).

Peanut proteins are of two types, the non storage and storage proteins: non storage proteins consist of enzymatic and structural proteins governing normal cellular activities, which include synthesis of storage proteins that are synthesized during seed development (Millerd 1975). Following seed germination and subsequent to hydrolytic breakdown, storage proteins serve as a source of nitrogen and carbon skeletons for the developing seedling. Eighteen of the 32 peanut proteins have been identified as allergens by binding to allergen-specific IgE antibodies (Dean 1998; Scurlock 2004). Most peanut allergens characterized thus far are seed storage proteins—class I allergens, which are water-soluble glycoproteins, and are resistant to heat, acid and enzymatic digestion (Pele 2010). These allergens belong to a few protein families and super families (Breitender and Radauser 2004):

1. The cupin superfamily, which are 7S and 11S seed storage proteins, i.e., vicilins and legumins, respectively;
2. The prolaminin superfamily such as 2S albumins, which are related to the protein family conglutin, nonspecific Lipid Transfer Proteins (nsLTPs);
3. Pathogenesis-related proteins (PRs), which are involved in providing resistance to pathogens and adverse environmental conditions;
4. Profilins, 2- to 15-kd cytosolic, actin-binding proteins that regulate polymerization of actin filaments in eukaryotic cells;
5. Oleosins, 16- to 24-kd protein components of plant lipid storage bodies known as oil bodies.

To date, the Allergen Nomenclature Subcommittee of the International Union of Immunological Societies has identified 11 such proteins, Ara h 1 to Ara h 11, from peanuts as allergenic (Table 8-1). Ara h 1 and Ara h 2 are major peanut proteins that accounts for 12–16% and 5.9–9.3% of the total peanut proteins, respectively (Koppelman et al. 2001). Individual content of these allergens in peanut varieties is similar regardless of where such peanuts were grown. Therefore, it would be useful to reduce the level of allergens in peanuts that are mixed with other food ingredients to protect consumers from potential life threatening allergic reactions. Elucidation of these allergens would make it possible to identify peanut germplasm with lower levels of allergens or genetically engineer the proteins so that they are less or no longer allergenic, but retain their beneficial functions.

Table 8-1 Peanut Allergens Identified Until August 2012 (Allergome and IUIS).

Allergen	Protein Family	MW (kDa)	Isoallergens[1]
Ara h1	Cupin: 7S Vicilin-like Globulins	64.0	Ara h 1.0101
Ara h2	Prolamin: 2S Albumin, Conglutin	17.0	Ara h 2.0101, Ara h 2.0201
Ara h3[2]	Cupin: Legumin-like (11S globin, Glycinin)	60.0	Ara h 3.0101, Ara h 3.0201
Ara h4[2]	Cupin: Legumin-like (11S globin, Glycinin)	37.0	Ara h 4.0101
Ara h5	Profilin	15.0	Ara h 5.0101
Ara h6	Prolamin: 2S Albumin, Conglutin	15.0	Ara h 6.0101
Ara h7	Prolamin: 2S Albumin, Conglutin	15.0	Ara h 7.0101, Ara h 7.0201, Ara h 7.0202
Ara h8[3]	Pathogenesis-related protein, PR-10	17.0	Ara h 8.0101, Ara h 8.0101
Ara h9	Prolamin: Nonspecific lipid-transfer protein	9.8	Ara h 9.0101, Ara h 9.0201
Ara h10	Oleosin	16.0	Ara h 10.0101, Ara h 10.0102
Ara h11	Oleosin	14.0	Ara h 11.0101

[1]Isoallergens are variants which may be recognized differently by patient IgE which either increases or decreases the severity of the allergenic response (Christensen et al. 2010)
[2]Ara h 3 and Ara h 4 are 91% homologous and are regarded as isoforms of each other and are considered to be the same allergen (Allergome http://www.allergome.org, and Koppelman et al. 2003)
[3]Ara h8 is homologous to *Betula verrucosa* (birch) allergen Bet v 1 (Mittag et al. 2004)
Abbreviations: MW = Molecular Weight, kDa = kilodalton

8.2 Peanut Breeding and Genetics

8.2.1 Wide Array of Adaptive Responses to Various Stresses

Peanuts are particularly valuable as a model crop for examining abiotic stress because of the tropical origin of the species, which has led to selection for a diverse set of adaptive responses. Peanuts have been cultivated in many semiarid areas in the US, parts of South America, Africa and India, resulting in selection for drought and heat tolerance and some accessions tolerant to abiotic stress has been identified (Upadhyaya et al. 2001; Kottapalli et al. 2009). Peanut's adaptation to abiotic stress is unusual among legumes in that shoot, root and pod-based mechanisms have also been identified (Kottapalli et al. 2009).

Breeding peanuts for drought tolerance has been accepted as one of the strategies for developing aflatoxin-tolerant peanut cultivars, which would serve to extend the life of aquifers and reservoirs, improve water-use efficiency in water-limited production areas, and help expand peanut production in marginal and sub marginal soils. Success in this effort has been slow due to lack of information on the interaction between drought affected pathways and or pathogen invasion. Although some efforts have been made to identify the quantitative trait loci and genetic basis of drought tolerance in peanut the use of molecular genetic techniques is at an early stage in peanuts compared to many other plant species of economic importance (Varshney et al. 2009).

8.2.2 Abiotic Stress

Environmental stresses such as drought, high salinity and extreme temperatures have adverse effects on crop productivity. Tolerant plants respond and adapt to these stresses through various morphological, physiological and molecular processes (Ingram and Bartels 1996). Most molecular processes studied in response to drought to date, deal with transcriptional modification of gene expression (Bray 2002). Gene products are two types: those that regulate gene expression and signal transduction during stress response and those that directly protect cells against environmental stresses (Seki et al. 2001). Increases in protease activities have been observed in several species including tropical and temperate legumes in response to water deficit (Guo et al. 2006). Proteome-level differences between a Salinity-Tolerant (ST) and -sensitive (SS) callus cell lines of *A. hypogaea* were characterized with a view of further understanding the molecular differences between the two lines that may be responsible for the higher tolerance to sodium chloride (Jain et al. 2006).

Recently, the development of new tools for peanut functional analyses have established transcript and protein analytical tools on an -omics scale (Payton et al. 2009). Using these tools, key enzymes and structural proteins of several biological functions have been identified that may be associated with water-deficit and heat-stress tolerance in peanuts. More importantly, these lines show significantly contrasting yields under managed field trials in the semi-arid southwestern United States (Kottapalli et al. 2009). These lines now represent the parental genotypes in on-going breeding studies and will serve as the foundation for identifying the molecular mechanisms controlling abiotic stress responses in peanut and improving the stress tolerance phenotype of cultivated varieties.

8.2.3 Drought and Aflatoxin Production

Drought prevents proper peanut seed maturation, which has a significant effect on the synthesis of bio-competitive compounds such as phytoalexins and phenols, which can inhibit the growth of *Aspergillus* species, preventing aflatoxin synthesis. Aflatoxin contamination threat increases with increasing seed maturity under drought. As the seed moisture content decreases during drought, the capacity of seed to produce phytoalexins decreases resulting in *Aspergillus* invasion and aflatoxin production (Payne 1998). Several peanut cultivars with natural preharvest resistance to aflatoxin production have been identified through field screening (Guo et al. 2008a). In peanuts a significant number of Expressed Sequence Tag (EST)-cDNA libraries from leaf and pod tissues showing homology to genes of known function have been released into the public domain (Luo et al. 2005a). These ESTs were used to generate cDNA microarray containing 384 unigenes, which were further used to identify stress-responsive genes and for the development of EST-derived SSR markers (Luo et al. 2005b). Likewise, ESTs were also generated from developing seeds to identify resistance-related genes involved in defensive response against *Aspergillus* infection and aflatoxin production (Guo et al. 2008b). The potential of developing new peanut lines with both resistance to aflatoxin and bacterial wilt and high oil content was investigated through extensive characterization of recombinant inbred lines (Liao et al. 2010).

Although many advances have been made in understanding host resistance to *Aspergillus* infection and aflatoxin contamination (Guo et al. 2003) but still very little is known about the molecular mechanisms of drought stress-drought tolerance and resistance to *Aspergillus* infection and aflatoxin contamination. The cDNA libraries were constructed from immature pods of peanut line tolerant to drought stress and preharvest aflatoxin contamination. The expression patterns of the resistant genes or cDNAs in response to *A. parasiticus* infection and drought stress using microarray analysis were studied. A total of 42 upregulated genes in several functional categories were detected under both *A. parasiticus* challenge and drought stress. A total of 52 upregulated genes were detected in response to drought stress alone. There were 25 genes commonly expressed in both treatments. Twenty upregulated genes from *A. parasiticus* challenged and drought stress were selected for validation of their expression levels using real-time PCR have been characterized (Luo et al. 2005b).

8.3 Functional and Genomics Update

8.3.1 Transcriptome

Molecular studies in peanut leaves and seeds showed that drought stress significantly altered gene expression with several transcripts either upregulated or downregulated or newly synthesized. Transcripts induced in response to drought were identified and two full-length genes namely; *AhSrp* (*Arachis hypogaea* serine-rich protein) and *AhLrp* (*Arachis hypogaea* lLeucine-rich protein) were constructed based on their sequence data (Jain et al. 2001; Devaiah et al. 2007). Payton et al. (2009) reported a high-density peanut oligonucleotide microarray using approximately 40,000 publicly available ESTs, which were used to identify genotypes with contrasting stress tolerance phenotypes for protein and transcript profiling studies. A total of 368 unique transcripts were upregulated in the tolerant peanut line under stress, but unchanged or repressed in the sensitive line, which identified five candidate genes namely *Pn-HsfA2, Myb91, EREBP, WRKY52* and *TF EGL1* (enhancer of Glabra3) associated with drought/heat stress. The differences between DT and DS lines were related to phenylpropanoid and flavanoid biosynthesis, common pathways also involved in plant-microbe interactions (Kottapalli et al. 2009). The proteomics and transcriptomics profiling of peanut leaf tissues revealed significant differences between DT and DS germplasms, relative to their basal thermotolerance response.

8.3.2 QTL Mapping

The genetic base of cultivated peanut is narrow because of the bottlenecks associated with the evolution of peanut. There has been a very limited introgression of useful traits using wild *Arachis* species due to differences in ploidy levels and linkage drag often associated with beneficial traits. Eliminating the linkage drag involves a lengthy process that may also result in dilution of resistance (Varshney et al. 2006). *Arachis* genomics and some efforts have been initiated towards QTL mapping and molecular breeding for resistance/tolerance to biotic/abiotic stresses for peanut improvement (Varshney et al. 2012).

8.4 Proteomics in Peanut

8.4.1 Integrating Biochemical Pathways

In peanuts, about 87% of the seed protein is globulin consisting of two major fractions, arachin (glycinin) and conarachin (vicilin) (Mosse et al. 1983). The peanut allergens Ara h 1, and Ara h 3/Ara h 4 have been assigned to the vicilin, conglutin and glycinin families of seed storage proteins, respectively (Kleber-Janke et al. 1999). Ara h 2, and the minor allergens Ara h 6 and Ara h 7 have been shown to have high sequence homology to proteins of the conglutin family from various dicotyledonous plants. Rabjohn et al. (1999) cloned *Ara h 3* gene using degenerate cDNA probes, whose sequences they determined based on N-terminal sequence information, and the gene expressed as a 60 kDa IgE-binding protein containing both acidic and basic chains. Studies of this new peanut allergen, Ara h 3-im, indicate that it may have lower allergenicity compared to other known Ara h 3 proteins (Kang and Gallo 2006). The peanut allergens, one cDNA fragment encoding an arachin isoform and one encoding a conglutin isoform (Paik-Ro et al. 2002) were isolated. Proteins are the major components of cellular metabolic pathways. Marc Wilkins derived the term "proteome" in 1994 from a blending of the words "protein" and "genome" to describe the complete complement of proteins and their variants or proteins as result of time, distinct system requirements and/or stresses undergone by a cell or organism (Wilkins et al. 1996). Proteomics is the high-throughput identification and analysis of proteins. It is an emerging field of research, facilitated by numerous advances made in mass spectrometry, genome sequencing/annotation and protein search algorithms (Anderson and Anderson 1998; Blackstock and Weir 1999). There is an increasing interest in proteomics technologies because DNA sequences provide only a static snapshot of the various ways in which the cell might use its proteins, whereas the life of the cell is a dynamic process. The direct utility of genomics is often limited in its power to identify the genetic basis for inherited diseases and it cannot address phenotypic changes, which are the result of physiological processes altered by time, diet, environmental changes or encounters with pathogens. Likewise, transcriptomics, the study the transcriptome or RNA transcripts, i.e., mRNA, rRNA, tRNA and other noncoding RNAs produced by a cell or a population of cells (Claverie 2005; Frith et al. 2005) do not provide adequate explanation for the phenotypic changes. For instance, it is known that mRNA content does not always correspond with the protein content (Dhingraa et al. 2005). This lack of correspondence can be due to either not all mRNA is translated into protein or the amount of mRNA ultimately translated into protein is often dependent on the gene it transcribed from and/or the physiological state of the cell (Buckingham 2003; Rogers et al. 2008). Therefore, proteomics is regarded as the "next step" after genomics and transcriptomics as it confirms and quantifies a particular protein or a set of proteins, including any metabolic-specific and/or time-dependent modifications.

8.4.2 Metabolomics

Metabolites are known as the measurable molecules, which will represent the phenotypes. Technologies that allow the creation of new annotated databases and interaction with already accessible annotated databases of metabolite concentrations reflecting individuals with various phenotypes are also needed. Recent development of mass spectrometry aimed to enhance the basic understanding of biochemistry of the phenotype and association of these metabolites to specific traits (Matuszewski et al. 1998). The metabolomics also assist in discovering biomarkers for specific proteins and the structural characterization.

Metabolic pathways and mechanisms of differential accumulation of various proteins resulting in the diversity in seed quality, taste and allergenicity remain unknown. Since proteins are directly associated with function, proteomics approaches are being applied increasingly to address biochemical and physiological questions. For example, Liang et al. (2006) used two-DE-based proteomics approach to differentiate the four peanut cultivars (New Mexico Valencia C, Tamspan 90, Georgia Green, and NC-7) based on market types prevalent in USA. They detected 20 protein spots that differed in relative abundance among the four cultivars and identified 14 non redundant proteins by nano-electrospray ionization liquid chromatography tandem mass spectrometry. The majority of these proteins belonged to the globulin fraction of seed proteins. It is also likely that some of these proteins may be associated with the basic differences in sensory attributes and nutritional traits including allergens in these cultivars (Kottapalli 2008). In a preliminary study using gel electrophoresis (Cherry et al. 1972), "standard" protein and enzyme patterns of peanuts from a single cultivar (Virginia 56R) were compared with gel profiles of these components from seeds contaminated with *A. parasiticus* in an attempt to develop new genetic or agronomic techniques designed to prevent or decrease susceptibility in peanuts (Cherry et al. 1974).

8.4.3 Proteomics and Stress Responses

Redox proteomics is an emerging branch of proteomics aimed at investigating oxidative-stress induced modifications of proteins. Oxidative injuries to proteins are produced by chemically reactive species. Modifications could address oxygen species and thus generate Reactive Oxygen Species (ROS), such as hydroxyl, peroxide and superoxide radicals or produce mixed nitrogen-oxygen species (RNS) such as nitric-oxide and peroxynitrite. The ROS/RNS are inevitably generated in metabolic pathways in all cells and some of them might play important roles in cell signaling (Møller and Sweetlove 2010). However excessive levels of ROS from either the environment or aberrations in electron transport can produce such high levels of oxidative stress, which may irreparably alter the proteins (Stadtman 2001). ROS/RNS alter the whole proteome through side-chain modifications and covalent changes, which have repercussions on protein activity, unfolding, degradation and cell functioning (Sheehan et al. 2010).

Comparative studies involving drought-tolerant and -intolerant lines under drought-stressed conditions have shown a negative correlation between seed protein composition in peanut. Thus, drought-tolerant lines showed little or no change in seed protein composition, while a susceptible line had major changes in protein composition (Basha et al. 2007; Katam et al. 2007). For example, seed storage proteins such arachin, conglutin and 2S proteins (Q647G9, P43238, Q647H2 and Q6PSU2) in an intolerant line were highly suppressed, while in a tolerant line (Vemana) their expression remained unchanged, suggesting significant alteration in the metabolic activity and seed protein composition in intolerant lines following drought.

The leaf protein reference map derived from drought-tolerant should serve as the basis for further investigations of peanut physiology such as detection of changes in gene expression due to biotic and abiotic stresses, plant development and genetic differences (Kottapalli et al. 2009; Katam et al. 2010a). Katam et al. (2010b) reported low abundance of certain proteins in susceptible peanuts, which led them to believe that certain proteins induced following stress in tolerant lines have a role in a drought tolerance mechanism. Protein-protein interaction detected equilogs or orthologous proteins in *Arabidopsis* (Jian et al. 2008).

Analysis of proteins is critical to define the function of their gene/s and linking proteins to their genome sequence information is very useful for functional genomics. Differentially expressed proteins in the control and infected seeds provide information about the changes in seed protein composition between tolerant and intolerant lines under drought-stressed conditions and *Aspergillus* infection. Seed proteomics may thus demonstrate if these proteins have a role in either pathogen suppression or supporting *A. parasiticus* growth and aflatoxin contamination under drought-stressed conditions.

8.4.4 Peanut Allergens and Human Health

The three peanut allergens extensively studied are Ara h 1, a vicilin-like protein; Ara h 2, a conglutin-homolog protein; and Ara h 3/Ara h 4, glycinin proteins (Mills et al. 2003; Piersma et al. 2005). Ara h 1 and Ara h 2 accounts for approximately 20 and 10% of the total protein content, respectively (Van Hengel et al. 2007).

The two-step mild extraction method and different degrees of processing (blanching, mild roasting and strong roasting) has been used to compare the extraction efficiency of peanut allergens. The complementary ELISA and 2D PAGE methods were applied for the detection of a number of peanut proteins with a direct relevance to the allergenic potential of peanut (Kiening et al. 2005). The composition of the peanut extracts has been investigated by employing ELISA test kits specific to Ara h 1, Ara h 2 or to soluble peanut proteins. The relative intensity of the ELISA signal obtained for the different protein extracts has been calculated for comparative purposes. The storage proteins in peanut seed, 14S and 7.8S globulins, possess different physical and chemical properties, with the former referred as arachin (legumin) and the latter as conarachin (vicilin) (Mosse and Pernollet 1983). The arachin, a member of the legumin-like globulins, accumulate as hexameric complexes, which consists of four or six polypeptides. Two-dimensional PAGE analysis shows that the conarachin shares two main subunits with the same molecular weight of 65 kD but different pI values of 6.3 and 7.4 (Li et al. 1998). A gene encoding the 65 kD subunit of conarachin was isolated and cloned and the structure of this gene in relation to several transcription factor motifs was investigated (Li et al. 2005).

Identification of peanut germplasm with a lower antigenic profile could be used in classic cross-breeding experiments to create less allergenic peanut cultivars. The development of peanut cultivars that are less allergenic has already been undertaken, with GT-C9, whose 2-D protein profile showed proteins with significantly lower levels of IgE binding, has been identified (Guo et al. 2008c).

8.4.5 Protein-Protein Interaction in Peanut

A hierarchical cluster analysis is used to study the expression profile of each individual protein. Each protein clusters that interacted were estimated by the representative time course that was calculated at each time point using the median value. Every interaction between clusters is tested based on a goodness-of-fit, which indicates how well the S-system differential equation simulates the expression of the corresponding cluster using modified computational method (Tanaka et al. 2005). The interactions showing R (multiple correlation coefficients) >0.98 were considered as candidate interactions.

Regulation and execution of biological processes require specific interactions of proteins. Dynamic networks of Protein-Protein Interactions (PPI) regulate numerous cellular processes and determine the ability to respond appropriately to environmental stimuli, which direct the implementation of developmental programs. Expression profiling at the protein level represents the core of current proteomic approach. PPI plays an important role in deciphering the function of the proteins discovered through molecular maps and proteomics. Katam et al. (2011) detected 302 proteins in leaf tissue of a peanut genotype of which 86 showed differential expression to water stress. Thirty-eight of these stress-responsive proteins have been mapped through unique protein identifiers, with 22 were shown to have putative protein interactions (Fig. 8-1). Among them, 21 proteins were further mapped across *Arabidopsis thaliana* Protein Interactome Database (*At*PID) linked to gene identifiers. All the 21 proteins with viable interactors were confirmed and visualized. The PPI map suggests 65 nodes form a Protein Interaction Network (PIN), a bona-fide functional interaction network. It is recommended that the said protein in peanut would also have these functional partners since there might be interactions prevailing in orthologs of peanut genomes. Yu et al. (2004) predicted functional orthologs among water-stress responsive proteins from putative protein interaction networks using orthology mapping, phylogenetic profiling, gene fusion and text mining. Furthermore, the network analysis revealed novel candidates among them to have potential functional linkages. These functional linkages would complement the molecular evidences in determining the traits.

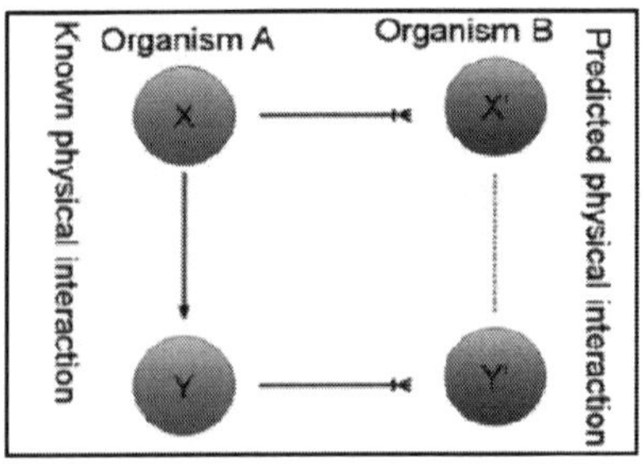

Figure 8-1 Illustration of interolog mapping mapped for those orthologous proteins known to have predicted interaction partners (Organism A: *Arabidopsis*; Organism B: Peanut).

Color image of this figure appears in the color plate section at the end of the book.

8.5 Future Outlook

It is important to note that candidate gene(s) or protein(s) associated with tolerance to drought and/or resistance to aflatoxin should be integrated into genetic maps. This will help identification of correlation/association between genes/QTLs conferring tolerance to aflatoxin and drought. It is necessary to identify peanut lines that possess tolerance to drought and resistance to aflatoxin for detailed understanding of the expression and functional characteristics of various proteins and their interactions associated with drought and aflatoxin resistance in peanut. Such a study is required to address the molecular and cellular components of both host plant and fungal pathogen to understand the events leading to *Aspergillus* entry into the drought-stressed plants, as well as discovering the molecular and cellular responses of peanut plant to drought stress, *Aspergillus* invasion and aflatoxin production. The research should focus to:

- Determine the interrelationship between drought tolerance, *Aspergillus* invasion and aflatoxin production;
- Monitor the effects of drought stress and *Aspergillus* invasion on the expression pattern of seed metabolites, amino acids, sugars and phytoalexins, genes and proteins;
- Identify key genes and pathways involved in pod/seed development, and tolerance to drought and aflatoxin production;
- Allele-specific sequencing of candidate genes/proteins in parents and mapping populations.

Acknowledgements

GB acknowledges the Department of Biological Sciences, Florida A&M University for financial support.

References

Anderson NL, Anderson NG (1998) Proteome and proteomics: new technologies, new concepts, and new words. Electrophoresis 19(11): 1853–1861.

Basha SM, Katam R, Naik KSS (2007) Identification of drought responsive proteins in peanut. Peanut Sci 34: 96–104.

Blackstock WP, Weir MP (1999) Proteomics: quantitative and physical mapping of cellular proteins. Trends Biotechnol 17(3): 121–7.

Bray EA (2002) Classification of genes differentially expressed during water-deficit stress in *Arabidopsis thaliana*: an analysis using microarray and differential expression data. Ann Bot 89: 803–811.

Breiteneder H, Radauer C (2004) A classification of plant food allergens. J Allergy Clin Immunol 113(5): 821–30.

Buckingham S (2003) The major world of microRNAs. Horizon Symposia, Nature Publishing Group: http://www.nature.com/horizon/rna/background/micrornas.html (Retrieved 14 June 2012).

Cherry JP, Ory RL, Mayne RY (1972) Proteins from peanut cultivars (*Arachis hypogaea*) grown in different areas. VI. Changes induced in gel electrophoretic patterns by *Aspergillus* contamination. J Am Peanut Res Edu Assoc 4: 32–40.

Cherry JP, Mayne RY, Ory RL (1974) Proteins and enzymes from seeds of *Arachis hypogaea* L. IX. Electrophoretically detected changes in 15 peanut cultivars grown in different areas after inoculation with *Aspergillus parasiticus*. Physiol Plant Pathol 4: 425–434.

Christensen LH, Riise E, Bang L, Zhang C, Lund K (2010) Isoallergen variations contribute to the overall complexity of effector cell degranulation: effect mediated through differentiated IgE affinity. J Immunol 184(9): 4966–4972.

Claverie JM (2005) Fewer genes, more noncoding RNA. Science 309: 1529–1530.

Cobb WY, Johnson BR (1973) Physicochemical properties of peanuts. In: Stillwater OK (ed) Peanuts: Culture and Uses. American Peanut Research of Education Association, USA, pp 209–256.

Dean TP (1998) Immunological responses in peanut allergy. [Editorial] Clin Exp Allergy 28(1): 7–9.

Devaiah KM, Geetha B, Athamaram TN, Basha SM (2007) Identification of two new genes from drought tolerant peanut up-regulated in response to drought. Plant Growth Regul 52: 249–258.

Dhingraa V, Gupta M, Andacht T, Fu ZF (2005) New frontiers in proteomics research: A perspective. Int J Pharmaceut 299 (1–2): 1–18.

Dorner JW, Cole RJ, Sanders TH, Blankenship PD (1989) Interrelationship of kernel water activity, soil temperature, maturity, and phytoalexin production in preharvest aflatoxin contamination of drought-stressed peanuts. Mycopathologia 105: 117–128.

Dwivedi SL, Nigam SN, Rao NRC, Singh U, Rao KVS (1996) Effect of drought on oil, fatty acids and protein contents of groundnut (*Arachis hypogaea* L.) seeds. Field Crops Res 48: 125–133.

Frith MC, Pheasant M, Mattick JS (2005) Genomics: The amazing complexity of the human transcriptome. Eu J Hum Genet 13: 894–897.

Guo B, Yu J, Holbrook CC, Lee RD, Lynch RE (2003) Application of differential display RT-PCR and EST/microarray technologies to the analysis of gene expression in response to drought stress and elimination of aflatoxin contamination in corn and peanut. J Toxicol Toxin Rev 22: 287–312.

Guo BZ, Xu G, Cao YG, Holbrook CC, Lynch RE (2006) Identification and characterization of phospholipase D and its association with drought susceptibilities in peanut (*Arachis hypogaea*). Planta 223: 512–520.

Guo B, Chen ZY, Lee RD, Scully BT (2008a) Drought stress and preharvest aflatoxin contamination in agricultural commodity: Genetics, genomics and proteomics. J Integr Plant Biol 50(10): 1281–1291.

Guo BZ, Chen S, Dang P, Scully BT, Liang X, Holbrook CC, Yu J, Culbreath AK (2008b) Peanut gene expression profiling in developing seeds at different reproduction stages during *Aspergillus parasiticus* infection. BMC Dev Biol 8: 12.

Guo B, Liang X, Chung S, Holbrook CC, Maleki SJ (2008c) Proteomic analysis of peanut seed storage proteins and genetic variation in a potential peanut allergen. Protein Pept Lett 15(6): 567–577.

Ingram J, Bartels D (1996) The molecular basis of dehydration tolerance in plants. Annu Rev Plant Physiol Plant Mol Biol 47: 377–403.

IUIS (International Union of Immunological Societies) Allergen Nomenclature: http://www.allergen.org (Accessed July 2012).

Jain AK, Basha SM, Holbrook CC (2001) Identification of drought responsive transcripts in peanut (*Arachis hypogaea* L.). Elec J Biotechnol 4: 59–67.

Jain S, Srivastava S, Sarin NB, Kav NN (2006) Proteomics reveals elevated levels of PR 10 proteins in saline-tolerant peanut (*Arachis hypogaea*) calli. Plant Physiol Biochem 44: 253–9.

Jian C, Peng L, Guang L, Feng X, Chen Z, Yuhua L, Zhongnan Y, Guang W, Qingbo Y, Yixue L, Tieliu S (2008) AtPID: Arabidopsis thaliana protein interactome database an integrative platform for plant systems biology. Nucleic Acids Research 2008, 36: D999–D1008.

Kang JH, Gallo M (2007) Cloning and characterization of a novel peanut allergen Arah 3 isoform displaying potentially decreased allergenicity. Plant Sci 172: 345–353.

Katam R, Vasanthaiah HKN, Basha SM (2007) Proteomic approach to screen peanut genotypes with enhanced nutritional qualities. In: Frontiers in the Convergence of Bioscience and Information Technologies. FBIT 171–178.

Katam R, Basha SM, Survajahala P, Pechan T (2010a) Analysis of peanut leaf proteome. J Proteome Res 9: 2236–2254.

Katam R, Basha SM, Sakata K (2010b) Leaf proteome analysis of water deficit stress responses in peanut. International Conference on Bioinformatics, Tokyo, Japan.

Katam R, Suravajhala P, William C, Gudimella R, Basha SM, Latinwo LM (2011) Protein–protein interaction analysis of water stress responsive proteins to reveal essential functional orthologues in peanut. International Conference on Bioinformatics, Malaysia, A. 54.

Kiening M, Niessner R, Drs E, Baumgartner S, Krska R, Bremer M, Tomkies V, Reece P, Danks C, Immer U, Weller MG (2005) Sandwich immunoassays for the determination of peanut and hazelnut traces in foods. J Agri Food Chem 53(9): 3321–3327.

Kleber-Janke T, Crameri R, Appenzeller U, Schlaak M, Becker WM (1999) Selective cloning of peanut allergens, including profiling and 2S albumins, by phage display technology. Int Arch Allergy Appl Immunol 119: 265–274.

Koppelman SJ, Vlooswijk RJ, Knippels LM, Hessing M, Knol EF, Reijsen FC, Bruijnzeel-Koomen CA (2001) Quantification of major allergens Ara h1 and Ara h2 in the peanut varieties Runner, Spanish, Virginia, and Valencia, bred in different parts of the world. Allergy 56: 132–137.

Koppelman SJ, Knol EF, Vlooswijk RA, Wensing M, Knulst AC, Hefle SL, Gruppen H, Piersma S (2003) Peanut allergen Ara h 3: isolation from peanuts and biochemical characterization. Allergy 58(11): 1144–51.

Kottapalli KR, Burow MD, Burow G, Burke J, Puppala N (2007) Molecular characterization of the Core of the US Peanut core collection using microsatellite markers. Crop Sci 47: 1718–1727.

Kottapalli KR, Payton P, Rakwal R, Agrawal GK, Shibato J, Burow M, Puppala N (2008) Proteomic analysis of mature seed of four peanut cultivars using two-dimensional gel electrophoresis reveals distinct differential expression of storage, anti-nutritional, and allergenic proteins. Plant Sci 175: 321–329.

Kottapalli KR, Rakwal R, Shibato J, Burow G, Tissue D, Burke J, Puppala N, Burow M, Payton P (2009) Physiology and proteomics of the water-deficit stress response in three contrasting peanut genotypes. Plant Cell Environ 32(4): 380–407.

Liang XQ, Luo M, Holbrook CC and BZ Guo (2006) Storage protein profiles in Spanish and runner market type peanuts and potential markers BMC Plant Biology 6: 24.

Li HG, Wang L, Zhang YS, Lin XD, Liao B, Yan YS, Huang SZ (2005) Cloning and sequencing of the gene Ahy-β encoding a subunit of peanut conarachin. Plant Sci 168: 1387–1392.

Li Y, Huang SZ, Fu JR (1998) One and two-dimensional polycrylamide gel electrophoretic analysis of seed polypeptide composition of peanut cultivars. Acta Bot Sin 406: 534–541.

Liao BS, Lei Y, Li D, Wang SY, Huang JQ, Ren XP, Jiang HF, Yan LY (2010) Novel Germplasm with high oil content and resistance to *Aspergillus flavus* and bacterial wilt developed from peanut recombinant Inbred lines. Acta Agron Sin 36(8): 1296–1301.

Luo M, Dang P, Guo BZ, He G, Holbrook CC, Bausher MG, Lee RD (2005a) Generation of expressed sequence tags (ESTs) for gene discovery and marker development in cultivated peanut. Crop Sci 45: 346–353.

Luo M, Liang XQ, Dang P, Holbrook CC, Bausher, Lee RD, Guo BZ (2005b) Microarray-based screening of differentially expressed genes in peanut in response to *Aspergillus parasiticus* infection and drought stress. Plant Sci 169(4): 695–703.

Matuszewski BK, Constanzer ML, Chavez-Eng CM (1998) Matrix effect in quantitative LC/MS/MS analyses of biological fluids: a method for determination of finasteride in human plasma at picogram per milliliter concentrations. Anal Chem 70(5): 882–889.

Millerd A (1975) Biochemistry of Legume Seed Proteins. Annu Rev Plant Physiol 26: 53–72.

Mills ENC, Madsen C, Shewry PR, Wichers HJ (2003) Food allergens of plant origin – Their molecular and evolutionary relationships. Trends Food Sci Technol 14: 145–156.

Mittag D, Akkerdaas J, Ballmer-Weber BK, Vogel L, Wensing M, Becker WM, Koppelman SJ, Knulst AC, Helbling A, Hefle SL, Van Ree R, Vieths S (2004) Ara h 8, a Bet v 1-homologous allergen from peanut, is a major allergen in patients with combined birch pollen and peanut allergy. J Allergy Clin Immunol 114: 1410–1417.

Møller IM, Sweetlove LJ (2010) ROS signaling-specificity is required. Trends Plant Sci 15(7): 370–4.

Mosse J, Pernollet JC (1983) Storage protein of legume seeds. In: Arora SK (ed) Chemistry and Biochemistry of Legumes. Edward Amold, Delhi, India, pp 111–193.

Paik-Ro OG, Seib JC, Smith RI (2002) Seed-specific, developmentally regulated genes of peanut. Theor Appl Genet 104: 236–240.

Payne GA, Brown MP (1998) Genetics and physiology of aflatoxin biosynthesis. Annu Rev Phytopathol 36: 329–362.

Payton P, Kottapalli KR, Rowland D, Faircloth W, Guo B, Burow M, Puppala N, Gallo M (2009) Gene expression profiling in peanut using high density oligonucleotide microarrays. BMC Genomics 10: 265.

Pele M (2010) Peanut Allergens [Review]. Romanian Biotechnological Letters 152: 5204–5212.

Piersma SSR, Gaspari M, Helfe SL, Koppelman SJ (2005) Proteolytic processing of the peanut allergen Ara h 3. Mol Nutri Food Res 49: 744–755.

Rabjohn P, Helm EM, Stanley JS, West CM, Sampson HA, Burks AW, Bannon GA (1999) Molecular cloning and epitope analysis of the peanut allergen Ara h 3. J Clin Invest 103: 535–542.

Rogers S, Girolami M, Kolch W, Waters KM, Liu T, Thrall B, Wiley HS (2008) Investigating the correspondence between transcriptomic and proteomic expression profiles using coupled cluster models. Bioinformatics 24(24): 2894–2900.

Sampson HA (2004) Update on food allergy. J Allergy Clin Immunol 113: 805–819.

Scurlock A, Burks AW (2004) Peanut allergenicity. Ann Allergy Asthma Immunol 93: S12–8.

Seki M, Narusaka M, Abe H (2001) Monitoring the expression pattern of 1300 *Arabidopsis* genes under drought and cold stress by using full-length cDNA microarray. Plant Cell 13: 61–72.

Sheehan D, McDonagh B, Bárcena JA (2010) Redox proteomics. Exp Rev Proteom 7(1): 1–4.

Sicherer SH, Sampson HA (2007) Peanut allergy: Emerging concepts and approaches for an apparent epidemic. J Allergy Clin Immunol 120: 491–503.

Simpson CE, Krapovickas A, Valls JFM (2001) History of Arachis including evidence of *A. hypogaea* L progenitors. Peanut Sci 28: 78–80.

Stadtman ER (2001) Protein oxidation in aging and age-related diseases. Ann NY Acad Sci 928: 22–38.

Talcott ST, Duncan CE, Del Pozo-Insfran D, Gorbet DW (2005) Polyphenolic and antioxidant changes during storage of normal, mid, and high oleic acid peanuts. Food Chem 89: 77–84.

Tanaka N, Mitsui S, Nobori H, Yanagi K, Komatsu S (2005) Expression and function of proteins during development of the basal region in rice seedlings. Mol Cell Proteomics 4: 796–808.

Tang K, Zhan JC, Yang HR, Huang WD (2010) Changes of resveratrol and antioxidant enzymes during UV-induced plant defense response in peanut seedlings. J Plant Physiol 167: 95–102.

Upadhyaya HD, Nigam SN, Mehan VK, Reddy AGS, Yellaiah N (2001) Registration of *Aspergillus flavus* Seed Infection Resistant Peanut Germplasm ICGV 91278, ICGV 91283, and ICGV 91284. Crop Sci 41: 559–600.

Van Hengel AJ, Anklam E, Taylor SL, Hefle SL (2007) Analysis of food allergens. Practical applications. In: Pico Y (ed) Food Toxicants Analysis. Elsevier, Amsterdam, Netherland, pp 189–229.

Varshney RK, Hoisington DA, Tyagi AK (2006) Advances in cereal genomics and applications in crop breeding. Trends Biotechnol 24: 490–499.

Varshney RK, Hoisington DA, Nayak SN, Graner A (2009) Molecular plant breeding: methodology and achievements. In: Somers D, Langridge P, Gustafson PJ (eds) Methods in Molecular Biology: Plant Genomics. The Humana Press, Totowa, NJ, USA, pp 283–304.

Varshney RK, Kudapa H, Roorkiwal M, Thudi M, Pandey MK, Saxena RK, Chamarthi SK, Mohan SM, Mallikarjuna N, Upadhyaya H, Gaur PM, Krishnamurthy L, Saxena KB, Nigam SN, Pande S (2012) Advances in genetics and molecular breeding of three legume crops of semi-arid tropics using next generation sequencing and high-throughput genotyping technologies. J Biosci 37(5): 811–820.

Wilkins MR, Pasquali C, Appel RD, Ou K, Golaz O, Sanchez JC, Yan JX, Gooley AA, Hughes G, Humphery-Smith I, Williams KL, Hochstrasser DF (1996) From proteins to proteomes: large scale protein identification by two-dimensional electrophoresis and amino acid analysis. Nat Biotechnol 14(1): 61–65.

Yen GC, Duh PD (1996) Antimutagenic effect of methanolic extracts from peanut hulls. Biosci Biotechnol Biochem 60: 1698–1700.

Yu H, Luscombe NM, Lu HX, Zhu X, Xia Y, Han JDJ, Bertin N, Chung S, Vidal M, Gerstein M (2004) Annotation transfer between genomes: protein–protein interologs and protein–DNA regulogs. Genome Res 14(6): 1107–18.

9

Transgenic Interventions in Peanut Crop Improvement: Progress and Prospects

Sowmini Sunkara, Pooja Bhatnagar-Mathur and *Kiran K Sharma**

ABSTRACT

Legumes rank third in world crop production in which the major constraint to crop productivity is attributed to biotic and abiotic stress. Peanut, also known as groundnut (*Arachis hypogaea* L.) is a major oilseed crop in the world, both for oil and as a protein source. Host plant resistance provides the most effective and economic option to manage stress tolerance in peanut which is also time consuming involving expensive agronomic practices. However, for many biotic and abiotic stresses, effective resistance gene(s) in cultivated peanut have not been identified. Success in breeding for better adapted varieties to biotic/ abiotic stresses depend upon the combined efforts of various research domains like plant and cell physiology, molecular biology, genetics and breeding. Moreover, availability of known genotypes with natural resistance to stresses is a prerequisite for the successful breeding program. With a few exceptions, crop improvement in peanut programs through conventional breeding has received little progress.

Over the years, biotechnology has emerged as a promising tool to overcome both biotic and abiotic stresses in plants. Biotechnology applications include potential approaches, especially where the existing germplasm lacks the required traits for conventional breeding and provide promising ways to increase peanut productivity, either through improved seed quality or stress resistance. However, the progress has

Genetic Transformation Laboratory, International Crops Research Institute for the Semi-Arid Tropics (ICRISAT), Patancheru, AP 502324, India.
*Corresponding author: *k.sharma@cgiar.org*

been very limited in legumes till date since these approaches require the identification of genes that control important agronomical traits, the understanding of gene regulation and metabolic pathways, along with methods of delivering genes or small RNAs into peanut plants. A new tool of engineering of multiple genes or regulatory machinery involving transcription factors has emerged for controlling the expression of different stress-responsive genes instead of inserting single genes for a single trait. Hence, researchers have focused their research on peanut functional genomics and biotechnology, and have achieved great strides during the recent decades. In this chapter, we discuss the recent progress and the current status of transgenic technology in peanut which offers the best option in host plant resistance breeding to combat various economically important biotic/abiotic stresses and its use in the crop improvement for stress tolerance.

Keywords: Transgenic breeding, Genetic transformation, Regeneration, Somatic embryogenesis, Genetically modified plants, Overexpression, Gene silencing, RNAi technology

9.1 Introduction

Legumes, rich sources of proteins and minerals, are referred to as "poor man's meat" in certain cultures. In order of importance, peanut, cowpea and beans represent about 80% of the production and cultivated area of food legumes, which are essential staples in the diets of millions. Peanuts share approximately 10% among production of 286.7 million metric tons of the world total of oilseeds behind soybeans (53%), rapeseed (15%) and cotton seeds (12%).

Peanut production process from planting to harvest is affected by different types of biotic and abiotic stresses that cause annual yield losses of over US$ 3.2 billion (Dwivedi et al. 2003). Since the mid-1970s, edible peanuts have increased in both domestic consumption and export trade in India. In contrast, production in Africa has declined by 17% over the last two decades. Acreage, production and productivity of peanut in India has shown large amount of fluctuations since 1993–94 to 2006–07. The productivity of peanut in India suffers mainly since 80% of the crop is grown under rainfed conditions by resource-poor farmers (Kaushik 1993). Lack of irrigation facilities to protect the crop from soil-moisture deficit during breaks in rainfall in the monsoon season affects germination. Rainfall pattern during the presowing months and availability of substitute high-value oilseed crops like soybean and sunflower with short durations requiring less water had a significant negative impact on acreage allocation decisions of the farmers (Patil et al. 2009). Resource-poor farmers who obtain low yields of 500–800 kg.ha^{-1} due to various biotic and abiotic constraints grow about 93.8% of the world's production of peanut. Moreover, a big gap exists between the realized yield and potential yield of peanut at both subsistence and commercial systems of production in Asia and Africa.

The decrease in peanut productivity is mainly affected by various biotic, abiotic and economic factors. The economic status of the small and marginal farmers restricts them to use poor quality local seed in addition to minimum or no fertilizer applied during cultivation, which is essential as peanut is mostly grown in marginal and poor soils of low fertility. Use of complex fertilizers may also add to deficiencies of nutrients such as calcium and sulfur affecting the yields.

The major abiotic factors affecting peanut production include drought, high temperature, low soil fertility, low soil pH and iron chlorosis. Among the biotic factors, fungal diseases, virus diseases, bacterial wilt disease, aflatoxin contamination, nematodes, foliar insect pests, and soil insect pests, pod borer (*Helicoverpa* spp.) play a significant role in yield reduction (Sharma and Oritz 2000; Dwivedi et al. 2003). The plant disease management technologies are greatly influenced by environmental pollution, deleterious effects of chemicals on nontarget organisms, resurgence of pesticide resistance among pathogens and outbreak of secondary pathogens. Hence, there is an urgent call for increased crop production to cater to the needs of the increasing population. In order to reconcile with the demands of intensive agriculture with maintenance of the ecosystem, pest control strategies employed in the future must be environmentally compatible and selective to target pests.

9.2 Rationale for Transgenic Peanut Breeding

From USDA estimates (FAS 2000), peanuts ranked third in production among oilseeds and 90% of world peanut production was accounted by developing countries (ERS 2001) with 2.5% increase annually. Though the world harvested area of peanut has changed very little since 1970s with an annual growth of only 0.1% (between 1972–1990) and 1.2% (between 1991–2000), the production has increased from 0.8 metric tons (during 1972) to 1.37 metric tons (during 2000), i.e., 1.9% increase per year (Revoredo and Fletcher 2002). It is generally accepted that the average yield of peanut is below its presumed potential, and efforts to improve the productivity of this crop by conventional breeding means have not been very effective. The major reason behind this is the lack of sufficient and satisfactory levels of genetic variability within the germplasm of cultivated peanut. Many wild annual *Arachis* species, which possess a wealth of agronomically desirable genes, are sexually incompatible with the cultivated varieties. Several advanced research institutes or groups are working with ICRISAT and other partners to apply modern biotechnology to the problems of peanut improvement in developing countries. Biotechnology tools such as marker-assisted breeding, tissue culture, *in vitro* mutagenesis, embryo rescue and genetic transformation have contributed to solve or reduce some of these constraints. Major yield increases could be achieved by development and use of cultivars addressing abiotic and biotic stresses. Comprehensive reviews on the history of molecular marker development in peanut were provided by Stalker and Mozingo (2001) and Dwivedi et al. (2003). However, only limited success has been achieved so far. The emergence of "omics" technologies and the establishment of model legume plants such as *Medicago truncatula, Glycine max* and *Lotus japonicus* (Cannon et al. 2009) are promising strategies for understanding the molecular genetic basis of stress resistance, which is an important bottleneck for molecular breeding. Understanding the mechanisms that regulate the expression of stress-related genes is a fundamental issue in plant biology and will be necessary for the genetic improvement of legumes (Bertioli et al. 2011).

Transgenic research has opened exciting opportunities in plant protection which result in prolonged benefit in sustainable agriculture with a high degree of safety which is also an important part of second green revolution. The techniques of genetic modification will allow breeders to access new gene pools, particularly those of wild *Arachis* species, bringing valuable traits into the modern cultivated peanut that cannot be addressed by conventional means. Development of transgenic peanut therefore has a good potential for its improvement. Advances in biotechnology have provided alternative pest control strategies that are based on natural biological processes. Tissue culture and genetic engineering have proven as important powerful tools in biotechnology that have been extensively used, either by taking advantage of naturally occurring defense mechanisms, which confer disease resistance of avoidance or by modifying plant genome to develop pest resistance.

9.3 Genetic Transformation in Peanut

Successful genetic transformation of plants, including peanut, generally requires a reproducible tissue culture system to regenerate whole fertile plants from single cells (totipotency) as well as a method to deliver the gene(s) of interest to those regenerating cells. Transformation frequencies are directly related to the tissue culture response, and therefore highly regenerative cultures are often transformation competent. The inefficient, inconsistent and genotype dependent published protocols for peanut regeneration have emboldened some researchers in adopting non tissue culture-based approaches, that do not depend on the regeneration of adventitious shoot buds for generating transgenic plants of peanut (Rohini and Rao 2000). *In vitro* regeneration of whole plants of economically important commercial cultivars of peanut from explants such as protoplasts, cell suspension cultures, callus tissue or organized tissue such as embryonic axes, mature and immature embryonic axes (Atreya et al. 1984; Hazra et al. 1989; Brar et al. 1994; Baker et al. 1995), cotyledons (Atreya et al. 1984; Ozias-Akins 1989) and leaves (Baker and Wetzstein 1992; Livingstone and Birch 1995) either by organogenesis or embryogenesis have been reported with different culture media containing different phytohormone combinations (Table 9-1, 9-2).

Regeneration by organogenesis occurs either by direct development of shoots from the surface of cultured explants (Hazra et al. 1989; McKently et al. 1991) or by an intervening callus phase (Bajaj et al. 1981; Bajaj and Gosal 1983, 1988). The reports of organogenesis from de-embryonated cotyledons, immature leaflets, seed explants, epicotyls, hypocotyls and anther-derived callus (Mroginski and Fernandez 1980; Mroginski et al. 1981; Narasimhulu and Reddy 1983; Pittman et al. 1983; McKently et al. 1990; Willcox et al. 1991; Li et al. 1994) had a very low frequency of transformation. However, not much success with genetic transformation of peanut genotypes was achieved until recently (Sharma and Anjaiah 2000) due to the lack of efficient protocols to obtain whole plants through *in vitro* regeneration of adventitious shoot buds from the transformed tissues. Direct regeneration systems favors easy accessibility for *Agrobacterium*-mediated genetic transformation because of advantages of *de novo* production of shoot primordia, synchronous with the period of cellular differentiation, rapidity of morphogenesis and lack of requirement for frequent subcultures. Sharma and Anjaiah (2000) obtained success of high-frequency direct shoot regeneration from mature cotyledon explants in various peanut genotypes. Shoot organogenesis and plants were also successfully obtained using immature leaflets (McKently et al. 1991; ICRISAT unpubl. data).

Regeneration via somatic embryogenesis also has been reported (Gill and Saxena 1992; Zhuang et al. 1999; Cucco and Jaume 2000) which has been used in transformation studies in peanut (Ozias-Akins et al. 1992; Sellars et al. 1990; Chengalrayan et al. 1994, 1997). However, conversion of somatic embryos into plants remains inefficient and limits the application of somatic embryogenesis in many systems, including genetic transformation (Wetzstein and Baker 1993).

Developments in genetic transformation for incorporation of novel genes into the peanut gene pool have emboldened researchers with new opportunities for crop improvement in this important legume to pursue the development of transgenic peanut plants resistant to various diseases, insect pests, enhanced nutritional quality and abiotic stresses (Sharma and Anjaiah 2000; Rohini and Rao 2001). Transformation of plants involves the stable introduction of desirable DNA/gene sequences into the nuclear genome of cells, which are capable of giving rise to a whole transformed plant.

Table 9-1 Responses of various explants and hormones on *in vitro* shoot regeneration in peanut.

Explant	Medium	Growth regulators	Morphogenic response	Genotype/cultivar	Reference
Ovaries	MS	BA (0.5 mg/l) + NAA (2 mg/l)		MK 374, M 13, TMV 2, Robut-33-1	Sastri et al. 1980
Ovules	MS	Kinetin+ GA_3	Shoots and roots		Martin 1970
Immature embryos	-	TDZ (10 mg/l)	-	New Mexico Valencia	Kanyand et al. 1994
	B5	Picloram (0.5–1 mg/l)	Shoots with roots	Several varieties	Ozias- Akins et al. 1992
Cotyledonary nodes	MS	NAA (1 mg/l) +BA (3 mg/l)	Multiple shoots		Banerjee et al. 1988
De-embryonated cotyledons	MS	Zeatin (4 mg/l) or kinetin (4 mg/l)	Multiple shoots	MK 374, M 13, TMV 2, Robut-33-1	Sastri et al. 1980
	MS	2,4-D (2 mg/l)+ kinetin (2 mg/l)	Multiple shoots	ICG 4367, US 48, TMV 2, TG 19B	Narasimhulu and Reddy 1983
	Moist cotton wool	BA (1 mg/l)	Multiple shoots	TG-17	Bhatia et al. 1985
Mature cotyledons	MS + B5 organics	BA(20 µM)+ 2,4-D (10 µM)	Multiple shoots	JL-24,J-11, ICGS-11, ICGS-44, Robut 33-1	Sharma and Anjaiah 2000
Embryo axis	MS	None	Shoots regenerated into plantlets		Atreya et al. 1984
Epicotyl	MS	Casein hydrolysate	Multiple shoots, roots		Bajaj 1982
	MS	BA (10 mg/l)+ NAA (1 mg/l)	Organogenesis	New Mexico Valencia	Cheng et al. 1992
	MS	None	9–28% shoots	ICG 4367, US 48, TMV 2, TG 19B	Narasimhulu and Reddy 1983
Mesocotyl	MS	IAA (11 µM)+ kinetin(2.3 µM)	Shoots with roots		Bajaj 1982
Hypocotyl	MS	IAA (2 mg/l)+ kinetin (2 mg/l)	Shoots	ICG 4367, US 48, TMV 2, TG 19B	Narasimhulu and Reddy 1983

Table 9-1 contd....

Table 9-1 contd.

Explant	Medium	Growth regulators	Morphogenic response	Genotype/cultivar	Reference
Apical meristem	MS +B5 vitamins	NAA (10 µM)+ BA (0.1 µM)	Single shoots with many roots		Kartha et al. 1981
		NAA (10 µM)+ BA (1 µM)	Shoots without any further development		Kartha et al. 1981
Plumule	MS	BA(30 µM)+ NAA(5 µM)+ brassin (1 µM)	Multiple shoots	Okrun	Ponsamuel et al. 1998
Immature leaflets	MS + Gamborg vitamins	NAA (1 mg/l)+ BA (1 mg/l)	50% shoots		Pitman et al. 1983
	MS	NAA (4 mg/l)+ BA (5 mg/l)		JL24	Chengalrayan et al. 1994
	MS	NAA (2 mg/l)+ BA (4 mg/l)	Shoots	NC-7	Utomo et al. 1996
Leaflets	MS	NAA (1 mg/l)+ BA (1 mg/l) BA (2 mg/l)+ NAA (0.5 mg/l)	Organogenic callus Shoot primordia	 TMV2	Mroginski et al. 1981 Venkatachalam et al. 1999

MS: Murashige and Skoog (1962)

Table 9-2 Update on Genetic transformation in peanut.

Explant	Gene delivery system	Gene introduced	Transformation frequency/status	Strain/Plasmid	Reference
Cotyledon	*Agrobacterium* *Agrobacterium* Biolistic Biolistic *Agrobacterium*	 *uidA, nptII* *uidA, hph* *uidA, hph* *gus, nptII*	3.30% 47% 1.6% 168 hygomycin resistant lines T2 generation viable seeds	 LBA4404, pBI121 pCAMBIA-1301 pMOG617/pxVGH	 Rohini and Rao 2000 Venkatachalam et al. 2000 Yang et al. 2001 Wang et al. 1998 ICRISAT 1994
Leaf	*Agrobacterium*	*Gus, nptII*	0.2–0.3%	pBI121	Cheng et al. 1997
Embryonic axis	*Agrobacterium* *Agrobacterium* Biolistic/Particle bombardment Biolistic	*uidA, nptII* *Bar* and *PSTV genome* *MerApe9, hph/ MerApe9*,mercuric ion reductase	9% Putative transformants 0.9–1%	EHA101/ pMON9793 pAC2MR/ pACH2MR	McKently et al. 1995 Cassidy and Ponsamuel 1996 Yang et al. 2003 Brar et al. 1994
Leaf, epicotyl	*Agrobacterium*	*uidA*	12–36% (leaves), 15–42% (epicotyls)	EHA 101	Egnin et al. 1998
Leaf Discs	*Agrobacterium*	*uidA, nptII*	6.7% putative shoots	pBI121	Eapen and George 1994
Embryonic axis, cotyledon, leaf, petiole explants, Mature cotyledons	*Agrobacterium* *Agrobacterium* *Agrobacterium*	*uidA, nptII* *IPCV (coat protein)* *H protein gene*	 55%	pTiBo542/pTiT37 pBI121/pROKII: IPCVcp	Lacorte et al. 1991 Sharma and Anjaiah 2000 Khandelwal et al. 2003
Seedling explants	*Agrobacterium*	*Gus, nptII*	Second generation callus colonies		Li et al. 1996
Protoplast	Electroporation Electroporation	*PstV coat protein* *Gus, nptII*	Protoplast derived callus colonies		Li et al. 1996 Li et al. 1996
Embryonic leaflets	Electroporation				Padua et al. 2000

Table 9-1 contd....

Table 9-1 contd.

Explant	Gene delivery system	Gene introduced	Transformation frequency/status	Strain/Plasmid	Reference
Epicotyl	Biolistic	*uidA, hph*		pKYLX80-N11 pTRA140	Magbanua et al. 2000
Embryonic callus	Biolistic	*Luc, hph*	54 independent transgenic lines	pDO432/pHygr/ pGIN	Livingstone and Birch 1995
	Biolistic	*hph*	1%		Ozias- Akins et al. 1993
Shoot meristem of embryonic axis	ACCELL (biolistic)	*Gus, bar, TSWV nucleocapsid protein*	Transgenic plants up to R2 generation		Brar et al. 1994
Somatic embryos	Biolistic	*hph gene, nucleocapsid protein gene of TSWV*	52 hygromycin resistant cell line	pCB13-N+ pCB13-N++	Yang et al. 1998
Immature cotyledons	Biolistic	*cry1AC*			Singsit et al. 1997
Mature Zygotic embryos	Biolistic	*GFP*		p524EGFP.1	Joshi et al. 2005

Gus/uidA: gene encoding glucoronidase activity; *hph*: gene conferring resistance to hygromycin; *nptII*: gene conferring resistance to neomycin and kanamycin; *TSWV*: tomato spotted wilt virus; *PStV*: peanut stripe virus; *PCV*: peanut clump virus; *bar*: gene conferring resistance to herbicide resistance

Transformation and regeneration are interdependent and the totipotency (i.e., single cell capable of giving rise to a whole plant *in vitro*) of the somatic plant cells via organogenesis or somatic embryogenesis under appropriate hormonal and nutritional conditions (Skoog and Miller 1957) is the essential feature for development of an efficient tissue culture techniques. Totipotent cells give rise to adventitious shoots or somatic embryos, which are both competent and accessible for gene transfer and will give rise directly to nonchimeric transformed plants. Development of an efficient transformation system for the introduction of genes into the crop plants also depends on the various factors such as development of reliable and reproducible tissue culture regeneration systems, selection and preparation of suitable gene constructs and vectors, recovery and multiplication of transgenic plants, molecular and genetic characterization of transgenic plants for stable and efficient gene expression, transfer of genes to elite cultivars by conventional breeding methods if required, evaluation of transgenic plants for their effectiveness in alleviating the biotic and abiotic stresses in the field condition, biosafety assessments including health, food and environmental safety and deployment of genetically modified plants.

A suitable system for selection of transgenic tissues and plants is one of the most important aspects of any transformation system. The utility of any particular gene construct as a transformation marker varies depending on the plant species and explant involved. Promoters are essential to control expression of the gene and also provide valuable insights about the overexpression or silencing of any gene in response to external stimuli. The most commonly developed transgenic plants use either the constitutive promoters like 35S of the Cauliflower Mosaic Virus (CaMV) or the maize ubiquitin or potato ubiquitin (Yang et al. 2003; Joshi et al. 2005) to drive expression of the gene of interest in their gene constructs. These promoters being constitutive in nature sometimes results in expression of the downstream transgenes in all organs and at all the developmental stages, which can be metabolically expensive leading to undesirable pleiotropic effects (Bhatnagar-Mathur et al. 2008). Hence, use of inducible or tissue-specific promoters is increasing in recent years for enhancing targeted gene expression, which also safeguards against biosafety and regulatory concerns to a certain extent. Use of these tissue-specific constructs is also important in RNAi technology to augment gene silencing strategies (Bhatnagar-Mathur et al. 2008).

The transformation and regeneration protocols for peanut are now well-established. Transformation techniques and plant regeneration from *in vitro* cultured tissues have been described for many species (Lindsey and Jones 1989; Dale et al. 1993; Birch 1997). There are numerous reports of tissue culture and transformation of peanut from various explants (Kartha et al. 1981; Sastri and Moss 1982; Kanyand et al. 1994). Regeneration via somatic embryogenesis has also been reported as one of the promising methods for transformation studies in peanut (Ozias-Akins et al. 1993; Sellars et al. 1990; Baker and Wetzstein 1995; Chengalrayan et al. 1994, 1997).

9.4 Transfer of Genetic Material

Different methods of DNA transfer have been developed for the production of transgenic peanut over the last few years. The most commonly used means of DNA delivery or transferring novel genes into either organogenic or embryogenic cultures of plant cells/peanut are either biologically by *Agrobacterium tumefaciens* or by direct gene transfer using microprojectile/particle bombardment or by electroporation (Table 9-2). Research is being carried out globally with single or multiple gene introductions to produce disease resistant, pest-resistant, healthier and high-quality peanuts. Peanut tissues are susceptible to infection by wild-type strains of *A. tumefaciens* (Lacorte et al. 1991). The choice between using microprojectile bombardment or *Agrobacterium* as the means by which to deliver DNA is determined by several factors including the laboratory facilities and technical skills available, the species and/or cultivar to be transformed (many monocots are still recalcitrant to transformation with *Agrobacterium*, although this is improving all the time), and the regeneration system.

9.4.1 Direct Gene Transfer

Direct DNA transfer methods can circumvent the genotype dependence of *Agrobacterium* infection. Direct gene transfer has been accomplished by several methods such as microprojectile bombardment, electroporation of protoplasts and intact tissues, microinjection of protoplasts or meristems and polyethylene glycol-mediated transformation of protoplasts. Among these, microprojectile bombardment is the most commonly used method for genotype-independent genetic transformation (Sharma et al. 2005).

Particle bombardment was developed by Sanford and coworkers (Sanford et al. 1987; Klein et al. 1988; Sanford 1990) and has been the most commonly used method for direct introduction of genes into a number of plant species including peanut. Transient expression (Li et al. 1995) was reported from cultures developed through bombardment of callus lines from immature peanut leaflet tissue (Clemente et al. 1992) and leaflets (Schnall and Weissinger 1995). However, bombardment of 1–2-year-old embryogenic callus derived from immature embryos followed by stepwise selection for resistance to hygromycin in semi-solid and liquid media produced transgenic shoots at a frequency of 1% (Ozias-Akins et al. 1993), while the shoot meristems of mature embryonic axis produced transgenic plants at a relatively low transformation frequency of 0.9–1.0% (Brar et al.

1994). Transgenic peanut plants using the somatic embryos were developed from immature cotyledons by transforming the *cry1Ac* gene for resistance to the cornstalk borer (*Elasmopalpus lignosellus*) (Singsit et al. 1997). Similarly, Livingstone and Birch (1995) obtained efficiently transformed Spanish and Virginia types of peanut by particle bombardment into embryogenic callus derived from mature seeds. More recently, cobombardment of embryogenic callus derived from mature seeds was used to develop peanut lines exhibiting high levels of resistance to Peanut Stripe Virus (PStV) (Higgins et al. 2004). Similarly, using particle bombardment transient expression of GUS and 2S albumin gene from Brazil nut was observed in peanut (Lacorte et al. 1997).

The advantages of particle bombardment system is that DNA may be transferred directly to cells by the introduction of multiple DNA fragments or multiple plasmids by cobombardment without using specialized or binary vectors, thus eliminating the necessity of constructing a single large plasmid containing multiple transforming sequences. However, the biolistic-based system is labor intensive since it requires bombardment of large number of explants for obtaining few stable transformation events. It may also result in the integration of multiple copies of the transgene, thereby leading to gene silencing which is the major drawback.

9.4.2 *Agrobacterium*-Mediated Genetic Transformation

The naturally-evolved unique system of *Agrobacterium* transfers the foreign DNA sequences precisely into plant cells using Ti plasmids. *Agrobacterium*-mediated transformation is the preferred method over microprojectile bombardment for gene delivery as it results in higher frequency of stable transformation with single or fewer integrated transgene copies, thus reducing the risk of gene silencing and transgene rearrangements. Moreover, when compared to direct DNA delivery system, *A. tumefaciens* infections are less complex and *Agrobacterium*-mediated transformation is generally precise in transferring and integration into the plant genome as it delivers long stretches of T-DNA between the right and left borders.

Several reports have been published for transforming peanut using *A. tumefaciens* method using hypocotyl explants (Dong et al. 1990; Lacorte et al. 1991; Mansur et al. 1993), leaf explants (Eapen and George 1994), and embryonic axes from mature seeds of peanut (McKently et al. 1995). High transformation frequency was reported by using precultured cotyledons as explants (Venkatachalam et al. 1998, 2000), or leaf segments with 0.3% frequency of fertile transgenic plants (Cheng et al. 1997), whereas stable 3% transformation frequency was reported using a nontissue-culture based *Agrobacterium* transformation involving direct cocultivation of cotyledon attached embryo axis supplemented with wounded tobacco leaf extract (Rohini and Rao 2000). Sharma and Anjaiah (2000) reported an efficient transformation system with >55% transformation frequency using cotyledon explants. Recently, promoter tagged peanut transgenics using the cotyledonary nodes as explants and a promoter-less fusion gene *nptII:gus* were produced (Anuradha et al. 2006).

9.5 Selection of Transformed Plants

Uptake of DNA transferred by either method only occurs in a minority of cells and selection of those cells is crucial. Most vectors used for the genetic transformation of plants carry marker genes that allow selection and screening of the transformed cells. More than 50 marker genes and molecular techniques were reported to screen for genetic transformation (Liang et al. 2010), which are divided into two categories: a) Selectable markers, and b) Screenable (scorable, reporter, visible) markers. Marker genes are usually co-introduced into a plant genome along with the transgenes in a single plasmid (Curtis et al. 1995), or as separate effector (for genetic transformation) and reporter (for screening) plasmids (Sakuma et al. 2006a). Protocols with selectable markers have yielded 10-fold higher frequency of recovered transgenic events compared to marker-free protocols (Birch 1997; de Vetten et al. 2003; Darbani et al. 2007) and so the use of marker genes is advantageous. Positive selectable marker genes promote the growth of transformed tissue whereas negative selectable marker genes inhibit growth or kill the nontransformed tissue (Liang et al. 2010).

Inclusion of selectable marker genes encoding resistance to an antibiotic such as kanamycin or hygromycin or to a herbicide such as phosphinothricin, glyphosate, bialaphos and several other chemicals (Wilmink and Dons 1993) in addition to the gene(s) of interest, allows the selection of such cells, by addition of the compound to the nutrient medium. Cells that express the resistance gene can proliferate while the untransformed cells die. Judicious choice of antibiotic and concentration levels may be an important criterion for the recovery of transformed cells, because too high a level would be deleterious even to the transformed cells at initial stages of screening. For peanut, hygromycin B is the most appropriate compound for the selection of transformed cells whereas kanamycin was also reported to be an effective selection agent to select stably transformed callus tissue obtained from immature leaflets of peanut (Clemente et al. 1992). The herbicide Basta® (active ingredient phosphinothricin) has also been used to select transgenic peanut tissue (Brar et al. 1994).

Screenable (reporter) genes have also been developed from bacterial genes, which encode proteins that are used for easy detection in a sensitive, specific, quantitative, reproducible and rapid manner, to measure transcriptional activity and are used to investigate promoters and enhancers of gene expression and their interactions. Some of the reporter genes reported include chloramphenicol acetyltransferase (CAT) (Herrera-Estrella et al. 1983), a bacterial enzyme that transfers radioactive acetyl groups to chloramphenicol; Luciferase (LUC/LUX) (Olsson et al. 1988), a firefly enzyme that oxidizes luciferin and emits photons; Green fluorescent protein (GFP) (Reichel et al. 1996), an autofluorescent jellyfish protein; β-galactosidase (GAL), a bacterial enzyme that hydrolyzes colorless galactosides to yield colored products; β-glucuronidase (GUS) (Beason 2003) (an enzyme that hydrolyzes colorless glucuronides to yield insoluble colored products) and nopaline synthase, and octopine synthase (Herrera-Estrella et al. 1988). ß-glucuronidase or GUS (Jefferson 1987) is the most commonly used reporter gene in plant genetic transformation studies including peanut. Assays for screenable markers can be destructive or nondestructive, in terms of the need to sacrifice the test material. GFP in peanut was reported as a nondestructive gene which requires no exogenous substrate to fluoresce by Joshi et al. (2005).

Identifying the small proportion of transformed cells in a large experimental cell population, using only screenable markers is tedious and time consuming. Hence, screenable markers are usually coupled with selectable markers in transformation systems as in almost all commercialized transgenic crops (Liang et al. 2010).

9.6 Future Roadmap for Transgenic Peanut

Genes for transformation can be broadly divided into those that will be used to overcome agronomic limitations (high yield potential, resistance to biotic and abiotic stresses) and ones that could be used to enhance value-added traits (Schnall and Weissinger 1995). Although major emphasis is currently being placed on improving the primary constraints, the manipulation of value-added traits, such as flavor and nutrition will be of much concern for peanut improvement using transgenic technology. Transgenic technology could conceivably be used in peanut for the introduction of disease and pest resistance as well as value-added traits such as improved vitamin, protein and oil quality, enhancing the crop product value, quality and safety. The genus *Arachis*, which itself is a repository for most of the valuable pest and disease resistance genes, could be used to transform cultivated peanut varieties (Bhatnagar-Mathur et al. 2008). Current efforts include incorporating immunity or very high resistance to several viral and fungal diseases through transformation of peanut cultivars that have very high demand for which no adapted resistant peanut genotypes are available. Improved crop protection through the transfer and expression of disease resistance genes will decrease or eliminate the usage of pesticides, which are costly to the grower and may be harmful to the environment.

9.6.1 Abiotic Stress Tolerance

Drought is the major cause for low and erratic pod yield in peanut that contributes to over 6.7 million t loss in annual world peanut production (Subbarao et al. 1995), resulting in estimated monetary losses of over US$ 520 million annually (Sharma and Lavanya 2002). Yield losses in peanut due to water deficits vary depending on timing, intensity and duration of the deficit, coupled with other location-specific environmental stress factors such as high irradiance and temperature (Nigam et al. 2001). Due to the scarcity of available water in semi-arid tropics regions, drought management strategies, whether agronomic or genetic, therefore need to focus on maximizing extraction of available soil moisture and the efficiency of its use in crop establishment, growth, biomass and grain yield (Serraj et al. 2005).

Many genes that display altered expression patterns in response to environmental stresses have been identified over the last 10 years (Bray 2004; Shinozaki and Yamaguchi-Shinozaki 2007) and the functions of some of these genes have been studied in detail (Vinocur and Altman 2005; Lemaux 2008, 2009; Mittler and Blumwald 2010). Several genes that confer drought tolerance have been tested in the field for many years (Yang et al. 2010) among which a few are waiting for the approval of commercial release at US federal regulatory agencies (Castiglioni et al. 2008; Yang et al. 2010).

Transgenic research using transcription factors has been the most widely used technology in developing drought-tolerant varieties (Dubouzet et al. 2003; Pellegrineschi et al. 2004; Oh et al. 2005; Behnam et al. 2006; Xiao et al. 2006; Wang et al. 2008; Morran et al. 2011). At ICRISAT, efforts for enhancing drought tolerance in peanut through genetic engineering was initiated as early as 2003 through *Agrobacterium*-mediated genetic transformation of drought sensitive cultivar of peanut, JL 24, using the transcription factor *At*DREB1A driven by constitutive CaMV35S promoter as well as a drought-responsive promoter rd29A, which resulted into ~18 35S:DREB1A and 50 rd29A: DREB1A T_0 transformants. Fourteen transgenic events showing high levels of stress tolerance were screened under contained greenhouse (Bhatnagar-Mathur et al. 2004, 2006) and field conditions (Bhatnagar-Mathur et al. 2013). Substantial yield improvement of at least 17% was observed under drought-stress conditions in a field trial across a wide range of vapor pressure deficits, where one of these transgenic events showed 40% higher transpiration efficiency than the control plants under water-limiting conditions (Bhatnagar-Mathur et al. 2007, 2009, 2013).

Another study revealed that transgenic plants having *AtNHX1* gene are more resistant to high concentration of salt and water deprivation than the wild type plants in which salt and proline level in the leaves of the transgenic plants were also much higher than that of wild type plants

(Asif et al. 2011). Similarly, regulated expression of isopentenyl transferase gene (*IPT*) in peanut significantly improved drought tolerance under both laboratory and field conditions (Qin et al. 2011).

9.6.2 Resistance to Biotic Stresses

Diseases attack by different pathogens which include primarily fungi, bacteria, viruses, mycoplasma, nematodes, insect pests and parasitic flowering plants are major constraints to peanut production throughout the world causing majority of economic losses of yield up to 40 to 60%. Although, many diseases infect the crop, only a few cause significant reduction in yields. Comparatively low annual yields have been reported in developing countries (~825 kg/ha) to developed countries (2,650 kg/ha). The major biotic stresses for peanut include the foliar fungal diseases, leaf spot (early and late) and rust. Seed and soil-borne diseases like collar rot, stem rot and dry root rot have also been identified as important. Among viral diseases, bud necrosis (BND), peanut mottle (PMV) and peanut clump (PCV) are important. With regard to insect pests, a wide range of pests like leaf miner, tobacco caterpillar, white grub, jassids, thrips, aphids, red hairy caterpillar and termite are known to cause serious damage to peanut crop (Ghewande et al. 1987; Basu 1995).

However, crop improvement by conventional breeding lacks to meet the demands of increasing population, especially in seed quality improvement and developing virus and insect-resistant varieties. Therefore, in peanut the Expressed Sequenced Tags (EST) would be a quick and economical approach to identify important peanut genes involved in defense response against fungal infections and also provide data on gene expression and regulation (Houde et al. 2006; Nelson and Shoemaker 2006). Utilizing genomic and proteomic tools, genes and proteins associated with *A. parasiticus* and drought stress were identified (Luo et al. 2005; Guo et al. 2006, 2008). Identified genes could be used for enhanced fungal disease resistance in peanut through marker-assisted selection in breeding or by direct up or down regulation of the target gene using genetic engineering. Identification of novel promoter and enhancer elements will also be critical to achieving efficacious expression of antifungal/anti-mycotoxin genes. The protocol for genetic modification is now standardized and available for routine applications (Sharma et al. 2000; Bhatnagar-Mathur and Sharma 2006). Hence the major focus lies on developing transgenic peanut varieties for resistance to insect pests/fungal pathogens/important viruses.

9.6.2.1 Fungal Diseases

Poor realization of potential yields has been mainly attributed to diseases in peanut (Ghuge et al. 1981; Chohan 1974). Fungal diseases in peanut are the most significant limiting factor causing more than 50% yield losses throughout the world. Among the foliar fungal diseases Early Leaf Spot (ELS) caused by *Cercospora arachidicola* S. Hori (*Mycosphaerella arachidis* Deighton), Late Leaf Spot (LLS) caused by *Phaeoisariopsis personata* Berk. & M.A. Curtis (*M. berkeleyi*), rust (*Puccinia arachidis*), crown rot (*Aspergillus niger* Teigh.), collar rot caused by *Aspergillus* spp., root rot caused by *Macrophomina phaseolina*, stem rot caused by *Sclerotium rolfsii* and Yellow mold (*Aspergillus flavus* and *A. parasiticus*) causing aflatoxin contamination are the major fungal diseases affecting peanut crop (Subrahmanyam et al. 1985; McDonald et al. 1985) (Table 9-3). Infection by these fungal

Table 9-3 Genetic Transformation of peanut against major fungal diseases/pathogens.

Disease/pathogen	Gene	Source	Reference
Late leaf spot by *Phaeoisariopsis personatum*	*Chitinase* *Chitinase* *Glucanase*	Tobacco Rice *Alfa alfa*	Rohini and Rao 2001 Chenault et al. 2005
Early Leaf spot by *Cercospora arachidicola*	*Glucanase* *Chitinase* *Chitinase* *Chitinase*	Tobacco Bacteria Rice Rice	Sundaresha et al. 2010 Iqbal et al. 2011 Iqbal et al. 2012 ICRISAT unpublished
A. flavus	*Glucanase* *mod1,* *D5C,* *anionic peroxidase* *synthetic peptide D4E1*	Tobacco Maize Tomato	Sundaresha et al.2010 Weissinger et al. 2003 Weissinger et al. 1999 Ozias-Akins et al. 2000
Cercospora arachidicola Hori. and *Phaeoisariopsis personata*	*SniOLP* *Rs-AFP2* *defensin*	*Solanum nigrum* Radish (*Raphanus sativus*) mustard	Vasavirama and Kirti 2010 Anuradha et al. 2008
Sclerotinia blight	*oxalate oxidase gene* *Chitinase* *Chitinase* *Glucanase*	barley Tobacco Rice *Alfa alfa*	Livingstone et al. 2005 Rohini and Rao 2001 Chenault et al. 2005
A. flavus and aflatoxin biosynthesis	*Loxl* *Nonheme chloroperoxidase gene(cpo)* *nonheme chloroperoxidase gene* *PnLOX3*	Soybean *Pseudomonas pyrrocinia* bacteria Peanut	Ozias-Akins et al. 2000 Niu et al. 2009 Ozias-Akins et al. 2003 ICRISAT Unpublished

pathogens results in severe yield losses and generates poor quality seeds (Pretorius 2005). The use of disease resistant peanut cultivars is the only means of controlling fungal diseases in peanut. Genetic enhancement in peanut through conventional breeding and chemical control has yielded only limited success (Nigam et al. 2012) and the narrow genetic base of the cultivated peanut *Arachis hypogaea* L. hampers the development of improved varieties through conventional breeding leaving with the development of transgenics as the only option.

9.6.2.1.1 Leaf spots: The annual economic losses caused by LLS and rust account for over US$ 599 m and US$ 467 m, respectively (FAO 2004) by causing yield loss of 50–70% (Gibbons 1980; Subrahmanyam et al. 1980a,b, 1984). These diseases damage the plant by reducing the green leaf area available for photosynthesis and by stimulating leaflet abscission leading to extensive defoliation (McDonald et al. 1985) which results in lower seed quality, reduced seed size and oil content besides affecting the haulm production and quality.

9.6.2.1.1.1 Early Leaf Spot: Early Leaf Spot, caused by *Cercospora arachidicola* was first reported from Japan in 1919 (Hemingway 1955). Interestingly, transgenic approaches using bacterial and rice chitinase genes for resistance to early leaf spot in peanut showed fairly good positive correlation between chitinase activity and fungal pathogen resistance (Iqbal et al. 2011, 2012) in which two lines transformed with bacterial chitinase gene showed 56–62% suppression of disease over the nontransgenic controls. Similarly, use of tobacco chitinase gene (Sundaresha et al. 2010) for developing transgenic peanuts against *Cercospora arachidicola* resulted in 16 plants which performed well against infection in the *in vitro* leaf bioassay against *Cercospora*, seven transgenic plants that showed the lowest percent disease index (i.e., 0–25% of leaf area was covered by spots) and delay in the onset of disease were considered to be resistant and were selected for analysis for further generations (Sundaresha et al. 2010).

9.6.2.1.1.2 Late Leaf Spot: Late Leaf Spot, caused by *Phaeoisariopsis personatum* was first described in the USA in 1885 (Jenkins 1938; Kolte 1985). Transgenic peanuts expressing tobacco chitinase gene (Rohini and Rao 2001), rice chitinase and an alfalfa glucanase gene (Chenault et al. 2005) have been shown to possess enhanced resistance to the late leaf spot. More recently, transgenic peanut plants carrying mustard defensin gene showed variable increased disease resistance to *Cercospora arachidicola* and *Phaeoisariopsis personata* in detached leaf assays and greenhouse evaluations using conidial suspensions (Anuradha et al. 2008). Similarly, over expression of SniOLP (osmotin like protein cloned from *Solanum nigrum*) and Rs-AFP2 (defensin gene from Radish (*Raphanus sativus*)) in a double construct resulted in enhanced resistance against *Cercospora arachidicola* and *Phaeoisariopsis personata* in transgenic peanut (Vasavirama and Kirti 2010). At ICRISAT efforts are carried out for developing peanut transgenics using rice chitinase gene which resulted at about >50% decrease in disease incidence (Prasad et al. 2012).

9.6.2.1.2 Rust: Rust, caused by *Puccinia arachidis* is another potential peanut disease of economic importance not only in India but also in Africa, Asia, Oceania and Australia (Hammons 1977; Mayee 1982, 1986, 1987a, 1989; Mayee et al. 1977). At ICRISAT efforts have been made to develop peanut transgenics using rice chitinase gene that resulted in over 50% decrease in disease incidence (Prasad et al. 2012).

9.6.2.1.3 Sclerotinia blight: Blight disease is caused by soil borne fungus *Sclerotinia minor* and *Sclerotinia sclerotiorum*. Transgenic peanut expressing a tobacco chitinase gene (Rohini and Rao 2001), rice chitinase and an alfalfa glucanase gene (Chenault et al. 2005) has been shown to possess enhanced resistance to Sclerotinia blight, respectively. Transgenic events developed using somatic embryos of the Okrun cultivar (Chenault et al. 2002, 2005) were tested over a 3 year period (2000–2002) under field conditions where 14 transgenic lines showed up to 43 to 100% reduction in disease incidence compared to their parent line Okrun showing increased resistance to Sclerotinia blight. Similarly, overexpression of barley oxalate oxidase gene in transgenic peanut developed from embryogenic cultures of Virginia peanut cultivars, showed enhanced resistance to oxalic acid producing fungi, *Sclerotinia minor* (Livingstone et al. 2005). Detached leaflet bioassays carried out under laboratory conditions indicated reduction in the lesion area ranging from 75 to 97% in these transformed plants when compared to their respective nontransformed control cultivars. These transgenic peanut lines identified with partial resistance to *Sclerotinia* blight might be useful in traditional breeding programs for fungal resistance.

9.6.2.1.4 Aflatoxin: Peanuts are susceptible to aflatoxin contaminations which are toxic, carcinogenic substances produced by fungi *Aspergillus flavus* and *Aspergillus parasiticus*. Since conventional breeding methods for controlling aflatoxin are only partially effective, novel biotechnological methods for enhancing host plant resistance to preharvest *A. flavus* invasion and aflatoxin contamination is considered to be the most cost-effective control measure. Besides, a complete knowledge of the resistance associated proteins/genes and their contribution to host plant resistance (comparative proteomics) is critical to harness their cumulative or complementary benefits in peanut for *A. flavus* infection and aflatoxin contamination.

Peanut produces stilbene phytoalexins in response to fungal infection. Organ-specific expression of multiple copies of a gene for stilbene synthesis (Stilbene synthase) has proven to inhibit fungal growth and spore germination of *Aspergillus* species and aflatoxin contamination. Hydrolytic enzymes such as chitinases and glucanases, which degrade the fungal cell wall, also pose as attractive candidates for development of disease-resistant peanut plants (Eapen 2003). Similarly, glucanase gene from tobacco introduced into peanut (PR protein from heterologous source) showed enhanced disease resistance to *in vitro* seed colonization (IVSC) and no accumulating aflatoxin (detected by HPLC) (Sundaresha et al. 2010). Maize and peanut transgenic expressing synthetic version of maize ribosome inhibiting protein gene, *mod1*, showed enhanced resistance to *A. flavus* and reduced aflatoxin contamination (Weissinger et al. 2003).

The aflatoxin biosynthetic pathway *in vitro* has been shown to be suppressed by enzyme encoded by soybean *loxl* gene that catalyzes the formation of a specific lipoxygenase metabolite of linoleic acid, (13S)-hydroperoxyoctadecadienoic acid ((13S)-HPODE). Transgenic peanut expressing soybean *loxl* gene under the control of carrot embryo specific promoter (DC3) (Ozias-Akins et al. 2000) resulted in reduction in the aflatoxin content. Efforts are being carried out at ICRISAT for generation of peanut transgenics with the rice chitinase gene (Prasad et al. 2012) and peanut lipoxygenase gene (*PnLOX3*). Work is being carried out at ICRISAT in developing construct for use in RNAi approach to suppress 9-hydroperoxide fatty acid producing lipoxygenases since incorporation of plant antisense genes for the 9-hydroperoxide fatty acid producing lipoxygenases also reduces mycotoxin contamination. Other antifungal genes such as D5C (Weissinger et al. 1999), tomato anionic peroxidase (tap 1), and synthetic peptide D4E (Ozias-Akins et al. 2000) are transformed into peanut and evaluated for antifungal activity against *A. flavus*. However, pure D5C showed strong activity against *A. flavus in vitro,* due to phytotoxicity of D5C, transgenic peanut callus showed poor recovery of plants. Expression of *cry1A(c)* (Ozias-Akins et al. 2002) in transgenic peanut lines could also be an effective means of inhibiting *A. flavus* infection by reducing the damage into peanut pods by lesser cornstalk borer (LCB) *Elasmopalpus lignosellus,* since it has been clearly reported that aflatoxin contamination can increase with insect damage (Lynch and Wilson 1991). Similarly, Ozias-Akins et al. (2003) reported 60–70% reduction in *A. flavus* colony growth in transgenic peanut lines expressing the bacterial chloroperoxidase gene (Rajasekaran et al. 2000). Niu et al. 2009 reported antifungal activity in transgenic peanut by transforming with a non-heme chloroperoxidase gene from *Pseudomonas pyrrocinia.*

9.6.2.2 Viral Diseases

Viruses pose a great threat to peanut production throughout the world. Viruses such as the Indian Peanut Clump Virus (IPCV), Peanut Bud Necrosis Virus (PBNV), Groundnut Rosette Assistor Virus (GRAV), Peanut Mottle Virus (PMV), Peanut Stripe Virus (PStV), Tobacco Streak Virus (TSV), and Tomato Spotted Wilt Virus (TSWV) cause considerable damage to the crop. The concept of pathogen-derived resistance (Sanford and Johnston 1985) has stimulated research on obtaining virus resistance through genetic engineering. Since, the insertion of genetic material from the virus had been shown to confer resistance to infection by preventing virus replication and spread in several crop species. Genetic transformation has been used to develop peanut varieties with total resistance and not just tolerance to these viral diseases. The development of new viral control strategies depends on the molecular mechanisms underlying the roles of both dominant and recessive resistance genes (Ritzenthaler 2005). In general, protein-mediated resistance provides moderate protection against a broad range of related viruses while RNA-mediated resistance has been shown to offer high levels of protection only against closely related strains of a virus (Pang et al. 1993; Lomonossoff 1995; Baulcombe 1996; Dawson 1996). Recent research indicates that pathogen-derived resistance to viruses is mediated, in most cases, by RNA-based Post-Transcriptional Gene Silencing (PTGS) mechanism (Baulcombe 2004) resulting in the degradation of mRNA produced both by the transgene and the virus. RNAi technology (RNA silencing or cosuppression of homologous genes) provides a significant tool for development of virus resistant peanut genotypes (Wang et al. 2000; Colbere-Garapin et al. 2005). The development of genetically transformed peanut cultivars with resistance to viruses and other biotic constraints potentially have tremendous impact on crop productivity, especially in the resource-poor agricultural systems of the semi arid tropics.

9.6.2.2.1 Groundnut rosette disease: Groundnut rosette disease is also one of the major destructive viral disease in sub-Saharan Africa (SSA) resulting in devastating losses to peanut production in Africa. The disease is caused by a complex of three casual agents such as Groundnut Rosette Assistor Virus (GRAV), Groundnut Rosette Virus (GRV) and a satellite RNA (satRNA) and is transmitted by an Aphid, *Aphis craccivora* (Naidu et al. 1998).

At ICRISAT Pathogen-Derived Resistance (PDR) for Groundnut Rosette Disease (GRD) by using GRAV*cp* gene has been exploited to induce host plant resistance to GRD for controlling GRD. Peanut transgenics for resistance to GRAV are being produced in ICRISAT (KK Sharma, unpubl. results) and the molecular characterized transgenic events have been transferred to South

Africa for phenotyping under greenhouse conditions. Introduction of GRAV or GRV genomic sequences or genes, or SatRNA–derived sequences that down regulate GRV replication (Taliansky et al. 1996) into suitable peanut cultivars is an ideal RNA-mediated/gene silencing approach.

9.6.2.2.2 Peanut Stem Necrosis Disease: PSND caused by Tobacco Streak Virus (TSV) was reported in India in 2000 (Reddy et al. 2002). TSV was reported as a frequent occurrence on peanuts in Brazil (Costa and Carvalho 1961), but it was first noticed on peanut in 1999 in South Africa (Cook et al. 1999).

At ICRISAT, work is being carried out on engineering TSV resistance through *A. tumefaciens*-mediated transformation of popular peanut variety JL 24 (Spanish type) with TSV coat protein gene (*TSV cp* gene), and recovery of transgenic plants that block systemic movement of TSV spread. The resistant transgenic events identified under greenhouse conditions will be evaluated under restricted field conditions in the TSV hot-spots in the near future. Similarly, transgenic peanut lines containing sense and antisense coat protein gene of TSV transformed through *Agrobacterium*-mediated transformation of de-embryonated cotyledons of cultivar JL 24 are under evaluation for their reaction to TSV (Bag et al. 2007).

9.6.2.2.3 Peanut Bud Necrosis: Peanut Bud Necrosis Disease (PBND) is caused by PBNV—transmitted by *Thrips palmi*. Strategies to combat peanut bud necrosis disease (PBND) include development of transgenic peanut plants expressing PBNV nucleocapsid gene at ICRISAT, which showed a modest tolerance to PBND (Chander Rao et al. 2006). Three selected transgenic peanut events of T_1 and T_2 generation showed a 40 to 67% decrease in disease incidence under greenhouse virus challenging experiments. However, under field conditions in a contained on-station trial only one event showed less than 25% disease incidence. The expression of symptoms in some plants was delayed by 40–60 days and 14–21 days under greenhouse conditions and contained on-station trial respectively as compared to the control plants. Because of the unexpected lower frequency of virus resistant events throughout the challenging experiments, an alternate strategy based on RNA interference (antisense and hairpin-RNA) mediated gene silencing is being used as a potential tool to address a complex constraint like PBNV. Currently, RNAi-mediated resistance approach to counter the effect of NSs gene in the PBNV genome is being pursued.

9.6.2.2.4 Tomato spotted wilt virus: Tomato spotted wilt virus (TSWV), first reported in Brazil (Costa 1941) is transmitted by thrips *Scirthothrips dorsalis* Hood (Mali and Patil 1979) and *Frankliniella schultzei* (Trybom) (Ghanekar et al. 1979).

Due to lack of availability of considerable levels of resistance in germplasm, development of transgenic plants through genetic engineering is the only effective approach for protection against TSWV which is carried over by both RNA and protein-mediated control (Pang et al. 1993). These approaches include using both sense and antisense TSWV nucleocapsid protein gene (*N* gene) expression. Nucleocapsid protein gene (*N* gene) was introduced into a runner and a Valencia type variety (Brar et al. 1994; Chenault and Payton 2003) whereas the *N* gene, was inserted into New Mexico Valencia A peanut, by Li et al. (1997). The field ratings from the study of Yang et al. (1998) indicated that there was a potential to combine nucleoprotein-mediated resistance in transgenic peanut with host-plant resistance that already had been identified in the peanut germplasm. Variety AT 120 transgenics with antisense nucleocapsid gene (Magbanua et al. 2000) and Marc 1 transgenics transformed with coat protein gene of TSWV (Ozias-Akins et al. 2002) showed lower disease incidence than respective nontransformed cultivar or than in moderately resistant cultivar Georgia Green. Transgenic progeny of Marc 1 peanut cultivar also showed lower incidence of spotted wilt in comparison to the nontransgenic controls in field evaluations and under controlled environmental conditions in the USA over years and locations (Yang et al. 2004), indicating its potential use in conventional breeding programs. Use of stable pathogen-derived resistance based on homology dependent RNA silencing for durable TSWV resistance was suggested by Bucher et al. (2003).

9.6.2.2.5 Peanut stripe virus (PStV): PStV is transmitted by seed and also by aphids (*Aphis craccivora, A. gossypii* and *Myzus persicae*). Transgenic plants of peanut varieties with high levels of RNA-mediated resistance to peanut stripe potyvirus (PStV) were obtained following cobombardment of embryogenic callus derived from mature seeds of the commercial cultivars, Gajah and NC 7, which were transformed with one of the two forms of PStV coat protein (*cp*) gene (an untranslatable, full-length sequence (*cp* 2) or a translatable gene encoding a *cp* with an N-terminal truncation (*cp* 4)) (Higgins et al. 2004). Resistance to PStV was stably inherited over at least five generations in these transgenic plants of Gajah variety (Dietzgen et al. 2004). From the study of Hapsoro et al. 2005, 2007, three different kinds of response to PStV infection were identified-resistant, recovery and susceptible, the transgenic peanut lines cv. Gajah proved stable up to seven generations of selfing and some pure lines were identified. Franklin et al. (1993) reported transformed callus expressing the PStV coat protein gene through *Agrobacterium*-mediated genetic transformation.

9.6.2.2.6 Peanut Clump Virus (PCV): The disease is soil borne and is caused by peanut clump virus (PCV) that is transmitted by a fungus, *Polymyxa* sp. living in the soil. ICRISAT has developed the first-ever transgenic peanut, resistant to the dreaded Indian Peanut Clump Virus (IPCV) by the introduction of *coat protein (cp)* gene and *replicase (rep)* genes of the target virus IPCV by using *Agrobacterium*-mediated transformation (Sharma and Anjaiah 2000). Field evaluations were carried out twice against IPCV under controlled conditions during the rainy season of 2002–2004 in an on-station sick plot at ICRISAT, Patancheru, India with 10 transgenic lines carrying single gene inserts (five each with *IPCVcp* and *IPCVrep* genes) of which four transgenic events (three with *IPCVcp* and one with *IPCVrep*) showed complete resistance to IPCV.

9.6.2.2.7 Bacterial wilt: It is a soil-borne disease caused by *Ralstonia solanacearum*. A novel approach of introducing microbial toxins (phytotoxins) such as tabtotoxin acetyl transferase and glucose oxidase into the plant has emerged as an efficient way to develop resistance in a wide range of host species (Eapen 2003). This approach can be conveniently used to impart resistance against bacterial wilt of peanut caused by *Burkholderia solanacearum*, formerly known as *Pseudomonas solanacearum*.

9.6.2.3 Insect Resistance

Among the insect pests *Spodoptera litura*, *Aproaerema modicella*, *Amsacta* spp., *Heliothis* spp., aphids, jassids, thrips and termites cause major yield losses. Though, a moderate level of resistance against specific pests was observed in wild relatives of peanut cultivars (Stalker and Moss 1987), but is often accompanied by undesirable agronomic features (low shelling and undesirable pod and kernel traits), interspecific reproduction barriers and linkage drag which impedes development of resistant cultivars using traditional breeding approaches. Hence the development of transgenic peanut for resistance to insects is gaining importance. The first transgenic peanut expressing *cry1EC* gene resistance to *S. litura* using de-embryonated cotyledon explants were developed by Tiwari et al. (2008). Leaf feeding bioassay was carried out twice under laboratory conditions on highly expressing transgenic lines, which showed 100% death of larvae at the 2nd instar stage of *S. litura*. Since, besides spodoptera, *Helicoverpa armigera* (Hubner) occasionally occurs on the peanut crop causing defoliation to a limited extent resulting in major crop loss, development of the peanut transgenics cv. TMV-2 expressing a chimeric *Bt* gene, *cry1X*, was reported (Entoori et al. 2008). *In vitro* detached leaf bioassays under laboratory conditions led to more than 50% mortality in 27 transgenic plants, showing not more than 10% damage against *H. armigera* and *S. litura*. Among the insect-pests, Lesser Cornstalk Borer (LCB), *Elasmopalpus lignosellus* (Zeller), is another major pest of peanut in the southern United States causing severe reduction in crop quality. Peanut transgenics against LCB using *cry1Ac* gene (Singsit et al. 1997) showed complete larval mortality to a 66% reduction in larval weight in insect feeding bioassay of transformed plants indicating various levels of resistance.

9.6.3 Biofortification and enhancing quality traits

Besides lysine, threonine and isoleucine, peanut is deficient in the essential amino acid methionine. The dietary and nutritional value of peanut can be improved by either raising the level of sulfur-containing amino acids of storage proteins or by changing the proportion of methionine-rich proteins already present in the peanut seed. Genetic transformation is an effective and an alternative approach for developing methionine-rich peanuts.

Efforts have been made to identify genes that play an important role in controlling the crucial and important regulatory biochemical steps whose constituents play a major role in determining the quality of peanuts. Attempts have been made to produce transgenic peanut plants with improved protein quality by transferring genes like the Brazil nut 2S albumin gene (Lacorte et al. 1997). Malnutrition due to vitamin A, zinc (Zn) and iron (Fe) deficiencies is a significant public health issue in most of the developing and undeveloped world involving one-third of the world's population (~1.02 billion people) (FAO 2009). Hence providing biofortified staple food with essential amino acids, vitamins and trace elements without imposing any additional cost to the consumer is an alternative and best solution to overcome the problem of vitamin and trace element deficiency for the poor in the population. The success in peanut transformation technology enabled researchers to address more complex and important aspects of biofortification in peanut for enhanced levels of beta-carotene (provitamin A). Work has been initiated at ICRISAT to develop genetically engineered groundnut having enhanced levels of ß-carotene (pro-vitamin A) to combat vitamin A deficiency. Owing to the high oil content >50% in peanut, targeting β-carotene to the oil bodies for enhanced bioavailability was thought to be critical. This has been achieved by using oleosin promoters for driving the carotenoid biosynthetic genes for targeting these to the oil bodies (Bhatnagar et al. 2010; Bhatnagar-Panwar et al. 2013), as has been previously reported in *Arabidopsis* and *Brassica napus* (Siloto et al. 2006; Hu et al. 2009). Over 200 primary transgenic events of groundnut have been developed by introducing the phytoene synthase gene (*psy1*) from maize that resulted in increased ß-carotone levels, in seed oil bodies to an extent of 20–25-folds when compared to the untransformed controls.

Table 9-4 Genes proposed for genetic transformation of peanut for nutritional enhancement.

Reason for modifications	**Gene/activity engineered**	**Modifications required**	**Success status of transgenic research**	**Reference**
Reduction in the risk for artherosclerosis	Antisense of stearoyl- CoA-β-ketoeicosanoyl CoA syhthetase	Reduction in long chain saturated fatty acids	Transgenic *Brassica* by antisense expression of stearoyl-ACP-desaturase gene	Knutzon et al. 1992
Reduction in aflatoxin load	Stilbene synthase	Increase in stilbenes	Transgenic tobacco	Hain et al. 1990
Improvement in nutritive value of protein	Gene encoding Brazil nut methionine-rich protein	Increase in polypeptides rich in S-containing amino acids	Transgenic tobacco	Altenbach et al. 1989
Reduction in flatus properties	Galactinol:sucrose-6-galactosyl transferase	Reduction in raffinose and stachyose	Not yet attempted	-
Prolongation of shelf-life	Stearoyl desaturase	Increase in oleic acid	Transgenic tobacco with yeast and rat genes	Polashock 1992; Garyburn 1992
Improve protein quality	Brazil nut 2S albumin gene	-	Transgenic peanut	Lacorte et al. 1997
Enhancement in carotenoid content	Maize *psy* gene, maize *lycopene cyclase* gene, bacterial *crtB*	Increase in β-carotene content	Transgenic peanut	Sharma Unpublished

9.6.4 Improvement in Quality of Oil

For peanut, oil content, oil quality and storage protein composition are major issues for quality improvement, and genes controlling these important agronomic traits have been the focus of peanut gene cloning. Currently efforts are carried over to increase stability and quality of peanut oil by hydrogenation to reduce the level of polyunsaturated fatty acids, which also has undesirable health and food quality consequences. Peanut's oils contain high levels of monounsaturated fatty acids that are prone to oxidation as compared to other oils with high levels of polyunsaturated fatty acids. Different genes for improving quality of oil have been proposed (Wang et al. 2011) that can be used for developing transgenic peanuts. For enhancing the shelf-life of peanut products, a higher oleic/linoleic (O/L) ratio is considered desirable. The introduction of the double bonds in the plant fatty acids occurs by the action of enzyme delta-12 fatty acid desaturase. Engineering a gene encoding for delta-12 fatty acid desaturase in peanut by antisense or RNAi strategies may help to reduce activity of this enzyme and hence produce oil with higher O/L ratio. Expression of additional copies of the gene for this enzyme may enhance the content of oleic acid and hence the O/L ratio. Several other reported genes which can be used for developing peanut transgenics for improving nutritional quality are listed in Table 9-4.

References

Altenbach SB, Pearson KW, Meeker G, Staraci LG, Sun SSM (1989) Enhancement of methionine content of seed proteins by the expression of a chimeric gene encoding a methionine-rich protein in transgenic plants. Plant Mol Biol 13: 513–522.

Anuradha ST, Jami SK, Datla RS, Kirti PB (2006) Genetic transformation of peanut (*Arachis hypogaea* L.) using cotyledonary node as explant and a promoterless *gus::npt*II fusion vector. J Biosci 31: 1–12.

Anuradha TS, Divya K, Jami SK, Kirti PB (2008) Transgenic tobacco and peanut plants expressing a mustard defensin show resistance to fungal pathogens. Plant Cell Rep 27: 1777–1786.

Asif MA, Zafar Y, Iqbal J, Iqbal MM, Rashid U, Ali GM, Arif A, Nazir F (2011) Enhanced expression of *AtNHX1*, in transgenic groundnut (*Arachis hypogaea* L.) improves salt and drought tolerence. Mol Biotechnol 49: 250–256.

Atreya CD, Rao JP, Subrahmanyam NC (1984) *In vitro* regeneration of peanut (*Arachis hypogaea* L.) plantlets from embryo axes and cotyledon segments. Plant Sci Lett 34: 379–383.

Bag S, Singh RS, Jain RK (2007) *Agrobacterium*-mediated transformation of groundnut with coat protein gene of Tobacco streak virus. Indian J Virol 18: 65–69.

Bajaj YPS (1982) Regeneration of plants from pollen embryos of *Arachis, Brassica* and *Triticum* species cryo-preserved for one year. Curr Sci 52: 484–486.

Bajaj YPS, Gosal SS (1983) Somatic hybridization and embryo culture studies on *Arachis hypogaea* and *Arachis villosa*. Am J Bot 70: 83–84.

Bajaj YPS, Gosal SS (1988) Isolation and fusion of protoplasts of *Arachis hypogaea* and *Arachis villosa*. Int *Arachis* Newslett 3: 13–14.

Bajaj YPS, Kumar P, Labana KS, Singh MM (1981) Regeneration of plants from seedling explants and callus cultures of *Arachis hypogaea* L. Indian J Exp Biol 19: 1026–1029.

Baker CM, Wetzstein HY (1992) Somatic embryogenesis and plant regeneration from leaflets of peanut (*Arachis hypogaea* L.). Plant Cell Rep 11: 71–75.

Baker CM, Wetzstein HY (1995) Repetitive somatic embryogenesis in peanut cotyledon cultures by continual exposure to 2, 4-D. Plant Cell Tiss Org Cult 40: 249–54.

Baker CM, Durham RE, Burns JA, Parrott WA, Wetzstein HY (1995) High frequency somatic embryogenesis in peanut (*Arachis hypogaea* L.) using mature, dry seed. Plant Cell Rep 15: 38–42.

Banerjee S, Bandyopadhyay S, Ghosh PD (1988) Cotyledonary node culture and multiple shoot formation in peanut: Evidences for somatic embryogenesis. Curr Sci 57: 252–257.

Basu MS (1995) Groundnut production technology for rainfed Kharif Workshop on Crop Production Management in Oilseeds and Pulses Based Cropping Systems, February 22–23, 1995, New Delhi Agricultural Science 43: 117–120.

Baulcombe D (1996) Mechanisms of pathogen-derived resistance to viruses in transgenic plants. Plant Cell 8: 1833–1844.

Baulcombe D (2004) RNA silencing in plants. Nature 431: 356–363.

Beason B (2003) GUS staining of transgenic *Arabidopsis* (updated March 2010). Rice University. http://www.owlnet.rice.edu/~bios311/bios311/bios413/413day5.html (accessed Dec 2, 2010).

Behnam B, Kikuchi A, Celebi-Toprak F, Yamanaka S, Kasuga M, Yamaguchi-Shinozaki K, Watanabe KN (2006) The *Arabidopsis DREB1A* gene driven by the stress-inducible *rd29A* promoter increases salt- stress tolerance in proportion to its copy number in tetrasomic tetraploid potato (*Solanum tuberosum*). Plant Biotechnol 23: 169–177.

Bertioli DJ, Seijo G, Freitas FO, Valls JFM, Leal-Bertioli SCM, Moretzsohn MC (2011) An overview of peanut and its wild relatives. Plant Genet Resour Characteriz Utiliz 9: 134–149.

Bhatia CR, Murty GSS, Mathews VH (1985) Regeneration from de-embryonated peanut (*Arachis hypogaea*) plants from elite cultivars using ACCELL technology. Plant J 5: 745–753.

Bhatnagar M, Kalyani P, Bhatnagar-Mathur P, Narasu ML, Waliyar F, Sharma KK (2010) An efficient method for the production of marker-free transgenic plants of peanut (*Arachis hypogaea* L.). Plant Cell Rep 29: 495–502.

Bhatnagar-Mathur P, Devi MJ, Reddy DS, Lavanya M, Vadez V, Serraj R, Yamaguchi-Shinozaki K, Sharma KK (2007) Stress- inducible expression of *Arabidopsis thaliana DREB1A* in transgenic peanut (*Arachis hypogaea* L.) increases transpiration efficiency under water-limiting conditions. Plant Cell 26: 2071–2082.

Bhatnagar-Mathur P, Devi MJ, Serraj R, Yamaguchi-Shinozaki K, Vadez V, Sharma KK (2004) Evaluation of transgenic groundnut lines under water limited conditions. Int *Arachis* Newslett 24: 33–34.

Bhatnagar-Mathur P, Devi MJ, Reddy DS, Vadez V, Yamaguchi-Shinozaki K, Sharma KK (2006) Over expression of *Arabidopsis thaliana* DREB1A in transgenic peanut (*Arachis hypogaea* L.) for improving tolerance to drought stress (poster presentation). In: Arthur M Sackler Colloquia (ed) From Functional Genomics of Model Organisms to Crop Plants for Global Health, April 3–5, 2006. National Academy of Sciences, Washington DC, USA.

Bhatnagar-Mathur P, Anjaiah V, Kirti PB, Sharma KK (2008) *Agrobacterium*-mediated genetic transformation of peanut. In: Kirti PB (ed) Handbook of New Technologies for Genetic Improvement of Legumes. CRC Press, Taylor & Francis Group, Boca Raton, New York, USA; London, UK, pp 227–251.

Bhatnagar-Mathur P, Devi MJ, Vadez V, Sharma KK (2009) Differential antioxidative responses in transgenic peanut bear no relationship to their superior transpiration efficiency under drought stress. J Plant Physiol 166: 1207–1217.

Bhatnagar-Mathur P, Rao JS, Vadez V, Reddy DS, Rathore A, Yamaguchi-Shinozaki K, Sharma KK (2013) Transgenic peanut overexpressing the DREB1A transcription factor have higher yields under drought stress. Mol Breed (in press). DOI: 10.1007/s11032-013-9952-7.

Bhatnagar-Panwar M, Bhatnagar-Mathur P, Bhaaskarla VV, Dumbala SR, Sharma KK (2013) Rapid, accurate and routine HPLC method for large-scale screening of pro-vitamin A carotenoids in oilseeds. J Plant Biochem Biotechnol (in press). DOI 10.1007/s13562-013-0239-1.

Birch RG (1997) Plant transformation: problems and strategies for practical application. Annu Rev Plant Physiol Plant Mol Biol 48: 297–326.

Brar GS, Cohen BA, Vick CL, Johnson GW (1994) Recovery of transgenic peanut (*Arachis hypogaea* L.) plants from elite cultivars utilizing ACCELL technology. Plant J 5: 745–753.

Bray EA (2004) Genes commonly regulated by water-deficit stress in *Arabidopsis thaliana*. J Exp Bot 55: 2331–2341.

Bucher E, Sijen T, De Haan P, Goldbach R, Prins M (2003) Negative-strand tospoviruses and tenuiviruses carry a gene for a suppressor of gene silencing at analogous genomic positions. J Virol 77: 1329–1336.

Cannon SB, May GD, Jackson SA (2009) Update on comparative genomics of legumes. Plant Physiol 151: 970–977.

Cassidy B, Ponsamuel J (1996) Update on groundnut transformation and evidence from mechanism of induced pathogen-derived resistance to peanut stripe virus in *Nicotiana benthamiana*. In: Groundnut Virus Diseases in the Asia-Pacific Region: Summary and Recommendations of the 4th Meeting of the International Working Group, 12–14, 1995, Khon Kaen University, Thailand: ICRISAT, Patancheru, Andhra Pradesh, India, p 14.

Castiglioni P, Warner D, Bensen RJ, Anstrom DC, Harrison J, Stoecker M (2008) Bacterial RNA chaperones confer abiotic stress tolerance in plants and improved grain yield in maize under water-limited conditions. Plant Physiol 147: 446–455.

Chander Rao S, Kumar PL, Reddy AS, Swamy Krishna T, Waliyar F, Nigam SN, Laxminarasu M, Sharma KK (2006) Evaluation of transgenic peanut plants against peanut bud necrosis disease (PBND) under greenhouse and field conditions. Indian J Virol 17: 135.

Chenault KD, Burns JA, Melouk HA, Payton ME (2002) Hydrolase activity in transgenic peanut. Peanut Sci 29: 89–95.

Chenault KD, Payton ME (2003) Genetic transformation of a runner-type peanut with the nucleocapsid gene of tomato spotted wilt virus. Peanut Sci 30: 112–115.

Chenault KD, Melouk HA, Payton ME (2005) Field reaction to *Sclerotinia* blight among transgenic peanut lines containing antifungal genes. Crop Sci 45: 511–515.

Cheng M, his DCH, Phillips GC (1992) *In vitro* regeneration of valencia-type peanut (*Arachis hypogaea* L.) from cultured petiolules, epicotyl sections and other seedling explants. Peanut Sci 19: 82–87.

Cheng M, Jarret RL, Li Z, Demski JW (1997) Expression and inheritance of foreign genes in transgenic peanut plants generated by *Agrobacterium* mediated transformation. Plant Cell Rep 16: 541–544.

Chengalrayan K, Sathaye SS, Hazra S (1994) Somatic embryogenesis from mature embryo-derived leaflets of peanut (*Arachis hypogaea* L.). Plant Cell Rep 13: 578–581.

Chengalrayan K, Mhaske VB, Hazra S (1997) High-frequency conversion of abnormal peanut somatic embryos. Plant Cell Rep 16: 783–786.

Chohan JA (1974) Recent advances in diseases of groundnut in India. In: Raychaudhury SP, Verma JP (eds) Current Trends in Plant Pathology. Lucknow University, Lucknow, UP, India, pp 171–184.

Clemente TE, Robertson D, Isleib TG, Beute MK, Weissinger AK (1992) Evaluation of peanut (*Arachis hypogaea* L.) leaflets from mature zygotic embryos as recipient tissue for biolistic gene transfer. Transgenic Res 1: 275–284.

Colbere-Garapin F, Blondel B, Saulnier A, Pelletier I, Labadie K (2005) Silencing viruses by RNA interference. Microbes Infect 7: 767–775.

Cook G, Miranda HR, Roossinck MJ, Pietersen G (1999) Tobacco streak ilarvirus detected on groundnut in South Africa. Afr Plant Prot 5: 13–19.

Costa AS (1941) Una molestia de virus de amendoim (*Arachis hypogaea* L.) A mancha anular. Biologico 7: 249–251.

Costa AS, Carvalho AMB (1961) Studies on Brazilian tobacco streak. Phytopathol Z 42: 113–138.

Cucco MF, Jaume ADR (2000) Protocol for regeneration *in vitro* of *Arachis hypogaea* L. Elec J Biotechnol 3: 154–160.

Curtis IS, Power JB, Davey MR (1995) NPTII assays for measuring gene expression and enzyme activity in transgenic plants. Methods Mol Biol 49: 149–159.

Dale PJ, Irwin JA, Scheffler JA (1993) The experimental and commercial release of transgenic crop plants. Plant Breed 111: 1–22.

Darbani B, Eimanifar A, Stewart CNJ, Camargo WN (2007) Methods to produce marker-free transgenic plants. J Biotechnol 2: 83–90.

Dawson WO (1996) Gene silencing and virus resistance: a common mechanism. Trends Plant Sci 1: 107–108.

de Vetten N, Wolters AM, Raemakers K, van der Meer I, der Stege R, Heeres E, Heeres P, Visser R (2003) A transformation method for obtaining marker-free plants of a cross-pollinating and vegetatively propagated crop. Nat Biotechnol 21: 439–442.

Dietzgen RG, Mitter N, Higgins CM, Hall R, Teycheney PY, Cruickshank A, Hapsoro D, Sudarsono (2004) Harnessing RNA silencing to protect peanuts from stripe disease. In new directions for a diverse planet (poster presentation). In: Proc 4th Int Crop Sci Congr, 26 Sept–1 Oct. Brisbane, Australia.

Dong JD, Bi YP, Xia LS, Sun SM, Song ZH, Guo BT, Jiang XC, Shao QQ (1990) Teratoma induction and nopaline synthase gene transfer in peanut. Acta Genet Sin 17: 13–16.

Dubouzet JG, Sakuma Y, Ito Y, Kasuga M, Dubouzet EG, Miura S, Seki M, Shinozaki K, Yamaguchi-Shinozaki K (2003) *OsDREB* genes in rice, *Oryza sativa* L., encode transcription activators that function in drought, high salt and cold responsive gene expression. Plant J 33: 751–763.

Dwivedi SI, Crouch JH, Nigam SN, Ferguson ME (2003) Molecular breeding of groundnut for enhanced productivity and food security in the semi-arid tropics: Opportunities and challenges. Adv Agron 80: 153–221.

Eapen S (2003) Regeneration and genetic transformation in peanut: current status and future prospects. In: Jaiswal PK, Singh RP (eds) Applied Genetics of Leguminosae Biotechnology. Kluwer Academic Publishers, Dordrecht, Netherlands, pp 165–186.

Eapen S, George L (1994) *Agrobacterium tumefaciens* mediated gene transfer in peanut (*Arachis hypogaea* L.). Plant Cell Rep 13: 582–586.

Economic Research Service (ERS) (2001) Production, Supply and Distribution (PS&D) database, USDA.

Egnin M, Mora A, Prakash CS (1998) Factors enhancing *Agrobacterium tumefaciens*-mediated gene transfer in peanut (*Arachis hypogaea* L.). *In vitro* Cell Dev Biol-Plant 34: 310–318.

Entoori K, Sreevathsa R, Manoj KA, Kumar PA, Kumar ARV, Madhusudhan B, Udayakumar M (2008) A chimeric *cry1X* gene imparts resistance to *Spodoptera litura* and *Helicoverpa armigera* in the transgenic groundnut. Eur Asian J Bio Sci 2: 53–65.

FAO (2004) The state of food and agricultural biotechnology, meeting the needs of the poor. Food and Agriculture Organization of the United Nations, Rome, Italy.

FAO (2009) State of Food Insecurity in the World. Food and Agricultural Organisation, Rome, Italy.

FAOSTAT (2004) http://faostat.fao.org.

Foreign Agriculture Service (FAS) (2000) Oilseeds Markets and Trade. Cotton, oilseeds, tobacco and seeds division. USDA.

Franklin CI, Shorrosh KM, Trieu AN, Cassidy BG, Nelson RS (1993) Stable transformation of peanut callus via *Agrobacterium*-mediated DNA transfer. Transgenic Res 2: 321–324.

Garyburn WS, Collins GB, Hilderbrand DF (1992) Fatty acid alteration by a δ-9 desaturase in transgenic tobacco tissue. Biotechnology 10: 675–678.

Ghanekar AM, Reddy DVR, Lizuka N, Amin PW, Gibbons RW (1979) Bud Necrosis of Groundnut (*Arachis hypogaea*) in India caused by tomato spotted wilt virus. Ann Appl Biol 93: 173–179.

Ghewande MP, Nandagopal V, Reddy PS (1987) Plant protection in Groundnut. Technical Bulletin no. 1: National Research Center for Groundnut, Junagarh, Gujarat, India, pp 35.

Ghuge SS, Mayee CD, Godbole GM (1981) Assessment of losses in groundnut due to rust and tikka leaf spots. Indian Phytopathol 34: 179–182.

Gibbons RW (1980) Groundnut improvement research technololgy for the semi-arid tropics. ICRISAT Proc Int Symp on Development and Transfer of Technology for Rainfed Agriculture and the SAT farmer, 28 Aug–Sept 1979. ICRISAT, Patancheru, Andhra Pradesh, India, pp 27–37.

Gill R, Saxena PK (1992) Direct somatic embryogenesis and regeneration of plants from seedling explants of peanut (*Arachis hypogaea* L.): Promotive role of TDZ. Can J Bot 70: 1186–192.

Guo BZ, Xu G, Cao YG, Holbrook CC, Lynch RE (2006) Identification and characterization of phospholipase D and its association with drought susceptibilities in peanut (*Arachis hypogaea*). Planta 223: 512–520.

Guo BZ, Chen X, Dang P, Scully BT, Liang X, Holbrook CC, Yu J, Culbreath AK (2008) Peanut gene expression profiling in developing seeds at different reproduction stages during *Aspergillus parasiticus* infection. BMC Dev Biol 8: 12.

Hain R, Bieseler B, Kindl H, Schroder G, Stocker R (1990) Expression of stilbene synthase gene in *Nicotiana tobacum* results in the synthesis of the phytoalexin resveratrol. Plant Mol Biol 15: 325–335.

Hammons RO (1977) Groundnut rust in the United States and the Caribbean. Proc Natl Acad Sci USA 23: 300–340.

Hapsoro D, Aswidinnoor H, Suseno R, Sudarsono J (2005) *Agrobacterium*-mediated transformation of peanuts (*Arachis hypogaea* L.) with PStV *cp* gene. J Trop Agri 10: 85–93.

Hapsoro D, Aswidinnoor H, Jumanto, Suseno R, Sudarsono J (2007) Resistance to peanut stripe virus (PStV) in transgenic peanuts (*Arachis hypogaea* L.) carrying PStV *cp* gene was stable up to seven generations of selfing. Biota 12: 83–91.

Hazra S, Sathaye SS, Mascarenhas AP (1989) Direct somatic embryogenesis in peanut (*Arachis hypogaea* L.). Nat Biotechnol 7: 949–951.

Hemmingsway JS (1955) The prevalence of two species of *Cercospora* on groundnuts. Trans Brit Mycol Soc 38: 243–246.

Herrera-Estrella L, Depicker A, Van Montagu M, Schell J (1983) Expression of chimeric genes transferred into plant cells using a Ti-plasmid-derived vector. Nature 303: 209–213.

Herrera-Estrella L, Teeri TH, Simpson J (1988) Use of reporter genes to study gene expression in plant cells. In: Gelvin SB, Schilperoort RA, Verma DPS (eds) Plant Molecular Biology Manual: B1, Kluwer Academic Publishers, Dordrecht, pp 1–22.

Higgins CM, Hall RM, Mitter N, Cruickshank A, Dietzgen RG (2004) Peanut stripe potyvirus resistance in peanut (*Arachis hypogaea* L.) plants carrying viral coat protein gene sequences. Transgenic Res 13: 59–67.

Houde M, Belcaid M, Ouellet F, Danyluk J, Monroy AF, Dryanova A, Gulick P, Bergeron A, Laroche A, Links MG, MacCarthy L, Crosby WL, Sarhan F (2006) Wheat EST resources for functional genomics of abiotic stress. BMC Genomics 7: 149.

Hu Z, Wang X, Zhan G, Liu G, Hua W, Wang H (2009) Unusually large oil bodies are highly correlated with lower oil content in *Brassica napus*. Plant Cell Rep 28: 541–549.

ICRISAT (1994) Legumes Program Annual Report 1993. Legumes Program. ICRISAT, Pattancheru, AP, India, p 260.

Iqbal MM, Zafar Y, Nazir F, Ali S, Iqbal J, Asif MA, Rashid O, Ali GM (2011) Over expression of bacterial chitinase gene in Pakistani peanut (*Arachis hypogaea* L.) cultivar GOLDEN. Afr J Biotechnol 10: 5838–5844.

Iqbal MM, Nazir F, Ali S, Asif MA, Zafar Y, Iqbal J, Ali GM (2012) Over expression of rice chitinase gene in transgenic peanut (*Arachis hypogaea* L.) improves resistance against leaf spot. Mol Biotechnol 50: 129–136.

Jefferson RA (1987) Assaying chimeric genes in plants: The Gus gene fusion system. Plant Mol Biol Rep 5: 387–405.

Jenkins WA (1938) Two fungi causing leaf spots of peanut. J Agri Res 56: 317–332.

Joshi M, Niu C, Fleming G, Hazra S, Chu Y, Nairn CJ, Yang H, Ozias-Akins P (2005) Use of green fluorescent protein as a non-destructive marker for peanut genetic transformation. *In vitro* Cell Dev Biol—Plant 41: 437–445.

Kanyand M, Desai AP, Prakash CS (1994) Thidiazuron promotes high frequency regeneration of peanut (*Arachis hypogaea*) plants *in vitro*. Plant Cell Rep 14: 1–5.

Kartha KK, Pahl K, Leung NL, Mroginski LA (1981) Plant regeneration from meristems of grain legumes, soybean, cowpea, peanut, chickpea and bean. Can J Bot 59: 1671–1679.

Kaushik KK (1993) Growth and instability of oilseed production in India. Indian J Agri Econ 48: 334–338.

Khandelwal A, Vally KJM, Geeta N, Venkatachalam P, Shaila MS, Lakshmi Sita G (2003) Engineering hemagglutinin (H) protein of rinder pest virus into peanut (*Arachis hypogaea* L.) as a possible source of vaccine. Plant Sci 165: 77–84.

Klein TM, Fromm ME, Gradziel T (1988) Factors influencing gene delivery into *Zea mays* cells by high velocity microprojectiles. Biotechnol 6: 923–926.

Knutzon DS, Thompson GA, Radke SE, Johnson WB, Knauf VC, Kridl JC (1992) Modification of *Brassica* seed oil by antisense expression of a stearoyl- acyl carrier protein desaturase gene. Proc Natl Acad Sci USA 89: 2624–2628.

Kolte SJ (1985) Diseases of Annual Edible Oilseeds Crops, Vol I. Groundnut, CRC Press, Boca Ration, FL, USA, p 155.

Lacorte C, Mansur E, Timmerman B, Cordeiro AR (1991) Gene transfer into peanut (*Arachis hypogaea* L.) by *Agrobacterium tumefaciens*. Plant Cell Rep 10: 354–357.

Lacorte C, Aragu FJL, Almeida ER, Mansur E, Rech EL (1997) Transient expression of GUS and 2S albumin gene from Brazil nut in peanut (*Arachis hypogaea* L.) seed explants using particle bombardment. Plant Cell Rep 16: 619–623.

Lemaux PG (2008) Genetically engineered plants and foods: a scientist's analysis of the issues (Part I). Annu Rev Plant Biol 59: 771–812.

Lemaux PG (2009) Genetically engineered plants and foods: a scientist's analysis of the issues (Part II). Annu Rev Plant Biol 60: 511–559.

Li Z, Jarret RL, Pittman RN, Demski JW (1994) Shoot organogenesis from cultured seed explants of peanut (*Arachis hypogaea* L.) using thidiazuron. *In vitro* Cell Dev Biol-Plant 30: 187–191.

Li Z, Jarret RL, Cheng M, Demski JW (1995) Improved electroporation buffer enhances transient gene expression in *Arachis hypogaea* protoplasts. Genome 38: 858–863.

Li Z, Cheng M, Xing A, Jarret RL, Pittman R, Demski JW (1996) Groundnut transformation research in Griffin, Georgia. In: Groundnut Virus Diseases in the Asia-Pacific Region: Summary and Recommendations of the 4th Meeting of the International Working Group, 12–14, 1995, Khon Kaen University, Thailand; ICRISAT, Patancheru, AP, India, pp 9–10.

Li ZJ, Jarret RL, Demski JW (1997) Engineered resistance to tomato spotted wilt virus in transgenic peanut expressing the viral nucleocapsid gene. Transgenic Res 6: 297–305.

Liang H, Kumar PA, Nain V, Powell WA, Carlson JE (2010) Selection and screening strategies. In: Kole C, Michler CH, Abbott AG, Hall TC (eds) Transgenic Plants, Vol 1: Principles and Development. Springer, Heidelberg, Germany, pp 85–144.

Lindsey K, Jones MGK (1989) Plant Biotechnology in Agriculture. Open University Press, Milton, UK.

Livingstone DM, Birch RJ (1995) Plant regeneration and microprojectile mediated gene transfer in embryonic leaflets of peanut (*Arachis hypogaea* L.). Aust J Plant Physiol 22: 585–591.

Livingstone DM, Hampton JL, Phipps PM, Elizabeth AG (2005) Enhancing resistance to *Sclerotinia minor* in peanut by expressing a barley oxalate oxidase gene. Plant Physiol 137: 1354–1362.

Lomonossoff GP (1995) Pathogen-derived resistance to plant viruses. Annu Rev Phytopathol 33: 323–343.

Luo M, Liang X, Dang P, Holbrook CC, Bausher MG, Lee RD, Guo BZ (2005) Microarray-based screening of differentially expressed genes in peanut in response to *Aspergillus parasiticus* infection and drought stress. Plant Sci 169: 695–703.

Lynch RE, Wilson DM (1991) Enhanced infection of peanut, *Arachis hypogaea* L. seeds with *Aspergillus flavus* group fungi due to external scarifycation of peanut pods by the lesser cornstalk borer, *Elasmopatpus lignosetlus* (Zeller). Peanut Sci 18: 110.

Magbanua ZV, Wilde HD, Roberts JK, Chowdhury K, Abad J, Moyer JW, Wetzstein HY, Parrott WA (2000) Field resistance to tomato spotted wilt virus in transgenic peanut (*Arachis hypogaea* L.) expressing an antisense nucleocapsid gene sequence. Mol Breed 6: 227–236.

Mali VR, Patil FS (1979) Occurrence of tomato spotted wilt virus on groundnut in Maharashtra. Indian Phytopathol 32: 193–197.

Mansur E, Lacorte C, De Freitas VG (1993) Regulation of transformation efficiency of peanut (*Arachis hypogaea* L.) explants by *Agrobacterium tumefaciens.* Plant Sci 99: 89–91.

Martin JP (1970) A contribution to the study of certain hereditary characters of agronomic importance in the groundnut. Oleagineux 22: 673–676.

Mayee CD, Godbole GM, Patil FS (1977) Appraisal of groundnut rust in India. Problems and Approach. Proc Natl Agri Sci 23: 162–165.

Mayee CD (1982) Groundnut rust, a review. Indian Bot Rep 1: 75–83.

Mayee CD (1986) Epidemiology and management of groundnut rust in vistas. In: Varma A, Verma JP (eds) Plant Pathology. Malhotra Publishing House, New Delhi, India, pp 305–309.

Mayee CD (1987a) Rust disease of groundnut in Maharashtra. In: Groundnut Rust Disease-Proceedings of Group Discussion meeting, 24–28th Sept 1984, ICRISAT, Patancheru, India, pp 81–89.

Mayee CD (1989) Dynamics of disease progress in groundnut; and epidemiological view. In: Tilak ST (eds) Recent Research in Ecology, Environment and Pollution, Vol 3, Today and Tomorrow's Printers and Publishers, New Delhi, pp 109–118.

McDonald D, Subrahmanyam P, Gibbons WR, Smith DH (1985) Early and Late leaf spots of groundnut. Info Bull no. 21. ICRISAT. Pattancheru, AP, India, pp 15–16.

McKently AH, Moore GA, Gardner FP (1990) *In vitro* plant regeneration of peanut from seed explants. Crop Sci 30: 192–196.

McKently AH, Moore GA, Gardner FP (1991) Regeneration of peanut and perennial peanut from cultured leaf tissue. Crop Sci 31: 833–837.

McKently AH, Moore GA, Doostdar H, Niedz RP (1995) *Agrobacterium*-mediated transformation of peanut (*Arachis hypogaea* L.) embryo axes and the development of transgenic plants. Plant Cell Rep 14: 699–703.

Milla SR (2003) Relationships and utilization of *Arachis* germplasm in peanut improvement. PhD Thesis, North Carolina State University, Reilph, NC, USA, pp 1–150.

Mittler R, Blumwald E (2010) Genetic engineering for modern agriculture: challenges and perspectives. Annu Rev Plant Biol 61: 443–462.

Morran S, Eini O, Pyvovarenko T, Parent B, Singh R, Ismagul A, Eliby S, Shirley N, Langridge P, Lopato S (2011) Improvement of stress tolerance of wheat and barley by modulation of expression of DREB/CBF factors. Plant Biotechnol J 9: 230–249.

Mroginski LA, Fernandez A (1980) Obtencion de plantulas por cultivo *in vitro* de anteras de escpecies silvestres *Arachis* (Leguminosae). Oleagineux 35: 89–92.

Mroginski LA, Kartha KK, Shyluk JP (1981) Regeneration of peanut plantlets by *in vitro* culture of immature leaves. Can J Bot 59: 826–830.

Murashige T, Skoog F (1962) A revised medium for rapid growth and bioassays with tobacco tissue culture. Plant Physiol 15: 473–497.

Naidu RA, Bottenberg H, Subrahmanyam P, Kimmins FM, Robinson DJ, Thresh JM (1998) Epidemiology of groundnut rosette virus disease: current status and future research needs. Ann Appl Biol 132: 525–548.

Narasimhulu SB, Reddy GM (1983) Plantlet regeneration from different callus cultures of *Arachis hypogaea* L. Plant Sci Lett 31: 157–163.

Nelson RT, Shoemaker R (2006) Identification and analysis of gene families from the duplicated genome of soybean using EST sequences. BMC Genomics 7: 204.

Nigam SN, Rao RCN, Wright GC (2001) Breeding for increased water-use efficiency in groundnut. In: New Millennium International Groundnut Workshop, Shandong Peanut Research Institute, Qingdao, China, pp 1–2.

Niu C, Akasaka-Kennedy Y, Faustinelli P, Joshi M, Rajasekaran K, Yang H, Chu Y, Cary J, Ozias-Akins P (2009) Antifungal activity in transgenic peanut (*Arachis hypogaea* L.) conferred by a nonheme chloroperoxidase gene. Peanut Sci 36: 126–132.

Oh SJ, Song SI, Kim YS, Jang HJ, Kim SY, Kim M, Kim YK, Nahm BH, Kim JK (2005) *Arabidopsis* CBF3/DREB1A and ABF3 in transgenic rice increased tolerance to abiotic stress without stunting growth. Plant Physiol 138: 341–351.

Olsson O, Koncz C, Szalay AA (1988) The use of the *luxA* gene of the bacterial luciferase operon as a reporter gene. Mol Gen Genet 215: 1–9.

Ozias-Akins P (1989) Plant regeneration from immature embryos of peanut. Plant Cell Rep 8: 217–218.

Ozias-Akins, P, Singsit C, Branch WD (1992) Interspecific hybrid inviability in crosses of *Arachis hypogaea x A. stenosperma* can be overcome by *in vitro* embryo maturation or somatic embryogenesis. J Plant Physiol 140: 207–212.

Ozias-Akins P, Anderson WF, Holbrook CC (1992) Somatic embryogenesis in *Arachis hypogaea* L.: Genotype comparison. Plant Sci 83: 103–111.

Ozias-Akins P, Schnall JA, Anderson WF, Singsit C, Clemente TE, Adang MJ, Weissinger AK (1993) Regeneration of transgenic peanut plants from stably transformed embryogenic callus. Plant Sci 93: 185–194.

Ozias-Akins P, Yang H, Roberson E, Akasaka Y, Lynch R (2000) Genetic engineering of peanut for reduction of aflatoxin contamination. In: Aflatoxin/Fumonisin Workshop, 25–27th Oct 2000, Yosemite, California, USA, pp 106–107.

Ozias-Akins P, Yang H, Gill R, Fan H, Lynch RE (2002) Reduction of aflatoxin contamination in peanut: A genetic engineering approach. In: Rajasekaran K, Jacks TJ, Finley JW (eds) Crop Biotechnology. American Chemical Society Symposium Series No. 829. Washington, DC, pp 151–160.

Ozias-Akins P, Yang P, Culbreath AK, Gorbet DW, Weeks JR (2002) Field resistance to *Tomato spotted wilt virus* in a transgenic peanut (*Arachis hypogaea* L.). Proc Amer Peanut Res Edu Socy 34: 70.

Ozias-Akins P, Niu C, Hazra S, Deng XY (2003) Genetic engineering for aflatoxin reduction in peanut: Utility of selectable markers and antifungal genes. Proc USDA-ARS Aflatoxin Elimination Workshop, Savannah, GA, p 99.

Padua VLM, Pestana MC, Margis-Pinheiro M, De Oliveira DE (2000) Electroporation of intact embryonic leaflet of peanut: Gene transfer and stimulation of regeneration capacity. *In vitro* Cell Dev Biol–Plant 36: 374–378.

Pang SZ, Slightom JL, Gonsalves D (1993) Different mechanisms protect transgenic tobacco against tomato spotted wilt and impatients necrotic spot tospoviruses. Biotechnol 11: 819–824.

Patil BN, Bhonde SR, Kandikar DN (2009) Trends in area, production and productivity of groundnut in Maharashtra. Financing Agriculture—A National Journal of Agriculture and Rural Development, pp 35–38.

Pellegrineschi A, Reynolds M, Pacheco M, Brito RM, Almeraya R, Yamaguchi-Shinozaki K, Hoisington D (2004) Stress-induced expression in wheat of the *Arabidopsis thaliana DREB1A* gene delays water stress symptoms under greenhouse conditions. Genome 47: 493–500.

Pittman RN, Banks DG, Kirby JS, Mitchell ED, Richardson PE (1983) *In vitro* culture of immature peanut leaves: Morphogenesis and plantlet regeneration. Peanut Sci 10: 21–25.

Polashock JJ, Chin CK, Martin CE (1992) Expression of the yeast δ-9 fatty acid desaturase in *Nicotiana tobacum*. Plant Physiol 100: 894–901.

Ponsamuel J, Huhman DV, Cassidy BG, Post-Beittenmiller D (1998) *In vitro* regeneration via caulogenesis and brassin-induced shoot conversion of dormant buds from plumular explants of peanut (*Arachis hypogaea* L. cv. 'Okrun'). Plant Cell Rep 11: 373–378.

Prasad K, Bhatnagar-Mathur, Waliyar F, Sharma KK (2012) Overexpression of a chitinase gene in transgenic peanut confers enhanced resistance to major soil borne and foliar fungal pathogens. J Plant Biochem Biot (in press) DOI: http://dx.doi.org/10.1007/s13562-012-0155-9.

Pretorius AE (2005) ARC-GCI Groundnut Department Progress Report, Potchefstroom, South Africa.

Qin H, Gu Q, Zhang J, Sun L, Kuppu S, Zhang Y, Burow M, Payton P, Blumwald E, Zhang H (2011) Regulated expression of an isopentenyltransferase gene (*IPT*) in peanut significantly improves drought tolerance and increases yield under field conditions. Plant Cell Physiol 52: 1904–1914.

Rajasekaran K, Cary JW, Jacks TJ, Stromberg K, Cleveland TE (2000) Inhibition of fungal growth in planta and *in vitro* by transgenic tobacco expressing a bacterial nonheme chloroperoxidase gene. Plant Cell Rep 19: 333–338.

Reddy AS, Prasada Rao RDVJ, Thirumala-Devi K, Reddy SV, Mayo MA, Roberts I, Satyanarayana T, Subramaniam K, Reddy DVR (2002) Occurrence of *Tobacco streak virus* on peanut (*Arachis hypogaea* L.) in India. Plant Dis 86: 173–178.

Reichel C, Mathur J, Eckes P, Langenkemper K, Koncz C, Schell J, Reiss B, Maas C (1996) Enhanced green fluorescence by the expression of an *Aequorea victoria* green fluorescent protein mutant in mono- and dicotyledonous plant cells. Proc Natl Acad Sci USA 93: 5888–5893.

Revoredo CL, Fletcher SM (2002) World Peanut Market: an overview of the past 30 years. Research Bulletin. Department of Agricultural and Applied Economics University of Georgia, Athens, GA, USA, p 437.

Ritzenthaler C (2005) Resistance to plant viruses: old issues, new answers? Curr Opin Biotechnol 16: 118–122.

Rohini VK, Rao KS (2000) Transformation of peanut (*Arachis hypogaea* L.): A non-tissue culture based approach for generating transgenic plants. Plant Sci 150: 41–49.

Rohini VK, Rao KS (2001) Transformation of peanut (*Arachis hypogaea* L.) with tobacco chitinase gene: variable response of transformants to leaf spot disease. Plant Sci 160: 889–898.

Sakuma Y, Maruyama K, Osakabe Y, Qin F, Seki M, Shinozaki K, Yamaguchi-Shinozaki K (2006a) Functional analysis of an *Arabidopsis* transcription factor, DREB2A, involved in drought-responsive gene expression. Plant Cell 18: 1292–1309.

Sanford JC, Johnston SA (1985) The concept of pathogen derived resistance. J Theor Biol 113: 395–405.

Sanford JC, Klein TM, Wolf ED, Allen N (1987) Delivery of substances to the cells and tissues using a particle gun bombardment process. Particle Gun Technol 5: 27–37.

Sanford JC (1990) Biolistic plant transformation. Physiol Planta 79: 206–209.

Sastri DC, Nalini M, Moss JP (1980) *Arachis* ovary and ovule culture *in vitro*. In: Rao PS, Heble MR, Chadah MS (eds) Proceedings of Symposium on Plant Tissue Culture, Genetic

Manipulation and Somatic Hybridization of Plant Cells. Bhabha Atomic Research Centre, Trombay, India, pp 366–374.
Sastri DC, Moss JP (1982) Effects of growth regulators on incompatible crosses in the genus *Arachis* L. J Exp Bot 53: 1293–1301.
Schnall JA, Weissinger AK (1995) Genetic transformation in *Arachis hypogaea* L. Vol 34, In: Bajaj YPS (ed) Biotechnology in Agriculture and Forestry. Springer-Verlag, Heidelberg, Germany, pp 135–144.
Sellars RM, Southward GM, Philips GC (1990) Adventitious somatic embryogenesis and cultured immature zygotic embryos of peanut and soybean. Crop Sci 30: 408–414.
Serraj R, Hash CT, Rizvi SMH, Sharma A, Yadav RS, Bidinger FR (2005) Recent advances in marker-assisted selection for drought tolerance in pearl millet. Plant Prod Sci 8: 334–337.
Sharma HC, Ortiz R (2000) Transgenics, pest management, and the environment. Curr Sci 79: 421–437.
Sharma HC, Sharma KK, Seetharama N, Ortiz R (2000) Prospects for using transgenic resistance to insect pests in crop improvement. Elec J Biotechnol 3: 76–96.
Sharma KK, Lavanya M (2002) Recent developments in transgenics for abiotic stress in legumes of the semi-arid tropics. In: Ivanaga M (ed) Genetic Engineering of Crop Plants for Abiotic Stress. JIRCAS Working Report No 23, JIRCAS:Tsukuba, Japan, pp 61–73.
Sharma KK, Ortiz R (2000) Program for the application of genetic transformation for crop improvement in the semi-arid tropics. *In vitro* Cell Dev Biol-Plant 36: 83–92.
Sharma KK, Anjaiah V (2000) An efficient method for the production of transgenic plants of peanut (*Arachis hypogaea* L.) through *Agrobacterium tumefaciens*-mediated genetic transformation. Plant Sci 159: 7–19.
Shinozaki K, Yamaguchi-Shinozaki K (2007) Gene networks involved in drought stress response and tolerance. J Exp Bot 58: 221–227.
Siloto RM, Findlay K, Lopez-Villalobos A, Yeung EC, Nykiforuk CL, Moloney MM (2006) The accumulation of oleosins determines the size of seed oil bodies in *Arabidopsis*. Plant Cell 18: 1961–1974.
Singsit C, Adang MJ, Lynch RE, Anderson WF, Wang A, Cardineau G, Ozias-Akins P (1997) Expression of a *Bacillus thuringiensis cryIA(c)* gene in transgenic peanut plants and its efficacy against lesser cornstalk borer. Transgenic Res 6: 169–176.
Skoog F, Miller CO (1957) Chemical regulation of growth and organ formation in plant tissues cultured *in vitro*. Symp Soc Exp Biol 11: 118–131.
Stalker HT, Moss JP (1987) Speciation, cytogenetics, and utilization of *Arachis* species. Adv Agron 41: 1–40.
Stalker HT, Mozingo LG (2001) Molecular genetics of *Arachis* and marker assisted selection. Peanut Sci 28: 117–123.
Subbarao GV, Johansen C, Slinkard AE, Rao RCN, Saxena NP, Chauhan YS (1995) Strategies for improving drought resistance in grain legumes. Crit Rev Plant Sci 14: 469–523.
Subrahmanyam P, Gibbons RW, Nigam SN, Rao VR (1980a) Screening methods and further sources of resistance to peanut rust. Peanut Sci 7: 10–12.
Subrahmanyam P, Mehan VK, Nevill DJ, McDonald D (1980b) Research on fungal diseases of groundnut at ICRISAT. Proceedings of the International Workshop on Groundnuts. 13–17 Oct 1980, ICRISAT Center, India. Patancheru, AP, 502324, pp 193–198.
Subrahmanyam P, Williams JH, McDonald D, Gibbons RW (1984) The influence of foliar diseases and their control by selective fungicides on a range of groundnut (*Arachis hypogoea* L.) genotypes. Ann Appl Biol 104: 467–476.
Subrahmanyam P, Moss JP, McDonald D, Subba Rao PV, Rao VR (1985) Resistance to late leaf spot caused by *Cercosporidium personatum* in wild *Arachis* species. Plant Dis 69: 951–954.
Sundaresha S, Manoj Kumar A, Rohini S, Math SA, Keshamma E, Chandrashekar SC, Udayakumar M (2010) Enhanced protection against two major fungal pathogens of

groundnut, *Cercospora arachidicola* and *Aspergillus flavus* in transgenic groundnut over-expressing a tobacco β-1, 3 glucanase. Eur J Plant Pathol 126: 497–508.

Taliansky ME, Robinson DJ, Murant AF (1996) Complete nucleotide sequence and organization of the RNA genome of groundnut rosette umbravirus. J Gen Virol 77: 2335–2345.

Tiwari S, Mishra DK, Singh A, Singh PK, Tuli R (2008) Expression of a synthetic *cry1EC* gene for resistance against, *Spodoptera litura* in transgenic peanut (*Arachis hypogaea* L.) Plant Cell Rep 27: 1017–1025.

Utomo SD, Weissinger AK, Isleib TG (1996) High efficiency peanut regeneration using a non-imbibed immature leaflet culture method. Peanut Sci 23: 71–75.

Vasavirama K, Kirti PB (2010) Generation of "tikka" resistant transgenic groundnut plants by introducing double gene construct. In: DNA 2010: 2nd International Conference on the science of DNA fingerprinting in crime investigation. 29–30th October 2010 Bioaxis DNA research centre (BDRC) private limited, Hyderabad, India.

Venkatachalam P, Geeta N, Jayabalan N (1998) Induction of somatic embryos and plantlet development in cell suspension cultures of *Arachis hypogaea* L. Breeding Sci 48: 231–236.

Venkatachalam P, Geeta N, Khandelwal A, Shaila MS, Sita GL (2000) *Agrobacterium*-mediated genetic transformation and regeneration of transgenic plants from cotyledon explants of groundnut (*Arachis hypogaea* L.) via somatic embryogenesis. Curr Sci 78: 1130–1136.

Venkatachalam P, Geetha N, Khandelwal A, ShailaMS, Lakshmi Sita G (1999) Induction of direct somatic embryogenesis and plant regeneration from mature cotyledon explants of *Arachis hypogaea* L. Curr Sci 77: 269–273.

Vinocur B, Altman A (2005) Recent advances in engineering plant tolerance to abiotic stress: achievements and limitations. Curr Opin Biotechnol 16: 123–132.

Wang A, Fan H, Singsit C, Ozias-Akins P (1998) Transformation of peanut with a soybean VspB promoter-*uidA* chimeric gene. 1. Optimization of a transformation system and analysis of GUS expression in primary transgenic tissues and plants. Physiol Plant 102: 38–48.

Wang M, Abbott D, Waterhouse PM (2000) A single copy of a virus derived transgene encoding hairpin RNA gives immunity to barley yellow dwarf virus. Mol Plant Pathol 1: 401–410.

Wang Q, Guan Y, Wu Y, Chen H, Chen F, Chu C (2008) Overexpression of a rice *OsDREB1F* gene increases salt, drought, and low temperature tolerance in both *Arabidopsis* and rice. Plant Mol Biol 67: 589–602.

Wang XJ, Liu ST, Xia H, Wan SB, Zhao CZ , Li AQ (2011) Peanut (*Arachis hypogaea* L.) Omics and biotechnology in China. Plant Omics J 4: 339–349.

Weissinger A, Liu YS, Scanlon S, Murray J, Cleveland TE, Jaynes J, Mirkoy E, Moonan F (1999) Transformation of peanut with the defensive peptidyl MIM D5C. In: USDA-ARS Proceedings of the USDA-ARS aflatoxin elimination workshop, 20-22, October 1999, Atlanta, GA, USA, pp 66–68.

Weissinger A, Wu M, Cleveland TE (2003) Expression in transgenic peanut of maize RIP 1, a protein with activity against *Aspergillus* spp. In: Proceedings of the USDA-ARS Aflatoxin Elimination Workshop, pp 100.

Wetzstein HY, Baker CM (1993) The relationship between somatic embryo morphology and conversion in peanut (*Arachis hypogaea* L.). Plant Sci 92: 81–89.

Willcox MC, Reed SM, Burns JA, Wynne JC (1991) Effect of microspore stage and media on anther culture of peanut (*Arachis hypogaea* L.). Plant Cell Tiss Org Cult 24: 25–28.

Wilmink A, Dons JJM (1993) Selective agents and marker genes for use in transformation of monocotyledonous plants. Plant Mol Biol Rep 11: 165–185.

Xiao H, Siddiqua M, Braybrook S, Nassuth A (2006) Three grape CBF/DREB1 genes respond to low temperature, drought and abscisic acid. Plant Cell Environ 29: 1410–1421.

Yang DX, Wei ZM, An HL (2001) Transgenic peanut plants obtained by particle bombardment via somatic embryogenesis regeneration system. Cell Res 11: 156–160.

Yang H, Singsit C, Wang A, Gonsalves D, Ozias-Akins P (1998) Transgenic peanut plants containing a nucleocapsid protein gene of tomato spotted wilt virus show divergent levels of gene expression. Plant Cell Rep 17: 693–699.

Yang H, Ozias-Akins P, Culbreath AK, Gorbet DW, Weeks JR, Mandal B, Pappu HR (2004) Field evaluation of Tomato spotted wilt virus resistance in transgenic peanut (*Arachis hypogaea*). Plant Dis 88: 259–264.

Yang HY, Nairn J, Ozias-Akins P (2003) Transformation of peanut using a modified bacterial mercuric ion reductase gene driven by an actin promoter from *Arabidopsis thaliana*. J Plant Physiol 160: 945–952.

Yang S, Vanderbeld B, Wang J, Huang Y (2010) Narrowing down the targets: towards successful genetic engineering of drought-tolerant crops. Mol Plant 3: 469–490.

Zhuang WJ, Zhang SB, Liu SH, Cai LL (1999) Somatic embryogenesis and plant regeneration from axes of peanut embryos. J Trop Subtrop Bot 7: 153–158.

10

Applications of Bioinformatics Tools to Genetic Mapping and Diversity in Peanut

K Venkatesh,[1] *A Roja Rani,*[1] *P Amareshwari*[1] and *Ramesh Katam*[2,*]

ABSTRACT

Peanut remains among the less-studied crops due to the limited genomic resources that have severely impeded molecular breeding activities. Tetraploid nature, low diversity, complicated and complex genome size hampers the progress in development of genome resources such as genetic linkage maps. Hence, more studies on genetic diversity among peanut cultivars are warranted for germplasm conservation and varietal development. Expressed sequence tags will play a vital role in identification of potential genetic markers and exhibit cross transferability that can serve as anchor markers for comparative mapping. Peanut database will facilitate genomic research and peanut proteomics is advancing with new challenges. We present an overview of available resources on peanut bioinformatics and their role in elucidating biological and genomic information on peanut. Also the high-resolution two-dimensional electrophoresis and high-sensitive mass spectrometry studies in peanut proteomics and the implications of these findings are reviewed.

Keywords: Expressed sequence tags, Genome Sequence, Cross-transferability, Comparative genomics, NGS technologies, Genotyping by sequencing (GBS), Gene discovery

[1]Department of Genetics and Biotechnology, Osmania University, Hyderabad, India.
[2]Department of Biological Sciences, College of Science and Technology, Florida A&M University, Tallahassee, Florida 32307 USA.
*Corresponding author: *ramesh.katam@gmail.com*

10.1 Introduction

Peanut (*Arachis hypogaea* L.) is an important principal commercial oilseed crop rich in high-quality edible oil (43–55%), digestible proteins (25–30%) and carbohydrates (10–20%) (Maiti 2002). Peanut is the third most important source of vegetable protein, fourth principal source of edible oil, which plays an important role in ameliorating livelihoods, nutrition and economy of many countries. Peanut is a good source of thiamin, riboflavin, nicotinic acid, niacin, vitamin E and B. Additionally it possesses significant amounts of phosphorus, calcium, magnesium, zinc, potassium and iron, which are important for human nutrition (Savage et al. 1994). Peanut with high-quality unsaturated fatty acid (oleic acid) and low saturated fat content along with antioxidants, nutraceuticals, such as resveratrol, exhibit superior oxidative stability and lowers serum LDL-cholesterol levels respectively (Guo et al. 2011). It fixes nitrogen and improves soil fertility by providing up to 60 kg/ha nitrogen to the soil (Sprent 1994).

Peanut is an economically important crop in arid and semi-arid tropics, mainly grown under rain-fed conditions. However, the yield of peanut is compromized due to the reduction in area and productivity in these areas due to various biotic factors, which include peanut stripe virus leaf spot, leaf miner and aflatoxin contamination. Furthermore, abiotic stresses drought, low pH, low temperatures and inadequate soil fertility (especially N and P) severely affect the crop yields. Cultivated groundnut is a self-pollinated, allotetraploid member of the section *Arachis*, classified into two subspecies: *hypogaea* and *fastigiata*. The subspecies *hypogaea* is further divided into *hypogaea* and *hirsute* varieties, while *fastigiata* into *fastigiata*, *vulgaris*, *aequatoriana* and *peruviana* based on morphology and growth habits (Krapovickas and Gregory 1994).

Drought is a multigenic, incompletely penetrant and quantitative trait in nature that makes difficult to breed for drought tolerance, thus adversely affecting crop productivity. In peanut limited empirical knowledge is available about the physiological and molecular events regulating gene expression under drought conditions (Jain et al. 2001). Breeding effort to introgress complex traits from drought tolerant germplasm met with limited success due to its complex polygenic nature, insufficient knowledge about genetics of tolerant components, no correlation of tolerance at different developmental stages and environmental effects (Jain et al. 2003).

10.2 Peanut Genome Sequencing

Cultivated peanut is an allotetraploid (2n = 4x (2A+2B) = 40) (Seijo et al. 2004) with a large genome size of 2,891 Mbp (Table 10-1; Feng et al. 2012). Complicated (duplicate sets of chromosomes) and large size genome makes

Table 10-1 Genome size comparison of Model Legumes.

Legume	*Cicer arietinum*	*Cajanus cajan*	*Arachis hypogaea*	*Glycine max*	*Medicago truncatula*
Ploidy	Diploid (2n = 2x = 16)	Diploid (2n = 2x = 22)	Allotetraploid (2n = 4x = 40)	Diploid (2n = 2x = 40)	Diploid (2n = 2x = 16)
Genome (Mb)	740	858	2,891	1,115	~454 to 526

the peanut difficult to sequence or assemble the whole genome and thus remains unexplored at the genomic level (Feng et al. 2012). Lack of genetic diversity as a result of polyploidization, hampered the genetic and genomic characterization (Guimarães et al. 2008). Expressed Sequence Tag (EST) sequencing generates functional genomic level data and is an alternative to whole genome sequencing for non-model organisms like peanut (Andersen and Thomas 2003; Bouck and Vision 2007).

Peanut Foundation and the American Peanut Council on behalf of international peanut research community and Peanut Genome Consortium (PGC) has been pursuing a peanut genome sequencing project on a collaborative approach with Chinese peanut collaborators for a possible joint sequencing project along with other international partners. Peanut Genome Consortium (PGC) was formed with specific goals: 1) a high-quality chromosome-scale draft of a tetraploid (cultivated species) as the reference genome sequence, plus high-density maps of both progenitor and synthetic amphidiploid genomes; 2) high-throughput transcriptome characterization of the reference tetraploid cultivar; 3) characterization of gene space in amphidiploid and diploid (progenitor species) germplasm; 4) phenotypic association with mapped genetic markers; and 5) interactive bioinformatics resources for data curation and application in a breeder's toolbox to enable molecular breeding approaches for enhancing peanut yielding ability, optimizing resistance to diseases and insects, tolerance to environmental stresses and improved quality traits. Beijing Genome Institute (BGI) proposed the sequencing and assembly strategies adopting integrated strategy whole-genome sequencing plus Bacterial Artificial Chromosome (BAC) by BAC sequencing with Hiseq 2000 technology.

10.3 Comparative Genomics

Comparative genomics has emerged as a new approach contributing to understand tetraploid peanut genome—origin, evolution, structure and function to promote its genetic improvement. In several crops, EST–Simple Sequence Repeat (SSR)'s cross-transferability has been reported across genera/species (Gale and Devos 1998; Bennetzen 2000; Paterson et al. 2000). Moreover, they can be used for comparative mapping as anchor markers as they are reported to have high level of transferability than the genomic SSRs

(Scott et al. 2000; Saha et al. 2004; Varshney et al. 2005). A comparison of SSR-based cultivated peanut map and legume anchor markers was made with maps from *Arachis*, Lotus, and *Medicago* (Moretzsohn et al. 2009). Percent transferability of EST-SSRs to peanut from *Medicago* is 20 and from soybean 34 along with 25% SSRs (Mullet et al. 2002; He et al. 2005; He et al. 2003; Wang et al. 2012). Recently, it rose to 39% from sorghum, thus prospects additional DNA markers and comparative mapping between peanut and sorghum as a possible monocot-dicot comparison (Savadi et al. 2012).

10.4 Genetic Resources and Their Utilization

Considerable variability for different traits has been exhibited by cultivated peanut (Kochert et al. 1996). Peanut core and minicore collections enabled to find new genes from the large number of accessions in gene banks (Table 10-2; Brown 1989). These germplasm collections assist in better germplasm exploitation by the breeders and enhanced use of genetic resources in crop improvement (Table 10-3; Upadhyaya et al. 2002; Holbrook and Dong 2005; Jiang et al. 2010). Both these germplasm collections identified sources of genetic variation and diversity in peanut for different traits (Jiang et al. 2010; Kottapalli 2011).

Table 10-2 Major gene banks holding peanut germplasm.

Cultivated genotypes	Wild genotypes	Total genotypes	Research Centers
14,968	477	15,445	International Crops Research Institute for the Semi-Arid Tropics (ICRSAT), India
-	-	14,585	National Bureau of Plant Genetic Resources (NBPGR), Indian Council of Agricultural Research (ICAR), India
9,310	607	9,917	Plant Genetic Resource Conservation Unit (PGRCU), Griffin, U.S. Department of Agriculture (USDA), USA
8,960	64	9,024	Directorate of Groundnut Research (DGR), ICAR, India
7,837	246	8,083	Oil Crops Research Institute (OCRI), Chinese Academy of Agricultural Sciences (CAAS), China
4,210	-	4,210	Crops Research Institute (CRI), Guangdong Academy of Agricultural
3,534	106	3,640	InstitutoNacional de TechnologiaAgropecuaria (INTA), Argentina
1,200	1,220	2,420	EMBRAPA and CENARGEN, Brazil
2,140	-	2,140	InstitutoAgronomico de Campinas, Brazil
-	1,200	1,200	Texas A&M University (TAMU), USA
740	406	1,146	North Carolina State University (NCSU), USA
-	472	472	Instituto de Botánica del Nordeste (IBONE), Argentina

Table 10-3 Peanut Core and Mini-core Collections.

Country	Core	Mini-core	Source
India	1,704 accessions	184	Upadhyaya et al. 2002; Upadhyaya et al. 2003
USA	831 accessions	112	Holbrook et al. 1993; Holbrook and Dong 2005
China	576 accessions	298	Jiang et al. 2008; Jiang et al. 2010

10.5 Construction and Improvement of Peanut Genetic Map

For several species molecular maps have been developed that are important for map-based cloning of potential genes with agronomic importance and comparative genomics. Genetic linkage maps of cultivated peanut have been developed for Quantitative Ttrait Locus (QTL) analysis and gene tagging thus finding application in breeding. The genetic information pertaining to the cultivated tetraploid was difficult to extract from the two Restriction Fragment Length Polymorphism (RFLP)-based genetic maps, the first one with 11 Linkage Groups (LGs) derived from a diploid interspecific hybridization of A-genome species (*A. stenosperma* x *A. cardenasii*) (Halward 1993) and the next from a synthetic interspecific tetraploid population with 23 LGs (Burow et al. 2001). Moretzsohn et al. (2005) constructed an SSR-based linkage map for *Arachis* from an A-genome with 170 SSR loci (from both genomic SSRs and EST-SSRs) on 11 LGs covering 1,231 cM and located five candidate genes for resistance on this map (Alves et al. 2008). Based on an F_2 population (93 lines) of *A. ipaensis* × *A. magna* a diploid B-genome map with 149 SSR loci covering 1,294.4 cM was been developed. A high level of synteny was revealed by comparative mapping of both these genomes (Gobbi et al. 2006; Moretzsohn et al. 2009). Peanut A-genome map (*A. duranenis* and *A. duranenis*) was constructed with 2,319 (971 SSRs, 1,127 SNPs and 221 Single Stranded Conformation Polymorphism (SSCP)) markers on 10 LGs (Nagy et al. 2010).

An integrated genetic linkage map of cultivated peanut was developed with 324 SSR loci into 21 LGs covering 1,352.1 cM derived from two Recombinant Inbred Line (RIL) populations ("Tifrunner"×"GT-C20" and "SunOleic 97R"×"NC94022"). The largest collection of 4,576 SSR markers was reported from three sources: SSR markers, EST-SSR markers and from BAC end-sequences for construction of a genetic linkage (Qin et al. 2012). EST-SSR's marker density is very low and Single Nucleotide Polymorphism (SNP) markers are yet to be integrated in the genetic maps of cultivated peanut. Furthermore, SNP makers facilitates the development of SNP-based genetic maps are in progress (Pandey et al. 2012b).

10.6 Transcriptome Sequencing

The first report of ESTs in cultivated peanut was developed from two cDNA libraries under biotic (leaf spot fungi/resistant response) and abiotic stress (hydric stress/tolerant response). A total of 1,825 ESTs with 1,305 unique ESTs were identified (Luo et al. 2005a). Currently, a total of 252,832 peanut ESTs (178,490 for *A. hypogaea* including 745 for subsp. *Fastigiata*; 35,291 for *A. duranensis*; 32,787 for *A. ipaensis*; and 6,264 for *A. stenosperma*) are available in the public domain in NCBI (as of January, 2013). There are a total of 2,83,566 nucleotide sequences for *A. hypogaea* in Gene Bank, including 90,001 nucleotide (core nucleotide) sequences, 1,77,745 EST, and 15,820 genome survey sequences (GSS). These EST resources have been used for novel gene discovery, gene sequence determination, microarray gene expression, marker development and genetic map construction (Luo et al. 2005b; Liang et al. 2009).

ESTs have been extensively used to monitor gene expression patterns to biotic stresses such as leaf spot caused by *Cercosporidium personatum*, aflatoxin by *Aspergillus parasiticus* (Luo et al. 2005a; Guo et al. 2008), virus infection by tomato spotted wilt virus (Guo et al. 2009) and bacterial wilt by *R. solanacearum* (Huang et al. 2012) have been analyzed via EST sequencing that serves as valuable resources (Table 10-4) for peanut genomic research. A total of 743,232 ESTs and functionally annotated 17,912 unigenes (singletons and contigs) from *A. stenosperma* and 21,714 unigenes from *A. duranensis* were generated by transcriptome analysis (Roche 454 GS FLX Titanium technology) of *A. stenosperma* and *A. duranensis* challenged with biotic and abiotic stress, respectively (Guimarães et al. 2011). A Golden-Gate assay for 1,536 SNPs has been developed, based on the SNPs identified after comparing the 454 sequence reads generated from 17 genotypes using 454/FLX transcript sequences (Uni. Georgia, Athens). Gene/transcript sequences derived markers referred as Genic Molecular Markers (GMMs) are used to develop the functional markers (Varshney et al. 2010a, 2012). On the contrary, transcript-based markers exhibit lower polymorphism than DNA

Table 10-4 ESTs from various tissues of peanut.

Tissue	ESTs	Source
1. Developing seeds	21,777	Guo et al. 2008
2. Leaf tissues	16,931	Guo et al. 2009
3. Diverse seed development stages	17,000	Bi et al. 2010
4. Different plant develop-mental stages	12,000	Song et al. 2010
5.Gynophores, roots, leaves and seedlings	10,102	Koilkonda et al. 2011
6. Root tissues	8,000	Proite et al. 2007
7. Roots, leaves and developing seeds	63,234	Huang et al. 2012

markers (Varshney et al. 2005, 2010c, 2010d). ESTs will play a vital role in identification of SSRs and SNPs as potential genetic markers and fill in the knowledge gaps after peanut whole-genome sequencing (Feng et al. 2012). For tetraploid peanut a consensus transcriptome assembly of 211,244 contigs has been developed. This transcriptome assembly can be used for aligning the transcript reads using NGS technologies (Varshney et al. 2009) and the alignments utilized for SNP discovery (Pandey et al. 2012b).

10.7 Molecular Markers in Peanut

Molecular markers being the most powerful genomic tool have revolutionized the entire scenario of biological sciences and increased the precision of selection. They instigated a revolution in the speed and quality of germplasm characterization, trait mapping and molecular breeding. In cultivated peanut markers that reveal polymorphism are available (Ferguson et al. 2004a). However, low level polymorphism has been detected using methods such as random amplified fragment DNA (RAPD), amplified fragment length polymorphism (AFLP), and RFLP (Halward et al. 1991; Kochert et al. 1991; Paik 1992; Subramanian et al. 2000; Gimenes 2002; Herselman 2004), Inter-Simple Sequence Repeat (ISSR) (Raina et al. 2001) and SSR markers (Mace et al. 2006). DNA polymorphism was revealed in *A. hypogaea* using SSRs and diversity arrays technology (DArT) markers (Varshney et al. 2006; Gupta et al. 2010). Polyploidization prevented gene flow to a certain extent from related species affecting the genetic diversity among the peanut cultivars, resulting in a narrow genetic base (Halward et al. 1991; Isleib et al. 1992; Young et al. 1996). EST-SSRs are PCR-based, genetically defined, codominant and multiallelic markers that are renowned for their versatility and high information content (Tautz 1989; Weber and May 1989; Powell et al. 1996). They can play a vital role in determining protein function, genetic development and regulation of gene expression (Lawson and Zhang 2006).

In recent years, many SSR markers were identified from the peanut genome (Hopkins et al. 1999; He et al. 2003; Ferguson et al. 2004b; Moretzsohn et al. 2005; Proite et al. 2007; Cuc et al. 2008) pertaining to resistance against rust, late leaf spot (Mace et al. 2006), *Ralstonia solanacearum* (Jiang et al. 2007) and *Sclerotinia minor* (Chenault et al. 2008). A new type of 138 (GGC) SSRs markers (Yuan et al. 2010) and 199 SSRs in addition to 946 novel SSR markers were identified (Pandey et al. 2012a). Allele-specific sequencing studies in cultivated genotype met with limited success compared to diploid and tetraploid species (Alves et al. 2008; Pandey et al. 2012b).

Efforts are underway worldwide on the development of novel SSR markers from SSR-enriched libraries, EST sequences, BAC-end sequences, DArT Arrays and transcript sequences produced by 454/FLX sequencing technology (Pandey et al. 2012b). Diversity Arrays Technology (DArT) is a cost-effective high-throughput whole-genome genotyping platform to screen a large number of polymorphic loci and has the potential for genome-wide diversity analysis and linkage mapping. DArT is a dominant marker that are sequence-independent and detects all types of DNA variation. Jaccoud et al. (2001) developed it for rice and it has been effectively deployed to genotype other species with large genomes. DArT markers have been developed in many species including several Triticeae crops for constructing genetic maps and diversity analysis (Wenzl et al. 2006; Neumann et al. 2011; Roy et al. 2011; Varshney et al. 2012). So far in legumes, the DArT arrays have been developed in pigeonpea (Yang et al. 2011), common bean (Briñez et al. 2011), chickpea (Thudi et al. 2011), soybean and mungbean (Hang et al. 2012). The first DArT-based framework linkage map for pigeonpea was generated based on an interspecific F_2 mapping population. Maternal and paternal genetic linkage map had 122 (270.0 cM) and 172 (451.6 cM) unique DArT loci, respectively (Yang et al. 2011). In common bean, a total of 2,501 polymorphic markers were found from a *Pst*I/*Bst*NI and *Pst*I/*Taq*I representation (Briñez et al. 2011). DArT arrays with 15,000 clones were developed from a *Pst*I/*Taq*I representation and 5,397 polymorphic clones were from 94 chickpea genotypes (Thudi et al. 2011). In soybean 1,500 polymorphic clones were identified. DArT marker transferability was 13.5% from soybean to mungbean and 3.1% from mungbean to soybean (Hang et al. 2012). More than 6,000 SSR markers and DArT arrays (15,360 features) have been developed in peanut. Very low and moderate level of polymorphism was exhibited by using DArT arrays in tetraploid (AABB) and diploid (AA, BB) genome species, respectively (Kilian 2008; Varshney et al. 2010b). Even DarT and SNP markers exhibited low-level polymorphism in cultivated genotypes. Perhaps DArT markers will be helpful in monitoring genome introgression from diploids into cultivated peanut lines and will empower advances in Genotyping By Sequencing (GBS) methods in addition to SNP markers (Pandey et al. 2012b).

Till date SSRs have proven to be the best choice of markers for genetics and breeding in cultivated peanut. Sequencing-based genotyping methods will reveal new SSR and SNP markers in the peanut genome. Efforts on development of allele-specific markers in peanut would make considerable advancement in the marker technology for modern breeding programs.

10.8 Advancing Peanut Biology through Proteomics

Mass Spectrometry (MS) has revolutionized the field of protein research and created a momentum in proteome analyses. Proteome analysis is significantly more challenging than analysis of genes and genomes. High-resolution two-dimensional electrophoresis (2-DE) along with mass spectrometry are the main technologies to study the organisms at the protein level (Wang et al. 2011). Peanut (*Arachis hypogaea* L.), at the genomic level, is limited due to the budding stage of peanut proteomics, numerous protein studies of peanut seed have emphasized on storage protein composition, nutritional value and immunological properties of the protein fractions (Viquez et al. 2003; Koppelman et al. 2004; Kang and Gallo 2007). Storage proteins analyzed from 12 different genotypes of cultivated peanut varieties discovered protein markers that distinguish the subspecies *hypogaea* and *fastigiata* (Liang et al. 2006). Methionine-rich proteins and arachin proteins in seed continued to express in drought-tolerant genotypes, while they were down regulated in drought-susceptible genotypes evident by MS analysis (Katam et al. 2007, 2010).

Overall, there is a need to identify proteins from peanuts leaves, seeds and their corresponding genes that can be incorporated in the marker-assisted breeding project to select for drought-tolerant peanut genotypes, which we see as a growing challenge. System-biology tools play an important role in analyzing the genome which aid in identification of useful traits in peanut at large. Protein-protein interactions, which cannot be identified by gene expression analysis, can be identified with bioinformatics tools. Peanut interaction database has to be developed similar to AtPID (Arabidopsis thaliana Protein Interactome Database) for better understanding the peanut proteomics. It has to be linked out with various other plant web servers and databases including Gramine, Phytozome for finding comparative mapping and synteny.

10.9 Peanut Database

PeanutMap is a new map database http://peanutgenetics.tamu.edu/cmap for providing a web-based interface for viewing specific LGs and compare multiple maps of a map set. The database allows identification of corresponding LG's from the results of different research projects by comparing and finding associations between LG's in multiple map sets. For cross-species comparisons the data is incorporated into Legume Information System website, http://www.comparative-legumes.org that facilitates comparison of synteny among different legume species (Arun and Mark 2006). Duan et al. (2012) generated peanut transcriptome assembly of 32,619 contigs with EC, KEGG and GO functional annotations. SSRs, SNPs along with other genetic polymorphisms were detected for each cintig. PeanutDB provides a comprehensive view of peanut transcriptome with all mRNA, Sanger ESTs and 454 Illumina reads generated by Institute of Plant Protection, Chinese Academy of Agricultural Science. The database is efficient to present a new perspective of peanut genome dynamics, incorporating the most comprehensive datasets of peanut transcriptome sequences.

10.10 Conclusions

Systems biology includes genomics, proteomics and metabolomics to understand the metabolic pathways involved in a specific function of the cell to improve the understanding of genetics and metabolism and also to assist in the selection of productive lines. The gene discovery phase is followed by the investigation of specific functions of the individual genes and, the definition of their structural characteristics. Eventually, the objective is to address "whole genome" analysis, through which the complete nucleotide sequence of a genome will be determined. For the investigation of a particular biological process a fundamental shift from hypothesis-based approaches is represented by functional genomics technologies. In this approach, only one or few genes or proteins are examined that involves the collection and analysis of data relating to large numbers of genes or proteins. It is well known that genes or proteins seldom act alone, and functional genomics therefore addresses the complexity of cellular processes. Most of the plant processes are mediated by large numbers of genes, therefore the technology must necessarily be focused on large-scale profiling of genes, mRNAs, proteins and metabolites that participate in cellular processes. This would allow the determination of the function of a maximum number of genes specifying the crop traits. Functional genomics studies on majority of organisms are in their early stages, because some species have genome duplications, self-incompatibilities and long generation time. In such cases, the proteomics approach is a powerful tool for analyzing the functions of the genes or proteins. Analysis of proteins is critical to define the function of their genes, and linking of proteins to genome sequence information is essential to follow sequential changes in protein expression in an organism. One of the focal points in this research is to use computational biology to understand the cellular processes regulated by protein-protein interactions, post-translational protein modifications and enzymatic activities that cannot be identified by gene expression studies.

Acknowledgements

The authors would like to thank Dr. Prashanth Suravajhala, Founder, Bioclues.org for his valuable suggestions and inputs in preparing the manuscript.

References

Alves DMT, Pereira RW, Leal-Bertioli SC, Moretzsohn MC, Guimarães PM, Bertioli DJ (2008) Development and use of single nucleotide polymorphism markers for candidate resistance genes in wild peanuts (*Arachis* spp). Genet Mol Res 7: 631–642.

Andersen JR, Thomas L (2003) Functional markers in plants. Trends Plant Sci 8: 554–560.

Arun MJ, Mark DB (2006) PeanutMap: An online genome database for comparative molecular maps of peanut. BMC Bioinformatics 7: 375.

Bennetzen JL (2000) Comparative sequence analysis of plant nuclear genomes: microcolinearity and its many exceptions. Plant Cell 12: 1021–1029.

Bi YP, Liu W, Xia H, Su L, Zhao CZ, Wan SB, Wang XJ (2010) EST sequencing and gene expression profiling of cultivated peanut (*Arachis hypogaea* L.). Genome 53: 832–839.

Bouck A, Vision T (2007) The molecular ecologist's guide to expressed sequence tags. Mol Ecol 16: 907–924.

Briñez B, Blair MW, Kilian A, Carbonell SAM, Chiorato AF, Rubiano LB (2011) A whole genome DArT assay to assess germplasm collection diversity in common beans. Mol Breed 30: 181–193.

Brown AHD (1989) Core collections: A practical approach to genetic resource management. Genome 31: 818–824.

Burow MD, Simpson CE, James L, Starr JL, Paterson AH (2001) Transmission genetics of chromatin from a synthetic amphiploid in cultivated peanut (*A. hypogaea* L.): Broadening the gene pool of a monophyletic polyploid species. Genetics 159: 823–837.

Chenault KD, Maas AL, Damicone JP, Payton ME, Melouk HA (2008) Discovery and characterization of a molecular marker for *Sclerotinia minor* (Jagger) resistance in peanut. Euphytica 66: 357–365.

Cuc LM, Mace ES, Crouch JH, Quang VD, Long TD, Varshney RK (2008) Isolation and characterization of novel microsatellite markers and their application for diversity assessment in cultivated groundnut (*Arachis hypogaea* L.). BMC Plant Biol 8: 55.

Duan X, Schmidt E, Li P, Lenox D, Liu L, Shu C, Zhang J, Liang C (2012) PeanutDB: an integrated bioinformatics web portal for *Arachis hypogaea* transcriptomics. BMC Plant Biol 12: 94.

Feng S, Wang X, Zhang X, Dang PM, Holbrook CC, Culbreath AK, Wu Y, Guo B (2012) Peanut (*Arachis hypogaea*) Expressed sequence tag project: Progress and application. Comp Funct Genom doi:10.1155/2012/373768.

Ferguson ME, Bramel PJ, Chandra S (2004a) Gene diversity among botanical varieties in peanut (*Arachis hypogaea* L.). Crop Sci 44: 1847–1854.

Ferguson ME, Burow MD, Schulze SR, Bramel PJ, Paterson AH, Kresovich S, Mitchell S (2004b) Microsatellite identification and characterization in peanut (*A. hypogaea* L.). Theor Appl Genet 108: 1064–1070.

Gale MD, Devos KM (1998) Comparative genetics in the grasses. Proc Nl Acad Sci USA 95: 1971–1974.

Gimenes MA, Lopes CR, Valls JFM (2002) Genetic relationships among Arachis species based on AFLP. Genet Mol Biol 25: 349–353.

Gobbi A, Teixeira C, Moretzsohn M, Guimarães P, Leal-Bertioli S, Bertioli D, Lopes CR, Gimenes M (2006) Development of a linkage map to species of B genome related to the peanut (*Arachis hypogaea*-AABB). In: Plant and Animal Genomes XIV Conference, San Diego, CA, USA.

Guimarães PM, Garsmeur O, Proite K, Leal-Bertioli SC, Seijo G, Chaine C, Bertioli DJ, Hont AD (2008) BAC libraries construction from the ancestral diploid genomes of the allotetraploid cultivated peanut. BMC Plant Biol 8: 14.

Guimarães PM, Brasileiro ACM, Leal-Bertioli SCM, Pappas G, Togawa R, Bonfim O et al. (2011) Comparative 454 pyrosequencing of transcripts from two wild Arachis genotypes under biotic and abiotic stress. In: Plant and Animal Genome XIX Conference, San Diego, CA, USA.

Guo B, Chen X, Hong Y, Liang X, Dang P, Brenneman T, Holbrook C, Culbreath A (2009) Analysis of gene expression profiles in leaf tissues of cultivated peanuts and development of EST-SSR markers and gene discovery. Int J Plant Genom doi:10.1155/2009/715605.

Guo BZ, Chen X, Dang P, Scully BT, Liang X, Holbrook CC, Yu J, Culbreath AK (2008) Peanut gene expression profiling in developing seeds at different reproduction stages during *Aspergillus parasiticus* infection. BMC Dev Biol 8: 12.

Guo B, Fedorova ND, Chen X, Wan CH, Wang W, Nierman WC, Bhatnagar D, Yu J (2011) Gene expression profiling and identification of resistance genes to *Aspergillus flavus* infection in peanut through EST and microarray strategies. Toxins 3: 737–753.

Gupta PK, Kumar J, Mir RR, Kumar A (2010) Marker-assisted selection as a component of conventional plant breeding. Plant Breed Rev 33: 145–217.

Hang VTT, Lawn RJ, Bielig LM, Molnar SJ, Xia L, Kilian A (2012) Development and initial evaluation of diversity array technology for soybean and mungbean. Euphytica 186: 741–754.

Halward TM (1993) Development of an RFLP linkage map in diploid peanut species. Theor Appl Genet 87: 379–384.

Halward TM, Stalker HT, Larue EA, Kochert G (1991) Genetic variation detectable with molecular markers among unadapted germplasm resources of cultivated peanut and related wild species. Genome 34: 1013–1020.

He G, Meng R, Newman M, Gao G, Pittman RN, Prakash CS (2003) Microsatellites as DNA markers in cultivated peanut (*Arachis hypogaea* L.). BMC Plant Biol 3: 3.

He G, Meng RH, Gao H, Guo B, Gao G, Newman M, Pittman RN, Prakash CS (2005) Simple Sequence Repeat Markers for Botanical Varieties of Cultivated Peanut (*Arachis hypogaea* L.). Euphytica 142: 131–136.

Herselman L, Thwaites R, Kimmins FM, Courtois B, VanDerMerwe PJ, Seal SE (2004) Identification and mapping of AFLP markers linked to peanut (*Arachis hypogaea* L.) resistance to the aphid vector of groundnut rosette disease. Theor Appl Genet 109: 1426–1433.

Holbrook CC, Dong W (2005) Development and evaluation of a mini core collection for the U.S. peanut germplasm collection. Crop Sci 45: 1540–1544.

Holbrook CC, Timper P, Xue HQ (1993) Selection of a core collection from the United States germplasm collection of peanut. Crop Sci 33: 859–861.

Hopkins MS, Casa AM, Wang T, Mitchell SE, Dean RE, Kochert GD, Stephen K (1999) Discovery and characterization of polymorphic simple sequence repeats (SSRs) in peanut. Crop Sci 39: 1243–1247.

Huang J, Yan L, Lei Y, Jiang H, Ren X, Liao B (2012) Expressed sequence tags in cultivated peanut (*Arachis hypogaea*): discovery of genes in seed development and response to *Ralstonia solanacearum* challenge. J Plant Res 125: 755–769.

Isleib TG, Wynne JC (1992) Use of plant introductions in peanut improvement. In: Shands HL, Wiesner LE (eds) Use of Plant Introductions in Cultivar Development. Crop Science Society of America, Madison, WI, USA, 2: 75–116.

Jaccoud D, Peng K, Feinstein D, Kilian A (2001) Diversity arrays: a solid-state technology for sequence information independent genotyping. Nucleic Acids Res 29: E25.

Jain AK, Basha SM, Holbrook CC (2001) Identification of drought-responsive transcripts in peanut (*Arachis hypogaea* L.). Elec J Biotechnol 4: 59–67.

Jain RK, Navinder S, Sunitha J, Randhir S (2003) Molecular strategies for developing salt tolerant crops. Indian J Biotechnology 2: 121–137.

Jiang HF, Ren XP, Liao BS, Huang JQ, Lei Y, Chen BY, Guo B, Holbrook CC, Upadhyaya HD (2008) Peanut core collection established in China and compared with ICRISAT mini core collection. Acta Agron Sin 34: 25–30.

Jiang HF, Ren XP, Zhang XJ, Huang JQ, Lei Y, Yan LY, Liao BS, Upadhyaya HD, Holbrook CC (2010) Comparison of genetic diversity based on SSR markers between peanut minicore collections from China and ICRISAT. Acta Agron Sin 36: 1084–1091.

Jiang H, Liao B, Ren X, Lei Y, Mace E, Fu T, Crouch JH (2007) Comparative assessment of genetic diversity of peanut (*Arachis hypogaea* L.) genotypes with various levels of resistance to bacterial wilt through SSR and AFLP analyses. J Genet Genom 34: 544–554.

Kang IH, Gallo M (2007) Cloning and characterization of a novel peanut allergen Ara h3 isoform displaying potentially decreased allergenicity. Plant Sci 172: 345–353.

Katam R, Basha SM, Prashanth S, Pechan T (2010) Analysis of peanut leaf proteome. J Proteome Res 9: 2236–54.

Katam R, Hemanth KNV, Basha SM (2007) Proteomic approach to screen peanut genotypes with enhanced nutritional qualities. In: Frontiers in the Convergence of Bioscience and Information Technologies, IEEE Computer Society, Jeju Island, Korea, pp 171–178. INSPEC Accession Number:

9979403 doi; 10.1109/FBIT.2007.94.

Kilian A (2008) DArT-based whole genome profiling and novel information technologies in support system of modern breeding of groundnut. 3rd International Conference for Peanut Genomics and Biotechnology on "Advances in *Arachis* through Genomics and Biotechnology (AAGB), India.

Kochert G, Halward T, Branch WD, Simpson CE (1991) RFLP variability in peanut (*Arachis hypogaea*) cultivars and wild species. Theor Appl Genet 81: 565–570.

Kochert G, Stalker HT, Gimenes M, Galgaro L, Lopes CR, Moore K (1996) RFLP and cytogenetic evidence on the origin and evolution of allotetraploid domesticated peanut, *Arachis hypogaea* (Leguminosae). Am J of Bot 83: 1282–1291.

Koilkonda P, Sato S, Tabata S, Shirasawa K, Hirakawa H, Sakai H, Sasamoto S, Watanabe A, Wada T, Kishida Y, Tsuruoka H, Fujishiro T, Yamada M, Kohara M, Suzuki S, Hasegawa S, Kiyoshima H, Isobe S (2011) Large-scale development of expressed sequence tag-derived simple sequence repeat markers and diversity analysis in *Arachis* spp. Mol Breed 30: 125–138.

Koppelman SJ, Wensing M, Ertmann M, Knulst AC, Knol EF (2004) Relevance of Ara h1 Ara h2 and Ara h3 in peanut-allergic patients, as determined by immunoglobulin E western blotting, basophil-histamine release and intracutaneous testing: Ara h2 is the most important peanut allergen. Clin Exp Allergy 34: 583–590.

Kottapalli P, Upadhyaya H, Kottapalli KR, Payton PR, Dwivedi S, Burow M, Puppala N (2011) Assessing genetic diversity in Valencia peanut germplasm using SSR markers. Crop Sci 51: 1089–1100.

Krapovickas A, Gregory WC (1994) Taxonomía del género *Arachis* (Leguminosae). Bonplandia. Facultad de Ciencias Agrarias. Corrientes Province, Argentina 8: 1–186.

Lawson MJ, Zhang L (2006) Distinct patterns of SSR distribution in the Arabidopsis thaliana and rice genomes. Genome Biol 7: R14.

Liang XQ, Luo M, Holbrook CC, Guo BZ (2006) Storage protein profiles in Spanish and runner market type peanuts and potential markers. BMC Plant Biol 6: 24.

Liang X, Zhou G, Hong Y, Chen X, Liu H, Li S (2009) Overview of research progress on peanut (*Arachis hypogaea* L.) host resistance to aflatoxin contamination and genomics at the Guangdong Academy of Agricultural Sciences. Peanut Sci 36: 29–34.

Luo M, Dang P, Guo B, He G, Holbrook CC, Bausher MG, Lee RD (2005a) Generation of expressed sequence tags (ESTs) for gene discovery and marker development in cultivated peanut. Crop Sci 45: 346–53.

Luo M, Liang XQ, Dang P (2005b) Microarray-based screening of differentially expressed genes in peanut in response to *Aspergillus parasiticus* infection and drought stress. Plant Sci 169: 695–703.

Mace ES, Phong DT, Upadhyaya HD, Chandra S, Crouch JH (2006) SSR analysis of cultivated groundnut (*Arachis hypogaea* L.) germplasm resistant to rust and late leaf spot diseases. Euphytica 152: 317–330.

Maiti RK (2002) About the peanut crop. In: Maiti R, Ebeling PW (eds) The Peanut (*Arachis hypogaea*) Crop. Science Publishers, New York, USA, pp 1–12.

Moretzsohn MC, Leoi L, Proite K, Guimarães PM, Leal-Bertioli SCM, Gimenes MA, Martins WS, Valls JFM, Grattapaglia D, Bertioli DJ (2005) A microsatellite-based, gene-rich linkage map for the AA genome of *Arachis* (Fabaceae). Theor Appl Genet 111: 1060–1071.

Moretzsohn MC, Barbosa AVG, Alves-Freitas DMT, Teixeira C, Leal-Bertioli SCM, Guimarães PM, Pereira RW, Lopes CR, Cavallari MM, Valls JFM, Bertioli DJ, Gimenes MA (2009) A linkage map for the B-genome of Arachis (Fabaceae) and its synteny to the A-genome. BMC Plant Biol 9: 40.

Mullet JE, Klein RR, Klein PE (2002) Sorghum bicolor—An important species for comparative grass genomics and a source of beneficial genes for agriculture. Curr Opin in Plant Biol 5: 118–121.

Nagy E, Guo Y, Khanal S, Taylor C, Knapp S, Oziasakins P, Stalker HT, Nielsen N (2010) Developing a high-density molecular map of the A-genome species *A.* duranensis. In: Starr JL, Nickels I (eds) The Proceedings American Peanut Research and Education Society (APRES), Florida, USA, 25 p.

Neumann K, Kobiljski B, Dencic S, Varshney RK, Börner A (2011) Genome-wide association mapping: a case study in bread wheat (*Triticum aestivum* L.). Mol. Breed 27: 37–58.

Paik-Ro OG, Smith RL, Knauft DA (1992) Restriction fragment length polymorphism evaluation of six peanut species within the Arachis section. Theor Appl Genet 84: 201–208.

Pandey MK, Gautami B, Jayakumar T, Sriswathi M, Upadhyaya HD, Gowda MVC, Radhakrishnan T, Bertioli DJ, Knapp SJ, Cook DR, Varshney RK (2012a) Highly informative genic and genomic SSR markers to facilitate molecular breeding in cultivated groundnut (*Arachis hypogaea*). Plant Breed 131: 139–147.

Pandey MK, Monyo E, Ozias-Akins P, Liang X, Guimarães P, Nigam SN, Upadhyaya HD, Janila P, Zhang X, Guo B, Cook DR, Bertioli DJ, Michelmore R, Varshney RK (2012b) Advances in *Arachis* genomics for peanut improvement. Biotechnology Adv 30: 639–651.

Paterson AH, Bowers JE, Burow MD, Draye X, Elsik CG, Jiang CX, Katsar CS, Lan TH, Lin YR, Ming R, Robert JW (2000) Comparative genomics of plant chromosomes. Plant Cell 12: 1523–1540.

Powell W, Morgante M, Andre C, Hanafey M, Vogel J, Tingey S, Rafalski A (1996) The comparision of RFLP, RAPD, AFLP and SSR (microsatellite) markers for germplasm analysis. Mol Breed 2: 225–238.

Proite K, Leal-Bertioli SCM, Bertioli DJ, Moretzsohn MC, Silva FR, Martins NF, Guimarães PM (2007) ESTs from a wild Arachis species for gene discovery and marker development. BMC Plant Biol 7: 7.

Qin H, Feng S, Chen C, Guo Y, Knapp S, Culbreath A, He G, Wang ML, Zhang X, Holbrook CC, Ozias-Akins P, Guo B (2012) An Integrated genetic linkage map of cultivated peanut (*Arachis hypogaea* L.) constructed from two RIL populations. Theor Appl Genet 124: 653–664.

Raina SN, Rani V, Kojima T, Ogihara Y, Singh KP, Devarumath RM (2001) RAPD and ISSR fingerprints as useful genetic markers for analysis of genetic diversity, varietal identification, and phylogenetic relationships in peanut (*Arachis hypogaea* L.) cultivars and wild species. Genome 44(5): 763–772.

Roy SJ, Tucker EJ, Tester M (2011) Genetic analysis of abiotic stress tolerance in crops. Curr Opin Plant Biol 14(3): 232–239.

Saha MC, Mian MA, Eujayl I, Zwonitzer JC, Wangnd L, May GD (2004) Tall fescue EST-SSR markers with transferability across several grass species. Theor Appl Genet 109: 783–791.

Savadi SB, Fakrudin B, Nadaf HL, Gowda MVC (2012) Transferability of Sorghum Genic Microsatellite Markers to Peanut. Am J Plant Sci 3: 1169–1180.

Savage GP, Keenan JI (1994) The composition and nutritive value of groundnut kernels. In: Smartt J, Chapman (eds) The Groundnut Crop: A Scientific Basis for Improvement. Chapman and Hall. London, UK, pp 173–213.

Scott KD, Eggler P, Seaton G, Rossetto M, Ablett EM, Lee LS, Henry RJ (2000) Analysis of SSRs derived from grape ESTs. Theor Appl Genet 100: 723–726.

Seijo JG, Lavia GI, Fernandez A, Krapovickas A, Ducasse D, Moscone EA (2004) Physical mapping of 5S and 18S-25S rRNA genes as evidence that *Arachis duranensis* and *A. ipaensis* are the wild diploid progenitors of *A. hypogaea* (Leguminosae). Am J Bot 91: 1294–1303.

Song GQ, Li MJ, Xiao H, Wang XJ, Tang RH, Xia H, Zhao CZ, Bi YP (2010) EST sequencing and SSR marker development from cultivated peanut (*Arachis hypogaea* L.). Elec J Biotechnol 13: 3.

Sprent J (1994) Nitrogen fixation. In: Smartt J (ed) The Groundnut Crop: A Scientific Basis for Improvement. Chapman and Hall London, pp 255–280.

Subramanian VS, Gurtu RC, Rao RCN, Nigam SN (2000) Identification of DNA polymorphism in cultivated groundnut using random amplified polymorphic DNA (RAPD) assay. Genome 43: 656–660.

Tautz D (1989) Hypervariability of simple sequences of a general source for polymorphic DNA markers. Nucleic Acids Res 17(16): 6463–6471.

Thudi M, Bohra A, Nayak SN, Varghese N, Shah TM, Penmentsa RV, Thirunavukkarasu N, Gudipate S, Gaur PM, Kulwal PL, Upadhyaya HD, Karikishor PB, Winter P, Kahl G, Town CD, Kilian A, Cook DR, Varshney RK (2011) Novel SSR markers from BAC-end sequences, DArT Arrays and a comprehensive genetic map with 1,291 marker loci for chickpea (*Cicer arietinum* L.). PLoS ONE 6(11): e27275 doi:10.1371/journal.pone.0027275.

Upadhyaya HD (2002) Developing a mini core of peanut for utilization of genetic resources. Crop Sci 42: 2150–2156.

Upadhyaya HD (2003) Development of a groundnut core collection using taxonomical, geographical and morphological descriptors. Genetic Resour Crop Evol 50: 139–148.

Varshney RK, Graner A, Sorrells ME (2005) Genic microsatellite markers in plants: features and applications. Trends Biotechnol 23: 48–55.

Varshney RK, Hoisington DA, Tyagi AK (2006) Advances in cereal genomics and applications in crop breeding. Trends Biotechnol 24: 490–499.

Varshney RK, Nayak SN, May GD, Jackson SA (2009) Next-generation sequencing technologies and their implications for crop genetics and breeding. Trends Biotechnol 27: 522–30.

Varshney RK (2010a) Gene-based marker systems in plants: high throughput approaches for discovery and genotyping. In: Jain SM, Brar DS (eds) Molecular Techniques in Crop Improvement. Springer, Netherlands, pp 119–142.

Varshney RK, Gowda MVC, Radhakrishnan T, Pandey MK, Gautami B, Sujay V, Koppolu R, Senthilvel S, Vadez V, Nigam SN, Upadhyaya HD, Isobe S, He G, Bertioli DJ, Knapp SJ, Cook DR (2010b) Development and application of genomic resources for molecular breeding in groundnut (*Arachis hypogaea* L). In: Proc 3rd International Conference on Plant Molecular Breeding (ICPMB), China.

Varshney RK, Thudi M, May GD, Jackson SA (2010c) Legume genomics and breeding. Plant Breed Rev 33: 257–304.

Varshney RK, Glaszmann JC, Leung H, Ribaut JM (2010d) More genomic resources for less-studied crops. Trends Biotechnol 28: 452–60.

Varshney RK, Kudapa H, Roorkiwal M, Thudi M, Pandey MK, Saxena RK, Chamarthi SK, Mohan SM, Mallikarjuna N, Upadhyaya H, Gaur PM, Krishnamurthy L, Saxena KB, Nigam SN, Pande S (2012) Advances in genetics and molecular breeding of three legume crops of semi-arid tropics using next-generation sequencing and high-throughput genotyping technologies. J Biosci 37: 811–820.

Viquez OM, Konan KN, Dodo HW (2003) Structure and organization of the genomic clone of a major peanut allergen gene, Ara h1. Mol Immunol 40: 565–571.

Wang H, Penmetsa RV, Yuan M, Gong L, Zhao Y, Guo B, Farmer AD, Rosen BD, Gao J, Isobe S, Bertioli DJ, Varshney RK, Cook DR, He G (2012) Development and characterization of BAC-end sequence derived SSRs, and their incorporation into a new higher density genetic map for cultivated peanut (*Arachis hypogaea* L.). BMC Plant Biol 12: 10.

Wang XJ, Shuan-Tao LIU, Han Xia, Shu-Bo Wan, Chuan-Zhi Zhao, Ai-Qin Li (2011) Peanut (*Arachis hypogaea* L.) Omics and Biotechnology in China. Plant Omics Journal 4(6): 339–349.

Weber RD, May PE (1989) Abundant class of human DNA polymorphisms which can be typed using the polymerase chain reaction. Am J Hum Genet 44: 388–396.

Wenzl P, Li H, Carling J, Zhou M, Raman H, Paul E, Hearnden P, Maier C, Xia L, Caig V, Ovesná J, Cakir M, Poulsen D, Wang J, Raman R, Smith KP, Muehlbauer GJ, Chalmers KJ, Kleinhofs A, Huttner E, Kilian A (2006) A high-density consensus map of barley linking DArT markers to SSR, RFLP and STS loci and agricultural traits. BMC Genomics 7: 206.

Yang SY, Saxena RK, Kulwal P, Ash GJ, Dubey A, Harper JDI, Upadhyaya HD, Gothalwal R, Kilian A, Varshney RK (2011) The first genetic map of pigeonpea (Cajanus cajan) based on Diversity Arrays Technology (DArT) markers. J. Genet 90(1): 103–9.

Young ND, Weeden NF, Kochert G (1996) Genome mapping in legumes (Family Fabaceae). In: Paterson AH (ed) Genome Mapping in Plants. Landes Biomedical Press, Austin, TX, USA, pp 212–227.

Yuan M, Gong L, Meng R, Li S, Dang P, Guo B, He G (2010) Development of trinucleotide (GGC)n SSR markers in peanut (*Arachis hypogaea* L.). Elec J Biotechnol 13: 5–6.

Index

About the Editors

Dr. Nalini Mallikarjuna, Principal Scientist (ICRISAT, India), has contributed significantly in diversification of legume crops through wide hybridization. She successfully developed diverse breeding material by using different wild species in groundnut, chickpea and pigeonpea. Her successful effort in development of synthetic tetraploids by using diploid species in groundnut was much appreciated. These synthetics are being now used in different breeding programmes of the world. She has published > 75 publications in various reputed journals.

Dr. Rajeev K. Varshney, a Principal Scientist, is serving at International Crops Research Institute for the Semi-Arid Tropics (ICRISAT)—a CGIAR research institute, as Research Programme Director—Grain Legumes and Director—Center of Excellence in Genomics. He also led Comparative and Applied Genomics program for CGIAR Generation Challenge Program for six years. Before joining CGIAR system, he worked at Leibniz Institute of Plant Genetics and Crop Plant Research (IPK), Germany for five years. He has a basic background in molecular genetics and possesses about 20 years' research experience.

Color Plate Section

Chapter 3

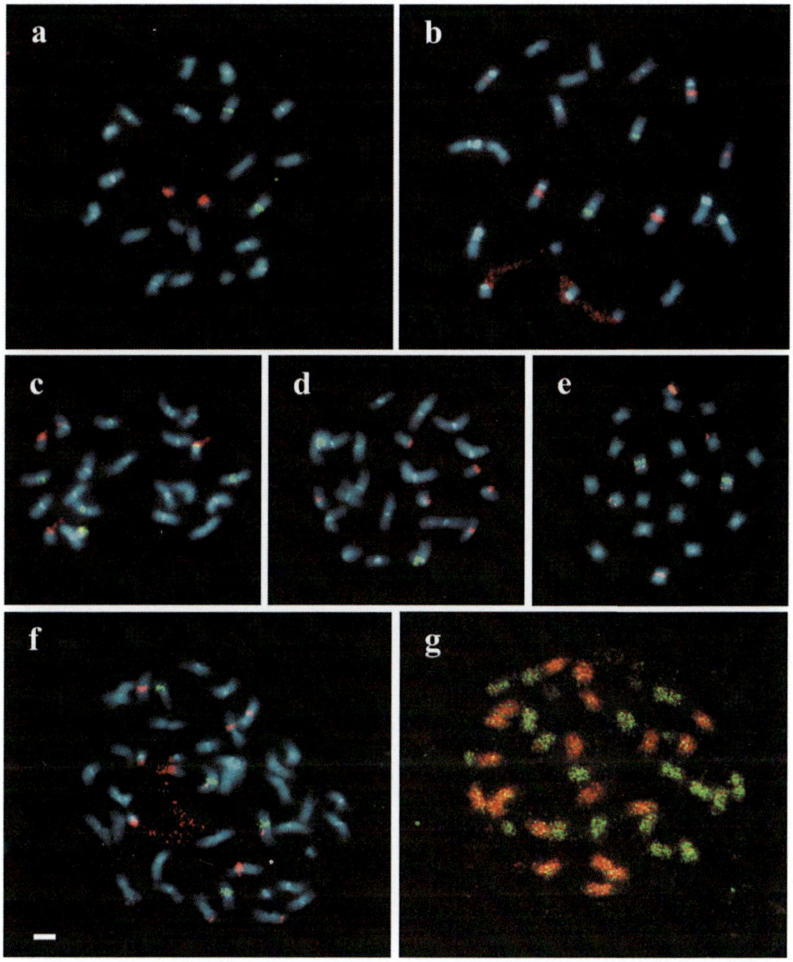

Figure 3-2 Somatic metaphases of *Arachis* species following double fluorescent *in situ* hybridization (a–f), showing yellow-green FITC signals from the 5S rDNA probe and red TRITC signals from the 18S–26S rDNA probe. DAPI counterstaining (light blue) subsequent to the FISH procedure was used to highlight the heterochromatin bands and to stain euchromatin. **(a)** *A. trinitensis* (F genome). **(b)** *A. glanduifera* (D genome). **(c)** *A. batizocoi* (K genome). **(d)** *A. duranensis* (A genome). **(e)** *A. ipaënsis* (B s.s. genome). **(f)** *A. hypogaea* ($2n = 40$). **(g)** Somatic metaphase of *Arachis hypogaea* after double genomic *in situ* hybridization (GISH) using total DNA probes of *A. ipaënsis* (red) and of *A. duranensis* (green). Scale bar = 3 µm for all the pictures.

Chapter 6

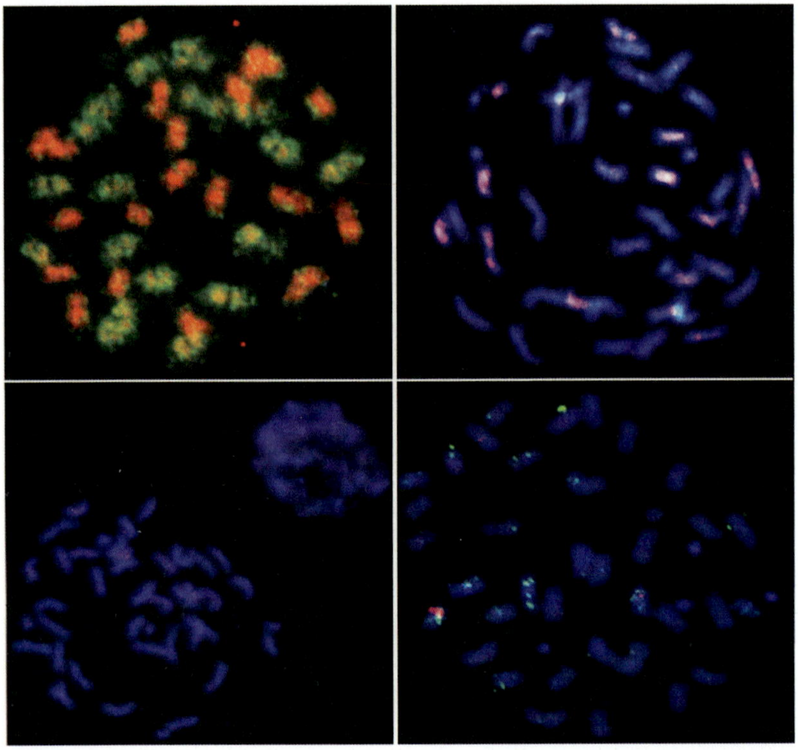

Figure 6-1 *In situ* hybridization on metaphase spreads of *Arachis* spp. with DAPI counterstaining. a) GISH on the synthetic amphidiploid (*Arachis duranensis* × *A. ipaënsis*)^{4x} with both parentals genomic DNA probes—green signals with *A. duranensis* on half of the chromosomes (A genome) and red signals with *A. ipaënsis* in the other half chromosomes (B genome) with signals overlapping part of some chromosomes. b) BAC-FISH with ADH79O23 (F12_Sl2_6OVER) probe with red signals over half of the chromosomes (A genome) and some dots on B genome chromosomes, more concentrated labeling but at different intensity depending on the chromosome. Hybridization signals were absent at centromere and telomere regions; c) BAC-FISH with ADH51I17 probe with diffused red signals in the pericentromere regions only on A genome chromosomes (F12_Sl4_5OVER); d) BAC-FISH with ADH179B13 probe with spotted green signals on A and B genome chromosomes but stronger on A chromosomes. Red signals correspond to the rDNA 5S sites (F17_Sl2_4OVER1). Scale bar: 5 µm.

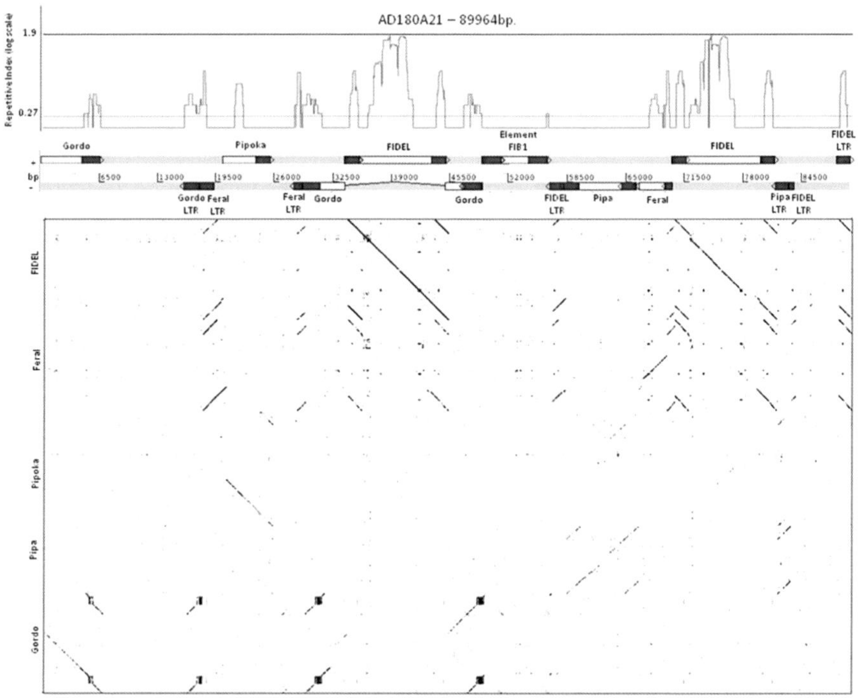

Figure 6-2 Representation of BAC clone AD180A21 consisting of two contigs, one of 84,046 bp positioned at left hand side, and one of 5,920 bp positioned at right. Top: Repetitive Index graph; Middle: Annotation scheme, and Bottom: dot plot. Repetitive Index is a score for repeat content based on BLASTN against 41,856 *A. duranensis* BAC-end sequences. The score is calculated using the formula Repetitive Index = log10(N+1), where N is the number of BLASTN homologies with an evalue of 1e-20 or less. The highest peak represented here is 1.9, which is equivalent to 88 BLASTN homologies; the lowest peak is 0.3, which is equivalent to a single BLASTN homology. The annotation scheme represents long terminal repeats (LTRs) in blue and internal regions of transposons in white. Transposons in positive orientation are represented on upper strand, and those in negative on the lower. The dot plot is of the BAC sequence (horizontal) against whole representative sequences of the transposons FIDEL, Feral, Pipoka, Pipa and Gordo (vertical).

This BAC clone consists almost entirely of LTR retrotransposons their solo elements and remnants, and does not contain any non-transposon gene. The sequence contains two complete FIDELs, one complete Pipa, and a complete Gordo interrupted by one of the FIDEL elements (names in bold type), a lower copy LTR transposon (Element-FIB1), plus retrotransposon fragments. All highly repetitive sequences in the BAC are derived from five retrotransposons: FIDEL, Feral, Pipoka, Pipa and Gordo.

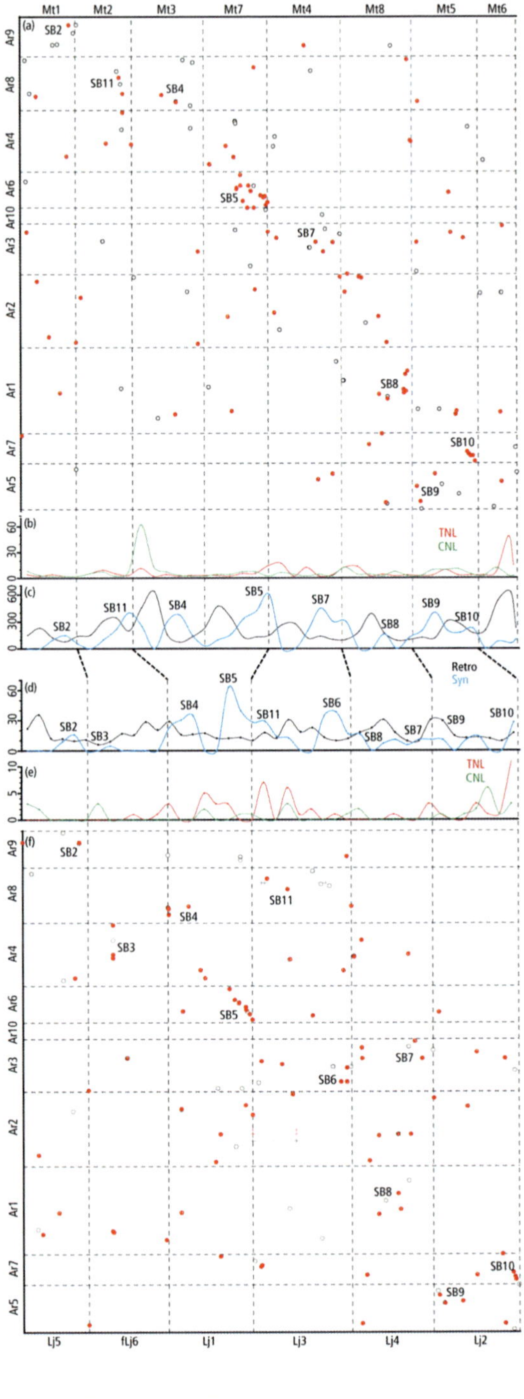

Figure 6-3 contd....

Chapter 7

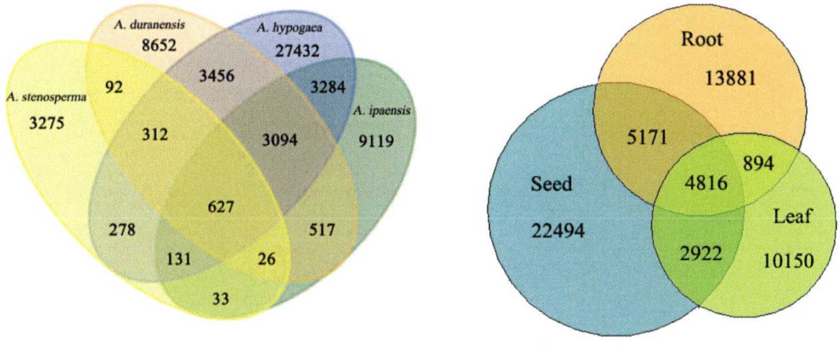

A B

Figure 7-1 Peanut transcriptome assembled from ESTs in dbEST. (A) Distribution of transcripts among cultivated species (*A. hypogaea*) and three wild relatives (*A. duranensis, A. ipaensis* and *A. stenosperma*). (B) Distribution of transcripts among seed, leaf and root tissues.

Figure 6-3 contd.

Figure 6-3 Genome plots of *Arachis* vs. *Medicago* and *Arachis* vs. *Lotus*, integrated with each other and graphs of synteny with *Arachis*, and retrotransposon, and resistance gene homolog distributions for *Medicago* and *Lotus* (original figure is from Bertioli et al. 2009). Chromosome orders and numbering of synteny blocks are the same as a *Medicago* vs *Lotus* plot in Cannon et al. 2006, allowing direct comparisons. Equivalent conserved regions (synteny blocks) and variable regions are present in all possible combinations of species comparisons *Arachis-Lotus-Medicago*. This shows that some genomic regions (synteny blocks) are consistently more stable during evolution than others.

(a) Genome Plot of *Arachis* vs. *Medicago*.
(b) Density of blast detected resistance gene homologs of the TNL (red line) and CNL (green line) subclasses plotted along the *Medicago* genome. High densities of resistance gene homologs and retrotransposons coincide.
(c) Black line: density of blast detected retrotransposons plotted along the *Medicago* genome. Cyan-blue line: scaled synteny score of *Medicago* with *Arachis*. Synteny blocks occur in regions of low retrotransposon density.
(d) Black line: percentage genome coverage of retrotransposons plotted along the *Lotus* genome. Cyan-blue line: scaled synteny score of *Lotus* with *Arachis*. Synteny blocks tend to occur in regions of low retrotransposon coverage.
(e) Density of resistance gene homolog encoding sequences, TNL (red) and CNL (green), plotted along the Lotus genome. Clusters of resistance gene homologs and retrotransposons coincide.
(f) Genome Plot of *Arachis* vs. *Lotus*. Markers mapped to intervals are plotted as horizontal lines.

Chapter 8

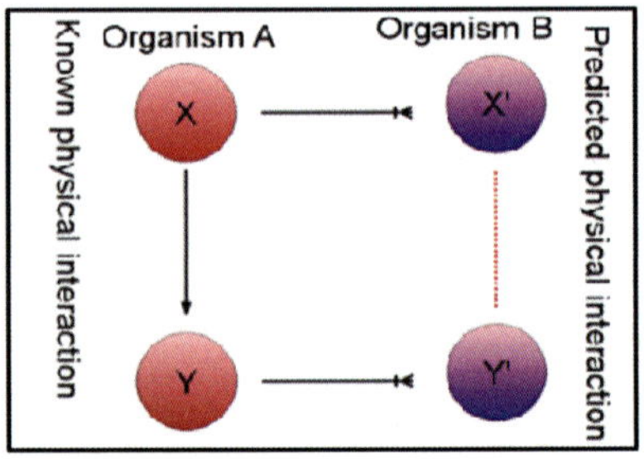

Figure 8-1 Illustration of interolog mapping mapped for those orthologous proteins known to have predicted interaction partners (Organism A: *Arabidopsis*; Organism B: Peanut).